Palaeolandscapes in Archaeology

What can we learn about the ancient landscapes of our world, and how can those lessons improve our future in the landscapes that we all inhabit? Those questions are addressed in this book, through a practical framework of concepts and methods, combined with detailed case studies around the world.

The chapters explore the range of physical and social attributes that have shaped and re-shaped our landscapes through time. International authors contributed the latest results of investigating ancient landscapes (or "palaeolandscapes") in diverse settings of tropical forests, deserts, river deltas, remote islands, coastal zones, and continental interiors. The case studies embrace a broadly accommodating approach of combining archaeological evidence with other avenues of research in earth sciences, biology, and social relations. Individually and in concert, the chapters offer new perspectives on what the world's palaeolandscapes looked like, how people lived in these places, and how communities have engaged with long-term change in their natural and cultural environments through successive centuries and millennia. The lessons are paramount for building responsible strategies and policies today and into the future, noting that many of these issues from the past have gained more urgency today.

This book reaches across archaeology, ecology, geography, and other studies of human-environment relations that will appeal to general readers. Specialists and students in these fields will find extra value in the primary datasets and in the new ideas and perspectives. Furthermore, this book provides unique examples from the past, toward understanding the workings of sustainable landscape systems.

Mike T. Carson (PhD in Anthropology, University of Hawai'i, 2002) has investigated archaeological landscapes throughout the Asia-Pacific region. He currently is Associate Professor of Archaeology at the Richard F. Taitano Micronesian Area Research Center at the University of Guam. He was author of *Archaeological Landscape Evolution: The Mariana Islands in the Asia-Pacific Region* (Springer, 2016) and *Archaeology of Pacific Oceania; Inhabiting a Sea of Islands* (Routledge, 2018), and he was editor of *Asian Perspectives: The Journal of Archaeology for Asia and the Pacific* (University of Hawai'i Press, 2014–2020).

Palaeolandscapes in Archaeology

Lessons for the Past and Future

Edited by Mike T. Carson

Routledge
Taylor & Francis Group

LONDON AND NEW YORK

First published 2022
by Routledge
2 Park Square, Milton Park, Abingdon, Oxon OX14 4RN

and by Routledge
605 Third Avenue, New York, NY 10158

Routledge is an imprint of the Taylor & Francis Group, an informa business

British Library Cataloguing-in-Publication Data
A catalogue record for this book is available from the British Library

Library of Congress Cataloguing-in-Publication Data
Names: Carson, Mike T., editor.
Title: Palaeolandscapes in archaeology : lessons for the past and future / edited by Mike T. Carson.
Other titles: Paleolandscapes in archaeology
Description: Abingdon, Oxon ; New York, NY : Routledge, 2022.
| Includes bibliographical references and index.
Identifiers: LCCN 2021032290 (print) | LCCN 2021032291 (ebook)
| ISBN 9780367689032 (hardback) | ISBN 9780367689056 (paperback)
| ISBN 9781003139553 (ebook)
Subjects: LCSH: Landscape archaeology--Case studies. | Landscape assessment--Case studies. | Paleolithic period. | Prehistoric peoples--Casestudies. | Human ecology--Case studies.
Classification: LCC CC77.L35 P35 2022 (print) | LCC CC77.L35 (ebook) | DDC 930.1--dc23
LC record available at https://lccn.loc.gov/2021032290
LC ebook record available at https://lccn.loc.gov/2021032291

ISBN: 978-0-367-68903-2 (hbk)
ISBN: 978-0-367-68905-6 (pbk)
ISBN: 978-1-003-13955-3 (ebk)

DOI: 10.4324/9781003139553

Typeset in Times New Roman
by MPS Limited, Dehradun

Contents

Figures

Tables

Contributors

Christopher J. H. Ames, University of Wollongong and University of Victoria, Canada

Jordan Brown, University of California, Berkeley

Ian Buvit, Oregon State University

Mereoni Camailakeba, Fiji Museum

Mike T. Carson, University of Guam

Lindsey E. Cochran, East Tennessee State University, USA

Shenglun Du, University College of London

Felicia De Peña, University of California, Berkeley

Andrea K. L. Freeman, University of Calgary

Kara A. Fulton, University of North Texas

Hsiao-chun Hung, The Australian National University

Masami Izuho, Tokyo Metropolitan University

Roselyn Kumar, University of the Sunshine Coast

Lisa A. Maher, University of California, Berkeley

David W. Mixter, Binghamton University

Elia Nakoro, Fiji Museum

Meli Nanuku, Fiji Museum

Patrick Nunn, University of the Sunshine Coast

Ben Shaw, The Australian National University and University of New South Wales

Glenn R. Summerhayes, University of Otago

Karisa Terry, Central Washington University

Victor D. Thompson, University of Georgia, USA

Bryan Tucker, New South Associates, Inc.

A. J. White, University of California, Berkeley

Yijie Zhuang, University College of London

Preface

This book originated in 2019, as an attempt to bring diverse researchers together in a shared theme of studying ancient landscapes. The book's editor organised a symposium for the 2020 annual meeting of the Society for American Archaeology, wherein colleagues could present their findings about ancient landscapes and consider how their results may or may not improve current and future issues of living through a changing world. Although the 2020 event was cancelled, due to the health and safety issues of a global pandemic, the contributors in this book continued their research and writing projects.

After several months of coping with a global pandemic, this book at last has been finalised. The experience with the pandemic has served to emphasise the importance of preparing for global change. In this respect, the current book may resonate with a growing international interest in looking at scientific research about our world's past in order to develop effective concepts, strategies, and policies on a global scale. If we can learn more about how people in the past have managed their challenging situations, then we can make more responsible decisions now and into the future.

Gratitude is extended to the many people who made this book possible. Matthew Gibbons, Manas Roy, and the teams at Routledge provided professional advice and support through the process of the book proposal, development of the contents, editorial reviews, and production details. Anonymous external reviewers offered constructive commentary and critical suggestions for improving the book, and their anonymous work is especially appreciated. The contributing authors persevered through the months of preparing their chapters and responding to review comments, editorial advice, and technical queries.

Ideally, this book can encourage more research in palaeolandscapes, not only for learning about the past but also for applying those lessons toward our future. This sentiment too often has been stated without actually delivering much substance. While the present book offers several data-rich and thought-provoking examples of palaeolandscape research in archaeology, numerous other examples have been produced outside this book and will continue to be performed by researchers worldwide. In this respect, hopefully, this book can attract broader interdisciplinary and international attention to the values of studying palaeolandscapes.

1 What can we learn from palaeolandscapes in archaeology?

Mike T. Carson

Archaeological studies of ancient landscapes or "palaeolandscapes" can operationalise knowledge about the past toward solving current and future problems of global concern. The natural and social landscapes of the world currently are imperilled by drastic change, thus making the lessons from palaeolandscapes notably urgent for informing wise policy decisions about adapting to complex and dynamic landscapes. Toward these goals, the current book embraces the challenge of making palaeolandscape research relevant and practical for modern and future issues.

Studies of palaeolandscapes can reveal what was most effective for people in the past when they lived through challenging circumstances of their ancient landscapes, for example, in terms of climate, sea level, water and food resources, social structure, population dynamics, and more. Those same challenges have become vital topics in the modern world, and therefore the lessons from palaeolandscape archaeology now have become more relevant than ever in the present. The real-life examples from the past can be instructive for managing through the future change of landscapes.

Palaeolandscapes encompass the varied physical, biotic, social, and cognitive aspects of landscapes that existed in the past and changed through time. Some of these lines of evidence have survived more clearly than others, for instance, in the physical remnants of archaeological sites, and these material findings necessarily provide an objective evidentiary record of ancient landscapes. While people today interpret their landscapes through their own subjective perspectives and personalised experiences, the surviving records of palaeolandscapes offer incomplete yet indisputably factual representations of ancient contexts.

Palaeolandscapes in archaeology illustrate the diverse natural and social environments where people have lived in the past, and furthermore, they can reveal uniquely valuable knowledge about the possibilities for modern and future landscapes. These studies expose the practices that were sustainable or unsustainable for the people who coped with the various challenges in their inhabited landscapes through different time scales and circumstances. The materially attested examples from the past provide direct parallels with modern landscape issues.

DOI: 10.4324/9781003139553-1

In order to learn as much as possible from the world's ancient landscapes or palaeolandscapes, this book embraces a broad view of the factors that contributed to the natural and cultural history of landscapes through time. Equally relevant approaches could examine components of geological formations, plant and animal communities, water sources, social settings, cognitive functions, and other fields of study. All of those aspects have been effective in palaeolandscape archaeology, as shown in the chapters of this book.

The global relevance of this book derives from the fact that palaeolandscape archaeology could be applied in any place and time period of the world, as proven in the diverse chapters of this book. For as long as human beings have existed, they have lived in landscapes that have shaped their lives, and likewise, people have shaped their landscapes. This chronological dimension opens the opportunities for learning how landscapes were configured in the past, how they changed through time, and how people lived with those changing conditions.

The present chapter introduces the core concepts and themes about palaeolandscapes in archaeology that are developed in the next chapters. In this fashion, the diverse case studies and perspectives of this book can become clarified and compatible in terms of: (1) the relationships between landscapes and palaeolandscapes; (2) the role of chronology in these studies; (3) the concurrent time scales of operation in complex landscape systems; and (4) the impact of any specific project within the larger scope of palaeolandscape studies.

This chapter concludes with an outline of the book structure, noting how all of the next chapters relate with each other and support the central goal of learning from palaeolandscapes. This research can offer lessons not only for the past but also for the future. Learning about the past is meaningful in its own right, and the results can create additional impact when they are applied for solving current and future problems.

Landscapes and palaeolandscapes

In the simplest terms, landscapes refer to currently functioning places in the world, and palaeolandscapes refer to ancient versions or conditions of landscapes that had existed in the past. Similar terminology may be applied for discussing the relationship between an environment and a palaeoenvironment, geography and palaeogeography, demography and palaeodemography, and so on. Studies in this book and elsewhere refer to past conditions of palaeohabitat, palaeobeach, palaeoshoreline, and other entities with modern-day parallels. Whatever physical phenomenon or cognitive notion can be experienced today, an ancient counterpart can be studied in the records from the past, thus enabling diverse perspectives in palaeolandscape research.

Modern and ancient contexts reflect each other in principle, but in practice, their available lines of evidence are fundamentally different. Modern contexts can be witnessed and experienced today in their entirety, including variable subjectivity and nuanced perspective, but ancient contexts are knowable only in their surviving material remnants. Those detectable vestiges of the past are inherently incomplete representations of their original contexts, yet they can be accepted as indisputable factual evidentiary records. Sociology, anthropology, and other fields of study can interpret modern living systems, while archaeology can document the physical assemblages of ancient artefacts, sites, and landscapes.

Studies of modern landscapes can be useful for building inferences about some of the intangible elements that have not survived intact in palaeolandscape records. Perspectives of cultural geography, landscape ecology, and social history, in particular, provide insights into the functional operations of landscape systems, and these studies could be applied toward interpreting ancient landscape records. In many cases, historical and modern documentation can contextualise the variable ways of living in landscapes, as well as the various interconnections among the natural and social aspects of landscapes.

Modern landscapes embody the results of palaeolandscapes, compounded through time in long chronological sequences of potentially several interrelated contexts. Logically, the modern versions of landscapes could not exist without all of the preceding millennia of the natural and cultural history of palaeolandscapes that shaped the outcomes that we can experience today. Furthermore, the events and processes of the modern era will continue to shape the future outcomes of landscapes.

Without this knowledge from the past, decisions in the present would be based on incomplete awareness of the long-term consequences of how landscapes can change in complex ways through long time scales. This book moves beyond offering cliché sentiments about learning from the past in order to understand the present and prepare for the future. Factual knowledge about palaeolandscapes can be applied toward responsible decisions by avoiding knowable mistakes and improving on prior outcomes that have been exposed through the long-term chronological records of palaeolandscapes.

Applying a chronological perspective

Palaeolandscape archaeology extends landscape ecology and historical ecology into a longer chronological view. Logically, every place has transformed through a chronological series of landscapes. Multiple palaeolandscapes existed in a measurable chronological order, potentially detectable through stratigraphic layers and other lines of evidence. In this way, palaeolandscape archaeology accounts for the complexities of social-ecological functioning that have occurred and changed through time. Moreover, the diverse components of a landscape can be specified, measured, and described in comparison with each other in a unified chronology.

The notion of chronological change must be clarified as qualitatively different from recognising the existence of time or history. A sense of history is acknowledged in most standard views of landscape ecology and historical ecology, but the specific forms of history in these cases do not necessarily involve dynamic or transformative qualities. Landscape ecology examines the complex interactive parts of landscape systems, wherein the component parts operate primarily in a synchronic sense of how they can affect each other. To be fair, landscape ecology could accommodate the role of historical time, yet history in this view tends to be portrayed as reflecting long-term stability and self-regulating continuity of an ecological system. This perspective does not necessarily account for a dynamically transformative or evolving landscape.

If a landscape has changed or transformed through time, then by definition, this place has involved more than one landscape during different time periods. If the full sequence can be revealed in its natural chronological order, then each discernible time period could refer to a separate portion of a palaeolandscape chronology. Within each measurable time period, the past conditions referred to different iterations of palaeolandscapes, and only the latest version can be described as directly connected with the knowledge of the present.

In a long-term view, multiple palaeolandscape contexts can be revealed in their natural chronological order, effectively illustrating how the components of landscapes have changed through time. For instance, a river during different times may have changed in its course, in its depth of cutting into the ground, and in its relation with adjoining landforms and biomes. Similarly, a hillslope may have undergone numerous iterations of erosional conditions, furthermore linked with re-deposition of slope-eroded sediments, differential conditions for vegetation growth, and potential for people to build houses, grow plant foods, or access mineral resources. Even if the complete details of a chronological sequence are not yet known, the notional framework, at a minimum, can accommodate the specified place and time period of any archaeological and palaeolandscape findings.

In a short-term view of a landscape, the records of the past are linked with historical and modern frames of reference. Most often, the past landscape and the present-day landscape are viewed as reflections of each other, stressing the continuity between them. In this view, stable landforms or landmarks, such as hilltops or coastlines, are noted as consistent through time in their physical conditions and in their cultural meaning. The same landscape components would be recognised as potentially changing through time in a long-term sequence.

Awkwardly, the concept of long-term chronological change only rarely has been consulted toward developing modern sustainable landscape systems. In principle, government agencies and independent research groups would welcome the findings from palaeolandscape studies, given the fact that policy-makers routinely have drawn on the last decades of observations of weather conditions, sea level, forest composition, and other lines of evidence. If these proximal historical records have been deemed relevant and useful, then the much deeper palaeolandscape evidence could be invaluable.

Partly, the potential of palaeolandscape research has been impeded by the limitations of vocabulary and concepts that can be translated into popular conventions about the past. Perhaps most frustrating, only a single word of a "palaeolandscape" unintentionally could reinforce the false notion of a monothetic past. Similarly, in the minds of many people today, archaeological sites 500 years ago, 2000 years ago, and 10,000 years ago are viewed as equally vague and distant from modern reality. Archaeologists routinely refer to numerous date ranges and time periods, and they may be disappointed when their findings are homogenised into summary views of ancient history, deep history, or "the past" as a singular entity.

Superficially in a modern view, the landscapes and events of recent history have retained more relevance today than would seem possible for ancient periods of prior centuries or millennia. This perspective tends to look at the past from the present, and therefore the more distant past seems to be less pertinent in the present day. In this view, the events of the middle through late AD 1900s may still affect business decisions and government policies today, but the societies of Medieval Europe or older periods have lost most of their influence by now. Continuing in this view, the last Ice Age could seem virtually incompatible with modern contexts, yet palaeolandscape studies can reveal the key points that are relevant for general knowledge.

Adding to the notion that ancient history has retained little or no relevance in the modern world, the older archaeological records do not point to the same kinds of details that can be seen in later written histories. A typical timeline of world history might begin with a few vague blocks of several thousands of years of pre-industrial or pre-written history, followed by increasingly refined detail during the last few centuries leading up to the present day. For those older time periods, broad time periods are constrained by the available stratigraphic layers and chronometric dating, such as when defining a palaeoshoreline of 1500 through 1100 BC. By comparison, the more recent periods with written records can refer to specific years or even to specific days, wherein historians of World War II would notice numerous unique contexts, all constrained within the few years between 1939 and 1945. Moreover, historians of World War II may look to the earlier decades of the 1900s in Europe and in Japan as potentially influential for the events of 1939 through 1945 in the Pacific Islands, but they probably would not consider the role of a palaeoshoreline of 1500 through 1100 BC beneath a World War II invasion beach.

In the view of landscape chronology, the past is viewed not from the present but rather through the perspective of the ancient contexts in their natural chronological order. In this view, the world has evolved continually through time, and each singular point or period in time has involved its own circumstances and conditions of landscapes or palaeolandscapes. For any of those identifiable portions of time, the surviving archaeological and environmental records could portray what a palaeolandscape looked like, how it functioned, and how it differed from the palaeolandscapes of older and younger time periods.

Concurrent time scales

Palaeolandscape chronologies can be complex due to the numerous natural and cultural components that act simultaneously yet at differing paces and magnitudes. Quite simply, some aspects of a palaeolandscape have changed more quickly or more dramatically than others. During a single measured time span, a palaeolandscape could exhibit stability or continuity in some of its parts, concurrent with transformation or change in other parts. For instance, the physical shapes of geological landforms may remain mostly stable through several thousands of years, while the forest composition, animal communities, and cultural land-use patterns in the same measured geographic space could change rapidly within decades or centuries.

With the notion of concurrent time scales, palaeolandscapes can be studied as complex systems, wherein longer time spans and more numerous lines of evidence can build more realistic portrayals of what happened through time. Logically, studies of longer time frames can account for more aspects of chronological change. Likewise, studies of more diverse material records can create more opportunities to notice the variations in patterns and trends.

Concurrent time scales have been relevant in most studies in palaeolandscape archaeology, as shown in the chapters of this book. Different intellectual perspectives and scopes of study, of course, have supported variable views of how concurrent time scales have operated. Despite the different viewpoints and examples, all of these studies can enhance our understanding of human-environment relations through time.

Scope of palaeolandscape studies

In this book, palaeolandscape studies involve diverse sets of substantive evidence, obtained directly from contexts of the past, within their measured geographic spaces and time periods of documented sites. The results can expose the various social-ecological components of the specified palaeolandscapes, effectively allowing the past contexts to speak for themselves. Those metaphorical voices are incomplete and fragmented today, but they can be deciphered through assembling the raw data of artefacts, food middens, preserved animal and plant remains, sedimentary layers, and other forms of evidence that have survived in palaeolandscape records.

Palaeolandscape research tends to be interdisciplinary, seeking whatever lines of evidence are available for learning about the past natural and cultural history of any given place of study. The broad scope of physical, biological, and social evidence may be compared with the diverse aspects of the historical and modern landscapes that people inhabit and experience today. Landscapes bear many meanings and implications, and palaeolandscape archaeology can explore the depth and breadth of this potential.

In terms of contextualising and interpreting the materials from ancient sites and palaeolandscapes, degrees of inference and subjectivity are often necessary, and the most convincing results always are based as closely as possible on the original primary evidence from tangible site records. For some rudimentary research goals, little or no interpretation may be needed. The basic factual evidence can depict what an ancient landscape looked like, where people lived, what foods they ate, and what durable tools and ornaments they used during the known time range of the study. For other research goals, steps of inference may be devised for interpreting the social and cognitive aspects of an ancient landscape, and in principle, these proposals could be tested against the reality of the supporting datasets.

The material basis of archaeology enables accurate descriptions of ancient sites and landscapes as a confident foundation for building meaningful interpretation or for testing theoretical notions. In this formulation, methodologies can clarify the logical links between the data-gathering techniques and the intellectual frameworks of testable hypotheses. The objects of study are the material evidentiary records of archaeology and palaeolandscapes, while the subjective qualities could involve any inferences or interpretations.

In this book's wide-ranging view of landscapes and palaeolandscapes, differing perspectives and approaches can be unified within a comprehensive framework of natural-cultural history. Most studies focus only on certain portions of this extensive field of potential, while all studies can be identified in terms of their specific contributions toward learning about palaeolandscapes in the broadest sense.

Structure and content of this book

This book aims to learn from palaeolandscapes in terms of clarifying what happened in the past and applying this knowledge into the future. Toward these goals, the book has been organised in three parts: (1) Two introductory chapters present the key themes and approaches in palaeolandscape research; (2) ten chapters offer the substantive details from intensive case studies; and (3) a conclusion chapter compares the case studies in a comprehensive framework, compatible with the key themes and approaches of the introductory chapters.

The two introductory chapters, including the present chapter, establish the scope of palaeolandscape research and the potential for learning from these studies of the past. Chapter 1 introduces palaeolandscapes in accessible terms for general readers, noting the operation of integrated natural-cultural histories through variable chronologies. Chapter 2 concentrates on unifying the diversity of evidence and approaches for studying palaeolandscapes toward identifying the potential contributions of research projects. Assorted aspects of these introductory chapters will be addressed, clarified, and expanded through the case studies in later chapters.

Ten chapters of case studies (Chapters 3 through 12) in total represent diverse geographic areas, time periods, cultural contexts, and aspects of natural and cultural histories in palaeolandscape research. While certainly more numerous case studies exist and could be included in further synthesis, the present examples offer a representation of the different perspectives and approaches that can be applied in palaeolandscape investigations.

Chapter 3 (by Andrea K. L. Freeman) reviews how palaeolandscape research has grown through the last decades of addressing questions about the migration routes of the first inhabitants of the American Continents. This long-enduring and broad-reaching topic necessarily has involved aspects of palaeolandscapes for conceptualising the possibilities of how people traversed waterways, coastlines, and environmental corridors near the end of the glacial conditions of the Pleistocene Epoch. Moreover, newer studies had begun to reveal substantive evidence from palaeolandscape contexts toward testing and refining the hypothetical models of what the ancient settings looked like, how people may have travelled through them, and more precisely when and where those events occurred.

Chapter 4 (by Ian Buvit, Karisa Terry, and Masami Izuho) draws from ancient stone tool assemblages and associated palaeolandscape records to depict the ancient contexts of the first communities who crossed Beringia into the Americas, just before 14,000 years ago. This approach compares the stone tool artefacts from relevant time periods in East Asia, Siberia, Beringia, and the Americas. The artefacts are considered in their measurable technical attributes, as well as in terms of their probable practical usage in ancient site contexts. The results enable a fuller sense of the patterns of past activities prior to 14,000 years ago, including the activities that substantiated at least one materially attested portion of the palaeolandscapes of the initial migration routes into the Americas.

Chapter 5 (by Lisa A. Maher, A. J. White, Jordan Brown, Felicia De Peña, and Christopher J. H. Ames) concentrates on the vital role of water sources and hydrology in supporting inhabitable landscapes, as shown in the case of people living in the Azraq Basin in eastern Jordan, since the end of the Last Glacial Maximum around 19,000 years ago. Several fluctuations can be traced in the water availability and degrees of aridification in different landforms and habitat areas of the Azraq Basin. Within this chronological narrative of the ancient water attributes, the archaeological evidence can be understood as reflecting how people adjusted their settlement locations and land-use traditions.

Chapter 6 (by Lindsey E. Cochran, Victor D. Thompson, and Bryan Tucker) develops multiple possible models for visualising the ancient landscapes and habitats where people may have lived along the coast of Georgia in North America during the last 5000 years. In the specified study area, the time range of the last 5000 years has disclosed material evidence from archaeological sites, associated terrain settings, and cultural practices of land-use patterns in variable areas of barrier islands, back-barrier islands, and estuaries that characterised the coastal context. The available lines of evidence supported different possible visualisation models, using approaches of a generalised linear model, a weighted overlay analysis, and suitability modelling based on fuzzy logic.

Chapter 7 (by Kara A. Fulton and David W. Mixter) considers the social and cultural factors that influenced how communities actively lived with their ancient landscapes of Actuncan in Belize during the last 3000 years. In this view, people continually have experienced landscapes through their thoughts and actions. This perspective can be applied for interpreting how people at different times during the last 3000 years made choices of where to live and how to relate with their landscapes in terms of their notions of self-sufficiency, perceived status, resilience, and other concepts.

Chapter 8 (by Yijie Zhuang and Shenglun Du) portrays the changing ecology and living environment of the Yangtze Delta in China over the last 10,000 years. This particular study highlights the role of a changing sea level in shaping and re-shaping the local environment in ways that created compound and complex results through time. Most notably, in this case study, people developed new ways of relating with their available landforms for varied economies of hunting, gathering, fishing, and farming at different times. The lessons from the Yangtze Delta are instructive for the many cases around the world and in all time periods that involved a change in sea level and subsequent effects in coastal morphology, habitat ecology, and other aspects of landscapes.

Chapter 9 (by Patrick D. Nunn, Elia Nakoro, Roselyn Kumar, Meli Nakunu, and Mereoni Camailakeba) examines how people adapted to a period of climate instability after AD 1300 specifically in the case of building and occupying hillfort settlements in Fiji. This period of unstable climate after AD 1300 offers important parallels with current and future global concerns of how to cope with the world's changing climate and sea level. Following an older context of predictable climate and weather patterns, people after AD 1300 witnessed significantly more dramatic and episodic weather events that interrupted their older traditions of sustainable land-use practices. Due to the pre-existing investment in optimal land-use patterns prior to AD 1300, the later reality of instability forced people to develop different modes of relating with their landscape, including the options of hillforts in Fiji.

Chapter 10 (by Ben Shaw) reviews the ancient landscape settings of the last several thousands of years in southern Papua New Guinea, in particular in the lowland coastal areas and nearby islands. Multiple lines of evidence refer to the ancient sea level, shapes of coastal landforms, forest composition, and placements of past cultural sites during key time periods, organised in chronological order. The archaeological records of the surrounding region extend nearly 50,000 years into the distant past when New Guinea was part of the larger landmass of Sahul that connected modern-day Australia and New Guinea. The later periods, generally within the last 5000 years, have revealed more numbers of sites in locations that can be accessed through traditional land surveys and exploratory excavations. A holistic and long-term view, though, supports a regional chronology of the changing palaeolandscape setting, and then the known archaeological evidence can be situated within this framework.

Chapter 11 (by Glenn R. Summerhayes) considers the ancient sites and landscapes along the north coast of New Guinea, New Ireland, and New Britain during the time of the region's initial pottery-bearing horizon following 3500 years ago. Within the span of a few centuries, people in this region created numerous sites in ancient coastal zones that since then have been transformed due to changes in sea level and coastal morphology, and therefore people may have inhabited several spots that have not yet been detected in the present-day archaeological record. Those few centuries coincided with the region's first pottery-bearing site horizons, including the distinctive Lapita style of dentate-stamped pottery among a diverse repertoire of artefacts, food remains, choice of settlement locations, and other cultural traits. A fuller understanding of the palaeolandscape context now can portray more realistically where people lived, what habitat ecologies they experienced, and generally how they related with their coastal environments.

Chapter 12 (by Mike T. Carson and Hsiao-chun Hung) investigates the changing palaeolandscape settings of the last 3500 years at the world-famous monumental site known as the House of Taga in Tinian of the Marianas, western Micronesia. Beneath the surface-visible stonework ruins of the House of Taga, multiple subsurface site layers extend back to the time when people first inhabited the remote-distance islands such as Tinian in the northwest tropical Pacific. The research here revealed a continuous chronological sequence of how people adapted to their changing coastal environment, from the time of first regional settlement through the time of the site's iconic stonework ruins of the AD 1600s.

The book concludes with Chapter 13, evaluating what has been learned from the ten case studies of palaeolandscape research toward improving the present and future relations between people and their inhabited environment. This cross-comparison approach begins with identifying the overall consistencies and inconsistencies in the case studies, and then the approach continues with reflexive evaluation and interpretation, based on the factual evidence as presented. Comparisons of the book's examples help to distinguish what was sustainable for people in the past when they faced challenges in their palaeolandscapes, with direct parallels in modern issues of managing water sources, food security, population densities, and investments in land-use policies. In this fashion, the book ends with a holistic evaluation of the lessons for the past and future, noting what further steps may yet be productive.

2 Potential contributions of palaeolandscape archaeology

Building strength through diversity

Mike T. Carson

Palaeolandscape archaeology can be approached through many perspectives, datasets, and interpretive lenses that all address questions about what ancient landscapes looked like, how people lived in these places, and how people and their environments mutually affected each other through time. If these different studies can be coordinated in a holistic framework, then the results may build a robust understanding of palaeolandscapes in the broadest sense. In contrast, if the studies are compartmentalised as separate specialisms, then the results can render palaeolandscape archaeology as a vague concept, lacking meaningful definition and applicability.

Palaeolandscapes, like their counterparts in historical and modern landscapes, could hold many different meanings. Throughout human evolutionary history, people have inhabited and interacted with landscapes, and therefore landscapes are fundamental in the human experience. Virtually every study of human society and behaviour can relate to landscapes in one way or another. By extension, almost every study of the past can relate to palaeolandscapes.

Landscapes and palaeolandscapes present innumerable open-ended possibilities, thereby prompting Gosden and Head (1994) to refer to a landscape as a "usefully ambiguous concept". In this view, the concept of a landscape is flexible, and no singular definition would be possible. Rather, multiple different and sometimes contradictory definitions co-exist about what landscapes are and how they can be studied. Nonetheless, the vagueness or ambiguity of a landscape may be regarded as useful for supporting a richness of approaches and interpretations of an inherently complex field of research.

A minor semantic nuance is offered here, noting that the concept of a landscape or palaeolandscape is broadly accepting rather than ambiguous. In this broadly accepting view, landscapes can be defined as encompassing any and all of the physical, ecological, and social aspects of places. Palaeolandscapes could extend this notion through long chronological sequences of the past, in essence accommodating diverse studies in the natural and cultural histories of landscapes through time.

In recognition of the capacity of archaeology to examine landscapes through multiple lines of evidence and across long time spans, Anschuetz

DOI: 10.4324/9781003139553-2

et al. (2001: 192) concluded that archaeology as a discipline was poised to "unify this truly interdisciplinary approach into a coherent whole". This goal of a unified approach would be effective for studies of ancient landscapes as complex natural-cultural systems, for instance, when posing questions about how and why landscapes have evolved through time. On the other hand, specific questions about social behaviours within landscapes or about cognitive perceptions of landscapes still would be open to variable philosophical perspectives and schools of thought. To be fair, Anschuetz et al. (2001: 192) aimed for a landscape paradigm that could understand "the anthropology of place" in the interpretive terms of social behaviour and cognitive aspects, beyond the physical limits of archaeological material evidence.

Differences of opinions about landscapes underscore the far-reaching potential of the concept, and most realistically, "landscapes must be understood in the broadest sense as inhabited social-ecological environments that can be studied in multiple ways" (Carson 2016: 3). Archaeology may provide a central focus for studies about ancient landscapes and about chronological sequences of palaeolandscapes, but ultimately these studies must involve interdisciplinary efforts that can account for the diverse lines of relevant evidence. One such unified approach was presented in an example of "archaeological landscape evolution" (Carson 2016), and much of the same approach applies in the present book about "palaeolandscape archaeology".

Toward building an inclusive framework for the diverse field of palaeolandscape archaeology, the present chapter considers two major questions. First, how can landscapes and palaeolandscapes be defined in accordance with different intellectual perspectives, approaches, and methodologies? Second, what are the major categories or lines of evidence that can reveal useful information about palaeolandscapes? After answering those questions, the potential contributions of palaeolandscape archaeology can be clarified more effectively.

Scope and perspectives of palaeolandscape archaeology

A discussion of palaeolandscape archaeology first requires clarification of what a landscape is and how it can be studied. Today, people can consider any of the details and individualised variables in the landscapes where they live, thus supporting numerous perspectives that potentially could be applied when studying the preserved archaeological remains of ancient landscapes.

A "landscape" blends objective physical properties with subjective experience and interpretation, such as happens whenever a person views the scenery of a place. The scenery can be perceived and sensed objectively, and this action can create opportunities for developing ideas and interpretations. This concept has enabled paintings, illustrations, and photographs of landscapes not only as descriptive records of those places but also as the basis for supporting further thought and contemplation about those places.

Variable degrees of objectivity and subjectivity co-occur and potentially could influence each other whenever a person perceives, experiences, or contemplates a landscape or anything else. Any of those views could be applied in studying a landscape, and the resulting approaches tend to be represented as versions of scientific objectivity, self-reflexive subjectivity, or combinations of objective and subjective aspects. Extreme views about pure objectivity or pure subjectivity can be rejected as unrealistic.

A landscape can be observed, recorded, experienced, and conceptualised. Accordingly, studies of landscapes may involve any of the possible ways of perceiving and interacting with a place. Moreover, philosophical issues may be considered about the human capacity for thinking objectively or subjectively about a landscape or about anything else. All of these approaches potentially can generate relevant information for clarifying what a landscape is, how it can be studied, and what it could signify or imply for different people.

When acknowledging landscapes as places where people actively live, potential research about those landscapes would need to accept social and behavioural sciences, including archaeology, among many other fields. The same landscape may be studied through differing perspectives, datasets, and interpretations of cultural geography, history, or sociology. Likewise, individual people can live in the same landscape with each other, but they can experience the place uniquely and develop variable new perspectives and relations with the landscape through their lifetimes.

The social and behavioural aspects of actively experienced landscapes have shaped much of the intellectual framing of how to define and study landscapes in archaeology, as reviewed by Julian Thomas (2012). In particular, perspectives in social anthropology have been greatly influential in developing frameworks about how people live in the world and interact with landscapes, for instance, as conveyed in a series of essays by Tim Ingold (2000). These ideas, along with many others, have supported a richness of approaches for examining landscapes in archaeology, as depicted in a collection of more than 60 essays and case studies (David and Thomas 2008).

The notion of a socially inhabited landscape can be most effective when recognising the mutual effects between the natural ecology and the social setting of a complex landscape. For instance, the given placement of water sources, usable landforms, and other factors of the natural environment can affect the choices of where people live, and simultaneously the cultural behaviours within those environments can influence the quality of water sources, erosion or purposeful re-shaping of landforms, and compositions of local plant and animal communities. These factors may continue to affect each other in repeated feedback cycles, extended through multiple generations and variable time spans.

Inter-connectedness within a landscape ecosystem has been a central principle of landscape ecology (Turner et al. 2001), wherein a change in one part can affect the other parts. All of the contributing components, therefore, are integral to the operation of a healthy ecosystem. This perspective could accommodate human beings and social groups as active parts of a social-ecological landscape, as has been stressed in fields of cultural ecology and human ecology.

The inter-connectedness between people and their natural worlds conceivably could develop in complex webs of interdependent relationships in a landscape system. In this view, the "natural world certainly has influenced much of human history, just as human actions have affected much of the natural world, inter-linked to such a degree as to create inter-dependence" (Carson 2016: 7). This process occurs, for example, when people and dogs influence each other in a co-domestication or co-evolutionary relationship. Similar symbiotic or synergistic processes could apply between people and their surroundings, forests, farmlands, and other components of their landscapes. Multiple such relationships can co-exist in a single landscape system.

If landscapes are recognised as integrated natural-cultural systems, then neither natural history nor cultural history alone can account for their operations through time. By extension of this logic, neither environmental determinism nor cultural determinism can account for the complexities of landscapes. Patrick Nunn (2003) clarified that physical environments present venues of possibilities for people to act, react, and adapt with the potential for affecting those same environments. In this view, the natural and cultural components of a landscape continually affect each other through time in overlapping feedback cycles, and therefore no single force can be isolated as a consistent determining factor for all of the processes that occur in a landscape system.

The prevailing schools of thought in social sciences have emphasised human agency to such a degree that cultural, historical, and archaeological landscape studies often have become reifications of human-caused impacts in landscapes. Many studies in environmental archaeology and historical ecology are designed to quantify human-caused impacts through proxies of counts of animal bones and other indicators, and the results by default advance a one-sided portrayal of people as the only meaningful factors in human-environment relations. These fallacies can be thwarted by rational thought, and they will not be illustrated in detail here, as a full discussion would distract from the key point of promoting an integrative approach.

In principle, palaeolandscape archaeology could apply a liberal scope of landscape ecology across a chronological dimension, but a long-term sense of chronology awkwardly has been missing or overlooked in most studies of modern landscape ecology. Typically, landscape ecology concentrates on the observable activities of living systems, documented within time spans that archaeologists would regard as narrow slices of time of a few years or decades. These studies clearly produce detailed information about the systemic operations of landscapes, yet palaeolandscape archaeology could extend the view across numerous consecutive centuries or millennia.

Within the temporal scopes of most landscape ecology studies, a landscape can change through processes that occur within defined geographic spaces (Cumming 2011), and all of a landscape's events and processes occur through measurable periods of time (Dearing et al. 2006a, 2006b). The detectable periods of time traditionally have been narrow or shallow in landscape ecology studies due to the practicalities of recording observations within living ecological systems, but nothing would exclude the possibility of accommodating datasets from older time periods. Any relevant datasets from more ancient time periods must derive from other fields of study, such as archaeology and palaeoenvironmental investigations, here described as palaeolandscape archaeology.

Longer time depths of study naturally can detect more individual instances of events and processes of chronological change in a landscape system. This obvious fact deserves mentioning because of the ability to produce more extensive raw datasets and to identify more instances of correlation, cause, and effect through long time scales. In shorter time scales and within some modern-day perspectives, landscapes may appear to be mostly stable or tending to seek internal equilibrium. In longer time scales of palaeolandscapes, however, chronological change through time can be a central research theme.

Within any slice of time, a landscape does not simply occur as a pre-existing entity, but instead, it already has evolved through a series of events and processes of chronological change. In this view, at every point in time, a landscape system operates as the result of the events and processes that already have occurred during prior time periods of natural and cultural history. Those prior time periods may be regarded as encompassing a series of palaeolandscapes, and each relevant window in time can be documented through the surviving physical evidence in archaeological sites, geological formations, and other records.

In order to learn the facts of what happened in a chronological series of palaeolandscape periods those individual time frames of palaeolandscapes need to be identified and recorded within the limits of their detectable contexts. Those contexts must have existed within measurable geographic areas and time periods, although the evidence from each such context can be variable and diminished by natural deterioration and other effects through time. In any case, though, the geographic and temporal boundaries of a context can be identified objectively, and next, those contexts can be populated with whatever lines of evidence have been preserved.

Documentation of an ancient context is not by any means a novel approach in archaeology, and indeed this scope of research continues the long-established practice of geoarchaeology as a form of extending human ecology into the distant past (Butzer 1982; Wilson 2011). While geoarchaeology applies principles and methods of geology toward addressing questions about archaeology, the results tend to situate archaeological findings within the contexts of ancient geological landforms and other aspects of ancient environments. Notably, in this view, an ancient context of human ecology is largely synonymous with an ancient context of landscape ecology, with an emphasis on the activities of people within a larger ecological setting.

Palaeolandscape archaeology shares much in common with broader perspectives and approaches about studying human-environment relations. Allied fields have been proposed as historical ecology (Crumley 1994), intersections of anthropology and environment (Crumley 2001), environmental history (Hornborg et al. 2007), and general human-environment interaction studies (Dearing et al. 2006a, 2006b). These fields of research do not necessarily involve archaeology as clearly as is the case with geoarchaeology, as noted, but they could be applied in archaeology.

Linkages between broad ecological approaches and archaeology clearly are viable, yet only rarely have they been formalised in terms of applying techniques, methods, and theories. In this respect, van der Leeuw and Redman 2002 outlined how archaeology could be situated within general studies of social-ecological settings. Similarly, Briggs et al. (2006) promoted the potential contributions that archaeology and ecology could offer for each other.

An intellectual framework of "niche construction" resonates with the current understanding of how landscape systems operate and continually shape or re-shape the possibilities into the future. "Niche construction" posits that people live in ecological niches, adapt with them, and thereby affect the outcomes of those niches for future generations (Odling-Smee et al. 2003). This framework has been examined formally in terms of how it can be applied in archaeology (Laland and O'Brien 2010; Smith 2015), and the founding principles mostly agree with the traditional approach of geoarchaeology for studying human ecology across long periods of time (Butzer 1982).

The current proposal of palaeolandscape archaeology aims for a widely accepting accommodation of perspectives and lines of evidence about ancient landscapes toward a unified approach. Accordingly, the chapters in this book offer diverse examples of studying ancient sites within their natural and cultural contexts, and they align with a range of studies in geoarchaeology, environmental archaeology, historical ecology, niche construction studies, and other fields.

Lines of evidence in palaeolandscape archaeology

Potential lines of evidence about palaeolandscapes first must be understood in terms of how they actually relate with the ancient contexts that no longer exist today as living entities. Useful pieces of evidence may have survived in archaeological sites and in various records of the past environment. While modern landscapes can be observed in their active use and operation, palaeolandscapes can be observed only through their preserved vestiges and relicts, and those findings next could be interpreted through variable perspectives and methodologies.

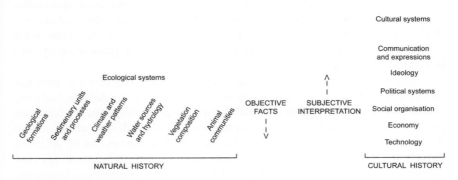

Figure 2.1 Schematic of lines of evidence in palaeolandscape archaeology.

Palaeolandscapes encompass all of the past time periods and contexts of natural and cultural histories of places (Figure 2.1). Many studies concentrate on illustrating the chronological order of palaeolandscape settings through time, owing in large part to the established traditions of combining archaeology with geology and other disciplines that obtain material datasets from specific contexts. Other studies concentrate on interpreting the social behaviours and cognitive perceptions of people who lived in ancient landscapes, for instance, when applying knowledge and experience of modern living landscapes into the research of ancient sites.

While the natural and cultural aspects of palaeolandscapes are deeply connected, they differ from each other in at least two fundamental ways that affect the ability to gather and interpret primary datasets. First, the natural world potentially can exist without people in it, but the cultural world cannot exist without a natural environment. Second and related to the first point, evidence from natural history tends to be more objective, while evidence from cultural history could vary in degrees of reasonable objectivity through extreme subjectivity.

Evidence about the past natural setting can be obtained not only from archaeological sites but also generally from palaeolandscape records. In this sense, evidence from natural history can be obtained from a much broader sampling of the ancient world, for example, from lake-bottom sediments and from minerals inside cave speleothems, and the resulting records from individual time periods can be coordinated with the findings from comparable ages of archaeological sites. Moreover, evidence from natural history can extend much farther back in time, for instance, revealing the shapes of landforms and the range of plant and animal species that existed prior to the appearance of anatomically modern humans in some particular regions of the world.

Natural history records of palaeoenvironments may relate with: (a) geological formations; (b) sedimentary units and processes; (c) climate and weather patterns; (d) water sources and hydrology; (e) vegetation composition; and (f) animal communities. In terms of linking these lines of evidence with archaeology and ancient landscapes, several compartmentalised specialities have developed as geoarchaeology, palaeoclimatology, palaeohydrology, archaeobotany, archaeozoology, and others. Additionally, the cross-connections among those lines of evidence may be pursued through palaeoecology, palaeogeography, and other fields.

Cultural history evidence in archaeology involves the durable material remnants that have survived through time in the forms of artefacts, features, and sites. The most straightforward and uncontested conclusions are based on physical evidence, such as when observing geological landforms or describing the technical qualities of stone tools. The technical properties can be observed directly or ascertained confidently, while other aspects of ancient life and society would require variable degrees of interpretive methodology. Questions about past social structure, politics, ideology, or cognitive functions tend to involve considerable intellectual frameworks as hypothetical models.

Based on the material records from palaeolandscapes, interpretations of those past contexts could range from basic descriptions of the physical environment through hypothetical models of cognitive perceptions. In this regard, Christopher Hawkes (1954) proposed a "ladder of inference", wherein the most convincing statements were about the physical formation processes of sites, and then increasing degrees of inference were needed for studies of past subsistence economy, followed by questions about the social and political institution, and reaching the limits of logical inference in topics of religious and spiritual life.

Numerous schematics have been proposed about the categories of research questions that can be examined through archaeological records of the past. Mostly, these schematics have built on the notions as outlined by Hawkes (1954), and here a general-utility framework may be proposed in categories of: (a) technology; (b) economy; (c) social organisation; (d) political systems; (e) ideology; and (f) communications and expressions. In this schematic, questions about technology and economy are the easiest to address through basic observable facts, and the next topics would require progressively more complicated interpretive methodology and proposals about testable hypotheses.

All of the components of natural and cultural history, as shown in Figure 2.1, could be studied individually or in their possible interconnections. Moreover, they all could exist in any given place and time period of a palaeolandscape or modern landscape. In these respects, the general-utility schematic is useful as a baseline reference for identifying what is being studied and how the scope of research relates to other possibilities.

Potential contributions of palaeolandscape archaeology

As defined here, a diverse scope of palaeolandscape archaeology can produce an impressive breadth of potential new contributions, both substantively and theoretically. If the field of study were to be restricted only to some lines of evidence while excluding others, then the potential contributions would be accordingly narrow. Similarly, if the intellectual scope were to focus only on particularistic material datasets or else only on new theoretical ideas, then again, the potential contributions would be limited. In all of these respects, diversity creates strength in research.

Palaeolandscape archaeology can clarify the factors that made some landscape systems sustain effectively through their defined periods, while other systems failed under certain conditions. Furthermore, the research results can generate new ideas and testable hypotheses. Toward these goals, a liberal view of the research field can be beneficial in at least three ways:

1. With more diverse lines of evidence, palaeolandscape research more confidently can portray what made a landscape system sustainable or unsustainable during specific circumstances.
2. With greater time depth of study, palaeolandscape research can account for more numerous instances of changing conditions in the natural and cultural environment through time.
3. With a broader sense of the relevant intellectual ideas and perspectives about landscapes generally, any site-specific findings can be interpreted through frameworks that accommodate the lines of evidence that happen to be discovered.

With the ability to study more of a landscape's natural and cultural variables through longer periods of time, the results can expose more confidently how those variables were interrelated. The inter-relations can be noted within single units of time, as well as across continuous chronological sequences of multiple time periods. The patterns in the datasets can be examined for positive and negative correlations, as well as for random associations. All of these findings would be informative about the functional relations among the diverse natural and cultural components of a palaeolandscape system.

Correlations among variables always are potentially instructive and interesting, but a correlation alone does not prove a causal relationship. For instance, a lowering of sea level in the Mariana Islands after 1100 BC was correlated with a change in artistic expressions in decorated pottery and shell ornaments, but this correlation did not substantiate a direct causal link between sea level and artistic output. Rather, other information was gathered about how people responded to the change in sea level generally (Carson 2016: 133–182), and this larger context of a cultural response accounted more holistically for the change in the styles of pottery and shell ornaments after 1100 BC.

Although palaeolandscape archaeology most directly produces knowledge about the past landscapes of the given site-specific or region-specific case studies, such as in the above-noted example in the Mariana Islands, those lessons could be applied toward issues of global relevance, such as in global concerns of changing sea level. The example in the Mariana Islands generated a new understanding of how people in this particular region of Pacific Oceania responded to a change in sea level after 1100 BC. In a broader perspective, however, a similar cultural response to the regional-scale of change in sea level was documented in numerous other cases of the west through central Pacific Oceanic islands over the next centuries (Nunn and Carson 2015). Moreover, the factor of a changing sea level clearly can be recognised as a driving determinant in the functioning of landscape systems generally around the world and in whatever time periods when these events might occur.

Palaeolandscape archaeology accomplishes much more than simply verifying that people tend to respond to any measurable change in sea level. More specifically, these studies reveal the degree and rapidity of change in the environment that triggered a cultural response in different situations around the world. Additionally, the cultural responses in each case had resulted in the continuation of some cultural traditions and use of resources, along with loss or alteration of other aspects of the larger social-ecological landscape system. In other words, aspects of both continuity and transformation co-occurred whenever people lived through a major environmental change, and all of this information in total can clarify what happened in each case under different circumstances.

By leveraging the lessons from real-life examples in the past, palaeolandscape archaeology can identify the past actions and choices that supported adaptations during and after periods of known environmental change, such as in the case of changing sea level. As a practical matter, greater numbers and diversity of case studies can strengthen the ability to identify the overall consistencies versus inconsistencies in the worldwide past records of cultural response.

The example about change in sea level can be understood as just one avenue for palaeolandscape archaeology to provide lessons for the past and for the future. Several other avenues could be pursued, such as concerning the management of water resources, the ability to cope with changing climate, and other issues as explored in the next chapters of this book. Lessons from the past are informative in their own right for learning about ancient time periods and contexts, and this knowledge can be applied toward solving current and future management of landscape issues in the broadest sense.

References

Anschuetz, Kurt F., Richard H. Wilshusen, and Cherie L. Scheick, 2001. An archaeology of landscapes: Perspectives and directions. *Journal of Archaeological Research* 9: 157–211.

Briggs, John M., Katherine A. Spielmann, Hoski Schaafsma, Keith W. Kintigh, Melissa Kruse, Kari Morehouse, and Karen Schollmeyer, 2006. Why ecology needs archaeologists and archaeology needs ecologists. *Frontiers in Ecology and the Environment* 4: 180–188.

Butzer, Karl W., 1982. *Archaeology as Human Ecology*. Cambridge University Press, Cambridge.

Carson, Mike T., 2016. *Archaeological Landscape Evolution: The Mariana Islands in the Asia-Pacific Region*. Springer International Publishing, Cham, Switzerland.

Crumley, Carole (editor), 1994. *Historical Ecology: Cultural Knowledge and Changing Landscapes*. School of American Research Advanced Seminar Series. School of American Research Press, Santa Fe.

Crumley, Carole (editor), 2001. *New Directions in Anthropology and Environment: Intersections*. AltaMira Press, Walnut Creek, California, USA.

Cumming, Graeme S., 2011. Spatial resilience: Integrating landscape ecology, resilience, and sustainability. *Landscape Ecology* 26: 899–909.

David, Bruno, and Julian Thomas (editors), 2008. *Handbook of Landscape Archaeology*. Left Coast Press, Walnut Creek, California, USA.

Dearing, J. A., R. W. Battarbee, R. Dikau, I. Larroque, and F. Oldfield, 2006a. Human-environment interactions: Learning from the past. *Regional Environmental Change* 6: 1–16.

Dearing, J. A., R. W. Battarbee, R. Dikau, I. Larroque, and F. Oldfield, 2006b. Human-environment interactions: Toward synthesis and simulation. *Regional Environmental Change* 6: 115–123.

Gosden, Chris, and Lesley Head, 1994. Landscape – a usefully ambiguous concept. *Archaeology in Oceania* 29: 113–116.

Hawkes, Christopher. 1954. Archaeological theory and method: Some suggestions from the Old World. American Anthropologist 56: 155–168.

Hornborg, Alf, J. R. McNeill, and Joan Martinez-Alier, (editors), 2007. *Rethinking Environmental History: World-system History and Global Environmental Change*. AltaMira Press, Latham, Maryland, USA.

Ingold, Tim, 2000. *The Perception of the Environment: Essays on Livelihood, Dwelling and Skill*. Routledge, London.

Laland, Kevin N., and Brien Michael J. O, 2010. Niche construction theory and archaeology. *Journal of Archaeological Method and Theory* 17: 303–322.

Nunn, Patrick D, 2003. Nature-society interactions in the Pacific Islands. Geografiska Annaler 85–B:219-229.

Nunn, Patrick D., and Mike T. Carson , 2015. Sea-level fall implicated in profound societal change about 2570 cal yr BP (620 BC) in western Pacific Island groups. *Geo, Geography and Environment* 2: 17–32.

Odling-Smee, F. John, Kevin N. Laland, and Marcus W. Feldman, 2003. *Niche Construction*. Monographs in Population Biology 37. Princeton University Press, Princeton, New Jersey, USA.

Smith, Bruce D., 2015. A comparison of niche construction theory and diet breadth models as explanatory frameworks for the initial domestication of plants and animals. *Journal of Archaeological Research* 23: 215–262.

Thomas, Julian, 2012. Archaeologies of place and landscape. In *Archaeological Theory Today*, 2nd Edition, edited by Ian Hodder, pp. 167–187. Polity Press, Cambridge.

Turner, Monica G., Robert H. Gardner, and Robert V. O'Neill, 2001. *Landscape Ecology in Theory and Practice: Pattern and Process*. Springer, New York.

van der Leeuw, Sander, and Charles L. Redman, 2002. Placing archaeology at the center of socio-natural studies. *American Antiquity* 67: 597–605.

Wilson, Lucy, 2011. The role of geoarchaeology in extending our perspective. In *Human Interactions with the Geosphere: The Geoarchaeological Perspective*, edited by Lucy Wilson, pp. 1–9. Special Publication 352. Geological Society of London, London.

3 25 years of geoarchaeological research on Paleoindian landscapes

A look back at the discipline

Andrea K. L. Freeman

An important theme in the conduct of Paleoindian archaeology is the convergence of archaeological, biological/environmental, and earth science data (see Holliday 2021 for a short but impactful discussion of this history). Landscapes that hold these three components provide rich sources of information about the past and help us construct sensible narratives about how the earliest occupants of North America entered, lived, and dispersed over that landscape. In 1995, Michael Faught and I held a symposium at the Society for American Archaeology Annual Meeting in Minneapolis titled "Paleoshorelines and Channel Margins". The session brought together people working principally in eastern North America to discuss what we understood about the context of Paleoindian sites and prospects in offshore, nearshore, and stream adjacent contexts (Freeman and Faught 1995). Having seen the benefits of decades of studying stratigraphy and chronology in the west, we thought that an examination of context would lead to new discoveries. This chapter examines what we have learned in the 25 years since that session. Looking broadly at geoarchaeological research of Palaeoindian sites across North America, the last quarter century has seen an expansion of contexts and the information that can be gathered from those contexts.

The dynamic geomorphic environments, many of them containing records extending into the Pleistocene, were often the source of stratified deposits, integral at separating Clovis deposits from both those that may precede them and those that followed. At the time, the record for Paleoindian sites in western North America seemed rather clear, however, one area lacking good stratigraphic and chronological control was eastern North America. There also appeared to be mounting evidence that offshore contexts in the eastern part of the continent held untapped resources (a reasonably thorough discussion of offshore archaeology at the time can be found in Garrison and Cook Hale 2021).

Although the 1995 symposium was focused on eastern North America, this paper will expand on how the general problems faced by Palaeoindian archaeology in eastern North America also were a symptom of barriers to knowledge that existed more pervasively. While fundamental geoarchaeological techniques are still being used, new methods and techniques have

DOI: 10.4324/9781003139553-3

emerged over the last 25 years that have allowed us to probe the same types of data with greater depth and precision. Two central themes emerged from the Minneapolis session: (1) We needed to expand methods in order to achieve the goals we were looking for; and (2) we needed to explore unconventional locations in order to access a greater diversity of data. My focus here lies in how addressing these two concerns over the last 25 years have advanced Palaeoindian archaeology, but the overall changes are something that we see across studies of palaeolandscapes from many different archaeological periods and areas (Killick and Goldberg 2009).

These two major themes could not have emerged were it not for the expansion of mental models or theories regarding the earliest peoples in the New World. Over the last 25 years, new theories have been advanced (Bradley and Stanford 2004; Stanford and Bradley 2000) and slightly older theories examined more intensively (Fedje et al. 2011; Freeman 2016; Pedersen et al., 2016; Potter et al. 2018). Although there is still a deep, underlying concern with when and from what route people came into North America, there has also been significant focus on finding new evidence and expanding understanding of old evidence. The last 25 years have seen fruitful research across North America, east to west and north to south. The Pleistocene – Holocene boundary is complex and so understanding the diversity of landscape data across that shift in the climate should be the first priority of all scientists working in the period. Hanging hope on any one theory of how people interacted with that landscape is immaterial without such evidence.

Methodological and technical advancements in palaeoindian landscape studies

When discussing palaeolandscapes from a physical or geoarchaeological point of view, there is a heavy reliance on converging data from biology, geology, and archaeology. Historically, progress is made when any limits to observational or analytical capacity are overcome. A very high priority was initially placed on refining chronologies in Palaeoindian archaeology (Haynes 1971; Taylor et al. 1996; for a history of Haynes' work at Palaeoindian chronology, see Holliday 2000), while large mammal hunting and stone tools also played a significant role in creating narratives about how people used the Palaeoindian landscape. The notable focus on chronology was largely due to the conflict over the earliest evidence for peopling the Americas, which began with Folsom and culminated in the visit to Monte Verde (Meltzer et al. 1997, Meltzer 1997). Intensive study of sites like Clovis and Folsom was followed with the scrutiny of the dates themselves (Waters et al. 2020), but this could not happen without greater stratigraphic scrutiny (which I will return to). Stratigraphic scrutiny would require that we understood the depositional and post-depositional environments from which samples were recovered a bit better. In this section, I focus on two themes that have emerged over the last 25 years: (1) The use of archaeological micromorphology to improve contextual information at the microscopic level; and (2) the employment or development of methods that could better inform about the environmental setting of the late Pleistocene through stable isotopes, phytoliths, phylogeography, and ancient or environmental DNA.

The emergence of archaeological micromorphology

Soil micromorphology provides information on the microscopic context of materials, using thin sections recovered from deposits collected in the field without disturbing their associative properties (Courty et al. 1989). When employed in archaeology (hereafter archaeological micromorphology, cf., Mentzer and Quade 2013), it can be used to demonstrate the relationships among archaeological features and microartifacts, sediments, pedogenic features, samples, etc., providing additional information on the depositional and post-depositional sequences of events that occurred at an archaeological site.

Soil micromorphology had been around for nearly 50 years before it became heavily applied in archaeology (Karkanas and Goldberg 2017). Goldberg and Aldeias (2018) blame both producers and consumers for the lacklustre interest in its application. Certainly, early issues with consistent terminology (as with the study of phytoliths, further explored below) created barriers requiring direct training that may be found beyond the boundaries of a graduate program (Holliday, 2021). However, the increase in both willing instructors (like Paul Goldberg and then Susan Mentzer) and willing students, as well as compelling problems in Palaeoindian archaeology, created enhanced interest in applying these techniques.

Some examples of the successful application of micromorphology include Goldberg and Arpin's (1999) study of potential sources of carbon contamination (or lack thereof) at Meadowcroft Rockshelter, Harris-Parks (2016) examination of variability in late Pleistocene-aged black mats across the American Southwest, and the examination of dark brown "stringers" that have often been described as weakly-developed soils in Alaskan loess (Kielhofer et al. 2020). Each of these studies demonstrates the power of micromorphology for resolving long-standing contextual, environmental, and chronological issues at Palaeoindian sites; these represent only a handful of studies that have been employed over the past 25 years. When micromorphology has been effectively used to describe and interpret important stratigraphic concerns in Palaeoindian archaeology, it is obvious that the technique not only holds promise but also should be employed more widely.

The power of microcontextual techniques does not end with the application of micromorphology. Importantly, such studies can not only be used to understand the context of environmental and chronological materials at archaeological sites but to precisely pinpoint measurements of such materials (Mentzer and Quade 2013). Goldberg and Aldeias (2018) list the other microcontextual techniques that have emerged in the last two decades, including FTIR, µXRF, etc. Rather than delve into these techniques in this paper, I focus on the use of two other techniques geared toward the biological environment: Stable carbon isotopes and phytoliths.

The use of stable isotopes

Although mass spectrometry was developed decades earlier as a scientific method, the demonstration that modern accelerators could accelerate and separate radioactive and stable isotopes of an element in a specimen (Muller 1977) revolutionised the use of both radioactive and stable isotopes in archaeology. The modern particle accelerator allowed scientists to directly measure radioactive carbon from smaller samples, enhancing both the precision and accuracy of age estimates. It also provided a method to detect the ratio of the stable isotopes of carbon in very small samples, and thus the metabolic pathway used by the plant specimen from which the radiocarbon age estimate was derived. In addition to providing a correction for the age of the sample specimen, the metabolic pathway could be used to reconstructing vegetation communities (Nordt 2001: see also Quade, Cerling, and Koch 1998).

While the use of other stable isotopes (H, S, N, O) can be beneficial in a variety of circumstances that also hold great promise for understanding paleoenvironments of the late Pleistocene and early Holocene (including offshore and other aquatic environments), I wish to focus on the common use of stable carbon isotopes as a way of understanding the terrestrial ecosystem. Although such determinations can also be achieved by examining the stable carbon isotope ratios from mammalian bone and teeth, my focus here is on the use of stable carbon isotopes preserved in soils at Palaeoindian sites.

The technique for examining changes in plant community composition works well in semi-arid to sub-humid climates exhibiting rapid and profound shifts (Nordt 2001). Because the ratio of C3 and C4 plant composition at the end of the Pleistocene exhibited significant variability, the technique has been particularly useful in the Great Plains (Fredlund and Tieszen 1997), where most of the classic Clovis sites have been located. Such techniques for reconstructing the environment at Palaeoindian sites are almost routinely a part of geoarchaeological sampling today (Cordova et al. 2011; Cyr et al. 2011; Mandel et al. 2014). These analyses need not be considered mere snapshots or windows of past landscapes. Shifts in plant community composition or the timing of those shifts can vary across different landscapes. In fact, Loren Davis (2001; Davis et al. 2002) employed stable carbon isotopes along the Lower Salmon River of Idaho at a number of archaeological sites, one of which is Cooper's Ferry, demonstrating that important asynchronous changes may occur in riparian zones.

Finally, the study of stable carbon isotopes in soils and archaeological deposits can help determine alternate pathways (carbonate deposition) for carbon entering the archaeological record as well as environmental effects on all plants, but in particular C3 plants (Fiorentino et al. 2014). Additionally, Mentzer and Quade (2013) have expanded the use of multiple proxies by developing better methods for pairing micromorphology and stable carbon isotopes. Although not typically considered microstratigraphic techniques, the employment of stable carbon isotopes alongside the analysis of phytoliths can be a powerful tool.

Use of phytoliths

For decades, environmental data for the Palaeoindian period has been primarily developed using pollen sequences. Rich, sometimes annual, records are produced from lakes and palaeolakes across the Americas. Pollen, a microscopic gametophyte of the plant, preserves well in these environments. Species-level identification can occur with pollen, and so the preservation of thousands of grains within lake sediments can provide plant community profiles for large regions. Both the breadth and specificity of data acquired from pollen is relatively well-known as scientists have accumulated decades of data on the deposition of each form of pollen, its preservation, and abundance. On the other hand, phytoliths form when silica replaces the cellular structure of parts of a plant and is incorporated into the soil. Because many different plants can have the same or similar cellular forms and because one plant can produce many different types of phytoliths, the specificity of identification of plants is limited. So why then would a record as taxonomically coarse-grained as that provided by phytoliths be important?

Like pollen, phytoliths can be preserved in deposits for long periods of time (Blinnikov et al. 2001), but unlike pollen, they are typically recovered in a wider variety of landscape contexts. Perhaps one of the most important traits of phytoliths when it comes to information on Palaeoindian landscapes, however, is that the locations in which they are the most useful cover an area that pollen covers more poorly (Fredlund 2005). Pollen grains representing grass are spherical, monoporate grains that distinguish themselves rather poorly, whereas, with grass phytoliths, panicoid, pooid, and chloridoid forms can be distinguished rather readily. While physical features obviously play a role here, the long history of the study of grass phytoliths is also a factor (Twiss 2001). Because grasses are well-distinguished and grasslands an important factor in where Clovis sites are recovered, the ability to provide paleoecological information on Palaeoindian grasslands is high. Finally, grass phytoliths preserve well because a fair proportion of the plant is below or at the ground surface.

An important concept in paleoecology is the fact that because some of the fossil records of plants are either differentially preserved and recovered or differentially represented. It is important to note that this is true of pollen data as well (Davis 2000), and while we know quite a bit about how certain types of pollen may be over- or under-represented in a sample, this places a limit on the confidence that we have that ancient plant communities re-constructed from palynological data are representative of the plant com-munity that existed in the past.

Replication using multiple proxy data is often important (Denham 2008; Twiss 2001) for providing greater information on past plant communities. Because phytoliths and especially grass phytoliths can be identified accordingly with reference to their metabolic pathways, and because they can be recovered in the same contexts as stable isotopes, they provide complementary data. This symbiotic pairing has become almost *de rigueur* sampling at Palaeoindian sites (Cordova et al. 2011, Cyr et al. 2011, Mandel et al. 2014) and in research of Quaternary landscapes (Baker et al. 2000). Their presence in slides extracted for micromorphological analysis also duplicates both the recovery and the context of plant structures that are preserved in soils.

The reconstruction of plant communities and the relationship to climate change during the Late Pleistocene is critically important for understanding the availability of plants and animals in any part of the landscape during the Palaeoindian occupation. However, a number of assumptions were made using only palynological analyses and the surviving osteological evidence from ani-mals (Burns 1996, Wilson 1996). For example, assumptions that the ice-free corridor could not support human migrants based on recovered fossil evidence or plant reconstruction (Mandryk 1993; White et al. 1985) has been challenged by recent geomorphologic evidence (Munyikwa, et al. 2017; Wolfe, et al. 2004) and compels us to look more closely at phylogeographic and ancient DNA evidence. The employment of these latest methods has not only allowed Palaeoindian archaeologists to consider a wider variety of options for human subsistence practices but also provides information for critically assessing less research accessible routes.

Advances in phylogeography and ancient DNA

Phylogeography attempts to trace the genetic footprint of an animal or plant species across ancient landscapes. Although the application of mtDNA to population genetics began in the 1970s, the rise of the discipline of phylo-geography occurred around 1991. During this time, the discipline saw a significant increase in papers studying various genetic molecular markers and also an increase in the types of markers studied (Avise 1998). For phylogeographers, the process of dispersal is somewhat more important than timing, but rates are also invoked in any phylogeographic study. This powerful tool can help to determine the presence and movement of plant and animal species across ancient landscapes, an important concept during a period in which both habitat loss and the emergence of new landscapes would have occurred.

Typically, phylogeographic studies rely on genetic information from ex-tant populations as well as the presence of species represented in fossil re-cords. Because the fossil evidence may be limited (to pollen or macrofossil evidence in the case of plants and fossil bone in the case of animals, there may be some unexplored limitations to understanding the evidence produced using phylogeography. However, it does provide evidence of the process of species dispersal, an important concept when considering the limitations of fossil plants and evidence in the ice-free corridor. I will return to this concept in the discussion. But first, it is important to comment on another piece of evidence that is important to reconstructing palaeolandscapes, DNA.

Ancient and/or environmental DNA has been used in the past decade or so to both reconstruct the phylogeographic distribution of species and to detect environmental DNA (eDNA) from deposits across palaeolandscapes. Both of these methods promise to provide more insight into the late Pleistocene environment and have been used to argue for and against coastal and ice-free corridor models for the peopling of North America. Here, I distinguish the use of mtDNA from bone (ancient or aDNA) from that recovered in geologic contexts (environmental or eDNA). Using ancient mtDNA extracted from fossil bison both within and outside the ice-free corridor, Heintzman et al. (2016) demonstrate the limitations on the spread of bison, both spatially and temporally. Around the same time, Pedersen, et al. (2016), using the University of Calgary Earth Sciences lab as an initial point of processing samples, extracted eDNA from several lake basins in Alberta and concluded that travel through the ice-free corridor would have met a "bottleneck", impeding the movement of people and also large fauna. Even more recently, researchers (da Silva Coelho et al. 2021; see also, Perri 2021) have argued through mtDNA from dog bones that a coastal route is more likely. My interest here is not which route was taken by Paleoindians nor whether the authors cited are right or wrong in their conclusions, but on how these types of studies can expand our knowledge of late glacial landscapes and the process of deglaciation. I find it curious that in order to understand palaeolandscapes more broadly, the focus becomes the detail.

Summary: methodological advancements

When considering the ice-aged landscape, microcontextual studies offer a resolution to many long-standing questions about the depositional and post-depositional processes, the context of human presence, and the context of samples recovered for paleoenvironmental and chronological sampling. When dealing with continental size questions, why are these microstratigraphic questions so important? Every bit of evidence is based on our interpretation of that stratigraphy (Denham 2008). These microstratigraphic analyses demonstrate that many of our earlier assumptions are simply false. For example, there is considerable variability in the formation of soils and sedimentary deposits at archaeological sites that have similar visible characteristics (Harris-Parks, 2016; Kielhofer et al. 2020). Soil or archaeological micromorphology should, therefore, be even more critical as we use evidence that provides higher resolution as well and should be combined with other methods to ensure that our landscape reconstructions are the highest quality, as attaining evidence from stable isotopes, phytoliths, and environmental DNA may destroy the samples recovered from these studies.

Stable isotopes and phytoliths provide evidence that is complementary to conventional pollen analysis and expands the variability that is expressed across palaeolandscapes. The ability to provide evidence across many different depositional contexts is an important benefit of using these methods; however, it is critically important that we continue to develop an understanding of the depositional, post-depositional, and taphonomic processes that may be involved in stable isotope (Cerling 1995; Tieszen and Boutton 1989) and phytolith (Fredlund and Tieszen 1997) studies.

Finally, ancient and environmental DNA, when used with phylogeographic studies, can help to provide insights into how palaeolandscapes changed during deglaciation and sea level rise and how newly exposed landscapes (covered in ice or water during glaciation) were populated by plants, animals, and ultimately people. These studies have the potential to expand the diversity that we see across that landscape. Importantly, we must be cautious about sampling and about the use of iterative data. Phylogeographic studies have demonstrated that some pollen are underrepresented in late glacial records (Gugger and Sugita 2010); the same could potentially be true for eDNA (only recovered sediments containing eDNA are preserved). While Beaudoin (2017) has demonstrated that eDNA can recover species that pollen has not, the reverse could also be true due to the locations where the samples are recovered.

More importantly, what we look for can create bias in the way in which we view phylogeographic evidence. There is no reason to suggest that Palaeoindians who may have travelled across rapidly changing landscapes subsisted only on megafauna, despite the fact that it may have been beneficial to do so once they entered southern North America (Surovell and Waguespack 2009). A wide variety of air, water, and land-based fauna was found at Charlie Lake Cave (Fladmark et al. 1988), and both I (Freeman 2016) and others (Fiedel 2005) have argued that Palaeoindians inhabiting a newly deglaciated landscape need not have relied on animals whose grazing habits would require extensive, well-developed grasslands. To search for bison or various plant-markers that provide pre-conceived notions of what a "habitable" landscape should look like would be akin to searching for phylogeographic evidence of mammoth, bison, or even kelp along the west coast; it belies the complexity of late glacial plant and animal communities. Moreover, if founding populations move in exponentially, while those that follow a move in logistically (Hewitt 2000), the evidence we seek may be far less abundant than we expect. We need to be open in our search of ancient landscapes and in the data that we recover from them.

Expanded access and site revisitation

A long-standing tradition in archaeology, particularly as new methods emerge, is the re-excavation of previously studied sites. The Clovis type site at Blackwater Draw Locality #1 has been studied off and on for nearly a century. We go back to old sites when the data is fruitful and when the site itself offers a bounty of information that can be newly studied with the techniques mentioned above. In addition to old sites, old regions can see a resurgence in interest when ideas or methods support the renewal of research interest. The past 25 years have seen efforts made to better understand the possible connection among early inhabitants of North America and the western stemmed tradition (Davis et al. 2019), and as described above, DNA evidence has been invoked to better understand the possibilities and limitations of evidence in the ice-free corridor. Yet underexplored areas include the Rocky Mountains (Pitblado 2017) and foothills both south of and within the ice-free corridor. In the foothills, Lanoë et al. (forthcoming) have made a convincing argument that we may be ignoring some subtle landscape features, such as kettles that may be fruitful contexts for discovering Palaeoindian sites and environmental data. Emboldened by a sense that the ice-free corridor was a closed route, researchers studying coastal margins have helped to uncover the possibilities offered by ancient environments there (Fedje and Christensen 1999; Fedje and Josenhans 2000) a striking gap from 25 years ago.

Apart from fortuitous discovery, the principal method of discovering new sites comes from the reinvestigation of under-studied areas or during work conducted through contract archaeology. The second reason for expanded knowledge emanates from contract archaeology. The very nature of CRM requires that unstudied areas are investigated, and the potential for new sites and new data emerge from us venturing into this expanded landscape. There is perhaps no single area of study that is more profoundly affected by this than the continental shelf. In 1995, Michael Faught and I had a very hard time finding scientists who had worked on underwater archaeological sites off the eastern coast of North America. There was the sense that late Pleistocene sites simply were not preserved there, though compelling evidence was emerging that the possibilities along all coastlines were not yet tapped (Faught 1996; Josenhans et al. 1997). New data supports the difficult, high capacity for recovering Palaeoindian sites in coastal settings (Gusick and Faught 2011; Joy 2019).

It is gratifying to see that over the last 25 years, research on the continental shelf has taken off (Bailey and Flemming 2008), both on the west and east coasts as well as the Gulf of Mexico. Of the more recent discoveries that provide hope for the coastal migration route into North America are discoveries in the Channel Islands, coastal California (Erlandson and Jew 2009; Erlandson et al. 2011; Rick et al. 2001, 2005), the possibilities offered for early sites on the Oregon coast (Punke and Davis 2006), renewed interest in submerged sites in Florida (Halligan 2012; Halligan et al. 2016) and the use of LIDAR and other methods on Haida Gwaii (Fedje and Christensen 1999; Mackie et al. 2011) and on the Beringian Coast (Dixon and Monteleone 2014). Coastal eastern North America has also been explored in more detail (Gingerich and Wagner 2017), including the protected Chesapeake Bay (Lowery 2010a, 2010b). These studies have opened a gate of unprospected land around North America, fenced in by sea level rise, thus greatly expanding our knowledge of underwater resources. An important aspect of the continued exploration of submerged sites in the Gulf of Mexico has been the influence of CRM projects there (Faught and Joy 2019), including projects related to the development of green energy (Halligan personal communication 2021; Hoffman et al. 2020).

Conclusion

Once the impediments brought about by a focus on the Clovis/pre-Clovis debate were brought to rest in 1996 with the validation of dates and materials from the Monte Verde site in Chile (Meltzer et al. 1997), slightly more focus on the process transpired as well as the exploration of new methods and new areas. Unfortunately, I would argue (Freeman 2016) that it led to a bit of a baby/bathwater situation with respect to the poorly investigated interior of Canada, particularly the ice-free corridor and adjacent mountain interiors of Alberta and British Columbia. Nevertheless, the confirmation of dates from Monte Verde led to increased study of alternate routes for entering North America, along with alternate timing.

In order to get past the mental models imposed by a Clovis/pre-Clovis debate, it became necessary to consider that people arriving in the Americas via coastal routes would need to have exploited a wider variety of animals than mammoth and bison and that those food resources exploited by Palaeoindians across the continent may inhabit a different landscape than we earlier expected. The application of new standards, such as routine stable isotope and phytolith sampling and the examination of micromorphology promise to provide exciting new data to old problems and refine our understanding of late Pleistocene palaeolandscapes in North America.

We must continue to be vigilant about how we discover, report, and analyse data, not shutting ourselves off to new possibilities. While DNA holds great potential for reconstructing ancient landscapes and the biota in them, our sights must not be limited by our expectations, nor should we overlook or ignore the biases that may be hidden in our sampling. Regardless of how and when early people arrived in North America, the investigation of landscapes that may be alternative routes must be investigated in order to fully understand the complex changes that occurred at the end of the last glaciation.

References

Avise, John C., 1998. The history and purview of phylogeography: A personal reflection. *Molecular Ecology* 7: 371–379.

Bailey, Geoffrey N., and Nicholas C. Flemming, 2008. Archaeology of the continental shelf: Marine resources, submerged landscapes and underwater archaeology. *Quaternary Science Reviews* 27: 2153–2165.

Baker, Richard G., Glen G. Fredlund, Rolfe D. Mandel, and E. Arthur Bettis, III, 2000. Holocene environments of the central Great Plains: Multi-proxy evidence from alluvial sequences, southeastern Nebraska. *Quaternary International 67*: 75–88.

Beaudoin, Alwynne B., 2017. How humans populated North America: Revisiting the ice-free corridor (and other places). Abstracts of the 21st Annual Symposium, Alberta Paleontological Society, Edmonton.

Blinnikov, Mikhail, Alan Busacca, and Cahty Whitlock, 2001. A new 100,000-year phytolith record from the Columbia Basin, Washington, USA. In *Phytoliths: Applications in Earth Sciences and Human History*, edited by M. Blinnikov, A. Busacca, and C. Whitlock, pp. 27–55. A. A. Balkema, Lisse, Netherlands.

Bradley, Bruce, and Dennis Stanford, 2004. The North Atlantic ice-edge corridor: A possible Palaeolithic route to the New World. *World Archaeology 36*: 459–478.

Burns, James A., 1996. Vertebrate paleontology and the alleged ice-free corridor: The meat of the matter. *Quaternary International 32*: 107–112.

Cerling, Thure E., 1995. Stable carbon isotopes in palaeosol carbonates. In *Palaeoweathering, Palaeosurfaces and Related Continental Deposits*, edited by Médard Thiry and Régine Simon-Coinçon, pp. 43–60. Wiley, New York.

Cordova, Carlos E., William C. Johnson, Rolfe D. Mandel, and Michael W. Palmer, 2011. Late Quaternary environmental change inferred from phytoliths and other soil-related proxies: Case studies from the central and southern Great Plains, USA. *Catena* 85: 87–108.

Courty, Marie-Agnes, Paul Goldberg, and Richard Macphail, 1989. *Soils and Micromorphology in Archaeology*. Cambridge University Press, Cambridge.

Cristescu, Melania E., and Paul D. N. Hebert, 2018. Uses and misuses of environmental DNA in biodiversity science and conservation. *Annual Review of Ecology, Evolution, and Systematics* 49: 209–230.

Cyr, Howard, Calla McNamee, Leslie Amundsen, and Andrea Freeman, 2011. Reconstructing landscape and vegetation through multiple proxy indicators: A geoarchaeological examination of the St. Louis Site, Saskatchewan, Canada. *Geoarchaeology* 26: 165–188.

da Silva Coelho, F. A., Stephanie, Gill, Crystal M., Tomlin, Timothy H., Heaton, and Charlotte, Lindqvist, 2021. An early dog from southeast Alaska supports a coastal route for the first dog migration into the Americas. *Proceedings of the Royal Society B* 288: Article 20203103.

Davis, Margaret B., 2000. Palynology after Y2K—understanding the source area of pollen in sediments. *Annual Review of Earth and Planetary Sciences* 28: 1–18.

Davis, Loren G., 2001. The coevolution of early hunter-gatherer culture and riparian ecosystems in the southern Columbia River Plateau. Unpublished doctoral dissertation, University of Alberta, Edmonton, Canada.

Davis, Loren G., Karlis Muehlenbachs, Charles E. Schweger, and Nathaniel W. Rutter, 2002. Differential response of vegetation to postglacial climate in the

Lower Salmon River Canyon, Idaho. *Palaeogeography, Palaeoclimatology, Palaeoecology* 185: 339–354.

Davis, Loren G., David B. Madsen, Lorena Becerra-Valdivia, Thomas Higham, David A. Sisson, Sarah M. Skinner, Daniel Stueber, Alexander J. Nyers, Amanda Keen-Zebert, Christina Neudorf, Melissa Cheyney, Masami Izuho, Fumie Iizuka, Samuel R. Burns, Clinton W. Epps, Samuel C. Willis, and Ian Buvit, 2019. Late Upper Paleolithic occupation at Cooper's Ferry, Idaho, USA,~ 16,000 years ago. *Science* 365: 891–897.

Denham, Tim, 2008. Environmental archaeology: Interpreting practices-in-the-landscape through geoarchaeology. *Handbook of Landscape Archaeology*, edited by Bruno David and Julian Thomas, pp. 468–481. Routledge, London.

Dixon, James E., and Kelly Monteleone, 2014. Gateway to the Americas: Underwater archeological survey in Beringia and the North Pacific. In *Prehistoric Archaeology on the Continental Shelf*, edited by Amanda M. Evans, Joe Flatman, and Nicholas Flemming, pp. 95–114. Springer, New York.

Erlandson, Jon M., and Nicholas Jew, 2009. An early maritime biface technology at Daisy Cave, San Miguel Island, California: Reflections on sample size, site function, and other issues. *North American Archaeologist* 30: 145–165.

Erlandson, Jon M., Torben C. Rick, Todd J. Braje, Molly Casperson, Brendan Culleton, Brian Fulfrost, Tracy Garcia, Daniel A. Guthrie, Nicholas Jew, Douglas J. Kennett, Madonna L. Moss, Leslie Reeder, Craig Skinner, Jack Watts, and Lauren Willis, 2011. Paleoindian seafaring, maritime technologies, and coastal foraging on California's Channel Islands. *Science* 331: 1181–1185.

Faught, Michael K., 1996. Clovis origins and underwater prehistoric archaeology in northwestern Florida. Unpublished doctoral dissertation, Department of Anthropology, University of Arizona, Tucson.

Faught, Michael K., and Shawn Joy, 2019. The Potential for offshore industry to enable discovery of paleo-landscapes and evidence for early people: Past, present, and an optimistic future. Presentation at the Offshore Technology Conference, Houston. https://doi.org/10.4043/29329-MS

Fedje, Daryl W., and Tina Christensen, 1999. Modeling paleoshorelines and locating early Holocene coastal sites in Haida Gwaii. *American Antiquity* 64: 635–652.

Fedje, Daryl W., and Heiner Josenhans, 2000. Drowned forests and archaeology on the continental shelf of British Columbia, Canada. *Geology* 28: 99–102.

Fedje, Daryl Quentin Mackie, Terri Lacourse, and Duncan McLaren, 2011. Younger Dryas environments and archaeology on the Northwest Coast of North America. *Quaternary International 242*: 452–462.

Fiedel, Stuart J., 2005. Man's best friend–mammoth's worst enemy? A speculative essay on the role of dogs in Paleoindian colonization and megafaunal extinction. *World Archaeology* 37: 11–25.

Fiorentino Girolamo, Juan Pedro Ferrio, Amy Bogaard, José Luis Araus, and Simone Riehl, 2014. Stable isotopes in archaeobotanical research. *Vegetation History and Archaeobotany* 24: 215–227.

Fladmark, Knut R. Jonathan C Driver, and Diana Alexander, 1988. The Paleoindian component at Charlie Lake Cave (HbRf 39), British Columbia. *American Antiquity* 53: 371–384.

Fredlund, Glen G., 2005. Grass phytolith analysis. *Plains Anthropologist* 50: 63–68.

Fredlund, Glen G., and Larry L. Tieszen, 1997. Phytolith and carbon isotope evidence for late Quaternary vegetation and climate change in the southern Black Hills, South Dakota. *Quaternary Research* 47: 206–217.

Freeman, Andrea K. L., 2016. Why the ice-free corridor is still relevant to the peopling of the New World. In *Stones, Bones, and Profiles: Exploring Archaeological Context, Early American Hunter-gatherers, and Bison*, edited by Marcel Kornfeld and Bruce B. Huckell, pp. 51–74. University Press of Colorado, Boulder.

Freeman, Andrea K. L., and Michael Faught. 1995. Pleistocene-Holocene shorelines and channel margins: The geoarchaeological context of human entry into eastern North America. Symposium organised and held at the 60th annual meeting of the Society for American Archaeology, 3–7 May 1995, Minneapolis, Minnesota.

Garrison, Ervan R., and Jessica W. Cook Hale, 2021. "The early days"–underwater prehistoric archaeology in the USA and Canada. *Journal of Island and Coastal Archaeology* 16: 27–45.

Gingerich, J. A., & Wagner, D. P. (2017). Terminal pleistocene depositional patterns and their relationship to the paleoindian occupation of drainage basins in the Middle Atlantic Region, USA. *PaleoAmerica, 3*(4), 383–394.

Goldberg, Paul, and Vera Aldeias, 2018. Why does (archaeological) micromorphology have such little traction in (geo) archaeology? *Archaeological and Anthropological Sciences* 10: 269–278.

Goldberg, Paul, and Trina L. Arpin, 1999. Micromorphological analysis of sediments from Meadowcroft Rockshelter, Pennsylvania: Implications for radiocarbon dating. *Journal of Field Archaeology* 26: 325–342.

Gugger, Paul F., and Shinya Sugita, 2010. Glacial populations and post-glacial migration of Douglas-fir based on fossil pollen and macrofossil evidence. *Quaternary Science Reviews* 29: 2052–2070.

Gusick, Amy E., and Michael K. Faught, 2011. Prehistoric archaeology underwater: A nascent subdiscipline critical to understanding early coastal occupations and migration routes. In *Trekking the Shore*, edited by Nuno F. Bicho, Jonathan A. Haws, and Loren G. Davis, pp. 27–50. Springer, New York.

Halligan, Jessi Jean, 2012. Geoarchaeological investigations into Paleoindian adaptations on the Aucilla River, Northwest Florida. Unpublished doctoral dissertation, Department of Anthropology, Texas A&M University, College Station, Texas.

Halligan, Jessi J., Michael R. Waters, A. Perrotti, I. J. Owens, J. M. Feinberg, M. D. Bourne, B. Fenerty, B. Winsborough, D. Carlson, D. C. Fisher, T. W. Stafford, and J. S. Dunbar, 2016. Pre-Clovis occupation 14,550 years ago at the Page-Ladson Site, Florida, and the peopling of the Americas. *Science Advances* 2: e1600375.

Harris-Parks, Erin, 2016. The micromorphology of Younger Dryas-aged black mats from Nevada, Arizona, Texas and New Mexico. *Quaternary Research* 85: 94–106.

Haynes, C. Vance, Jr., 1971. Time, environment, and early man. *Arctic Anthropology* 8(2): 3–14.

Heintzman, Peter D., Duane Froese, John W. Ives, André E. Soares, Grant D. Zazula, Brandon Letts, Thomas D. Andrews, Jonathan C. Driver, Elizabeth Hall, P. Gregory Hare, Christopher N. Jass, Glen MacKay, John R. Southon, Mathias

Stiller, Robin Woywitka, Marc A. Suchard, and Beth   Shapiro, 2016. Bison phylogeography constrains dispersal and viability of the Ice Free Corridor in western Canada. *Proceedings of the National Academy of Sciences of the United States of America* 113: 8057–8063.

Hewitt, Godfrey, 2000. The genetic legacy of the Quaternary ice ages. *Nature* 405: 907–913.

Hoffman, William, Joseph Hoyt, and William Sassorossi, 2020. North Carolina collaborative archaeological survey: Wilmington East and West Wind Energy Areas. Final Report to the US Department of the Interior, Bureau of Ocean Energy Management, Office of Renewable Energy Programs, Sterling, VA and US Department of Commerce, National Oceanic and Atmospheric Administration, Office of National Marine Sanctuaries, Silver Spring, MD. OCS Study BOEM, 16.

Holliday, Vance T., 2021. "Geoarchaeology in the Academic Trenches". Discussant contribution in forum titled: Teaching and training in geoarchaeology: pedagogical methods and curricular pathways. Organized by Carlos Cordova and Andrea Freeman, Saturday, 17 April 2021. Society for American Archaeology, 86th Annual Meeting (Online).

Holliday, Vance T., 2000. Vance Haynes and Paleoindian geoarchaeology and geochronology of the Great Plains. *Geoarchaeology* 15: 511–522.

Holliday, Vance T., 2021. Geoarchaeology and the search for the first North Americans. *Geoarchaeology* 36: 3–17.

Josenhans, Heiner, Daryl Fedje, Reinhard Pienitz, and John Southon, 1997. Early humans and rapidly changing Holocene sea levels in the Queen Charlotte Islands-Hecate Strait, British Columbia, Canada. *Science* 277: 71–74.

Joy, Shawn, 2019. The trouble with the curve: Reevaluating the Gulf of Mexico sea-level curve. *Quaternary International* 525: 103–113.

Karkanas, Panagiotis, and Paul Goldberg, 2017. Soil Micromorphology. In *Encyclopedia of Geoarchaeology*, edited by Allan S. Gilbert, pp. 830–841. Springer, New York.

Kielhofer, Jennifer, Christopher Miller, Joshua Reuther, Charles Holmes, Ben Potter, Ben, François Lanoë, Julie Esdale, and Barbara Crass, 2020. The micromorphology of loess-paleosol sequences in central Alaska: A new perspective on soil formation and landscape evolution since the Late Glacial period (c. 16,000 cal yr BP to present). *Geoarchaeology* 35: 701–728.

Killick, David, and Paul Goldberg, 2009. A quiet crisis in American archaeology. *SAA Archaeological Record* 9: 6–10.

Koch, Paul, 1998. Isotopic reconstruction of past continental environments. *Annual Review of Earth and Planetary Sciences* 26: 573–613.

Lanoë, François, M. Nieves Zedeño, Anna M. Jansson, Vance T. Holliday, and Joshua D. Reuther, forthcoming. Glacial kettles as archives of human settlement along the northern Rocky Mountain Front. *Quaternary Research* in press. doi:10.1017/qua.2021.40

Lowery, Darrin., 2010a. The Late Quaternary geology and archaeology of Mockhorn Island, Virginia: A summary of 2009 and 2010 research. Manuscript on file, Virginia Department of Historic Resources, Richmond.

Lowery, Darrin L, 2010b. Geoarchaeological investigations at selected coastal archaeological sites on the Delmarva Peninsula: The long term interrelationship between climate, geology, and culture. Unpublished doctoral dissertation, University of Delaware.

Mackie, Quentin, Daryl Fedje, Duncan McLaren, Nicole Smith, and Iain McKechnie, 2011. Early environments and archaeology of coastal British Columbia. In *Trekking the Shore*, edited by Nuno F. Bicho, Jonathan A. Haws, and Loren G. Davis, pp. 51–103. Springer, New York.

Mandel, Rolfe D., Laura R. Murphy, and Mark D. Mitchell, 2014. Geoarchaeology and paleoenvironmental context of the Beacon Island site, an Agate Basin (Paleoindian) bison kill in northwestern North Dakota, USA. *Quaternary International* 342: 91–113.

Mandryk, Carole S., 1993. Paleoecology as contextual archaeology: human viability of the late Quaternary ice-free corridor, Alberta, Canada. Unpublished doctoral dissertation, University of Alberta, Edmonton.

Meltzer, David J., 1997. Monte Verde and the Pleistocene peopling of the Americas. *Science* 276: 754–755.

Meltzer, David J., Donald K. Grayson, Gerardo Ardila, Alex W. Barker, Dena F. Dincauze, C. Vance Haynes, Francisco Mena, Lautaro Núñez, and Dennis J. Stanford, 1997. On the Pleistocene antiquity of Monte Verde, southern Chile. *American Antiquity* 62: 659–663.

Mentzer, Susan, and Jay Quade, 2013. Compositional and isotopic analytical methods in archaeological micromorphology. *Geoarchaeology* 28: 87–97.

Muller, Richard A., 1977. Radioisotope dating with a cyclotron *Science* 196: 489–494.

Munyikwa, Kennedy, Tammy M. Rittenour, and James K. Feathers, 2017. Temporal constraints for the Late Wisconsinan deglaciation of western Canada using eolian dune luminescence chronologies from Alberta. *Palaeogeography, Palaeoclimatology, Palaeoecology* 470: 147–165.

Nordt, Lee C., 2001. Stable carbon and oxygen isotopes in soils. In *Earth Sciences and Archaeology*, edited by Paul Goldberg, Vance T. Holliday, and C. Reid Ferring, pp. 419–448. Springer, New York.

Pedersen, Mikkel W., Anthony Ruter, Charles Schweger, Harvey Friebe, Richard A. Staff, Kristian K. Kjeldsen, Marie L. Zepeda Mendoza, Alwynne B. Beaudoin, Cynthia Zutter, Nicolaj K. Larsen, Ben A. Potter, Rasmus A. Nielsen, Rebecca A. Rainville, Ludovic Orlando, David J. Meltzer, Kurt H. Kjær, and Eske Willerslev, 2016. Postglacial viability and colonization in North America's ice-free corridor. *Nature* 537: 45–49.

Pedersen, Mikkel Winther Pedersen, Søren Overballe-Petersen, Luca Ermini, Clio Der Sarkissian, James Haile, Micaela Hellstrom, Johan Spens, Philip Francis Thomsen, Kristine Bohmann, Enrico Cappellini, Ida Bærholm Schnell, Nathan A. Wales, Christian Carøe, Paula F. Campos, Astrid M. Z. Schmidt, M. Thomas P. Gilbert, Anders J. Hansen, Ludovic Orlando, and Eske Willerslev. 2015. Ancient and modern environmental DNA. *Philosophical Transactions of the Royal Society B: Biological Sciences* 370(1660): article 20130383.

Perri, Angela R., Tatiana R. Feuerborn, Laurent A. F. Frantz, Geger Larson, Ripan S. Malhi, David J. Meltzer, and Kelsey W. Whitt, 2021. Dog domestication and the dual dispersal of people and dogs into the Americas. *Proceedings of the National Academy of Sciences of the United States of America* 118: article e2010083118.

Pitblado, Bonnie L., 2017. The role of the Rocky Mountains in the peopling of North America. *Quaternary International* 461: 54–79.

Potter, Ben A., Alwynne B. Beaudoin, C. Vance Haynes, Vance T. Holliday, Charles E. Holmes, John Ives, and Todd Surovell, 2018. Current evidence allows multiple models for the peopling of the Americas. *Science Advances* 4 (8). https://doi.org/1 0.1126/sciadv.aat5473

Punke, Michele L., and Loren G. Davis, 2006. Problems and prospects in the preservation of late Pleistocene cultural sites in southern Oregon coastal river valleys: Implications for evaluating coastal migration routes. *Geoarchaeology* 21: 333–350.

Rick, Torben C., Jon M. Erlandson, and René L. Vellanoweth, 2001. Paleocoastal marine fishing on the Pacific Coast of the Americas: Perspectives from Daisy Cave, California. *American Antiquity* 66: 595–613.

Rick, Torben, Jon Erlandson, René L. Vellanoweth, and Todd Braje, 2005. From Pleistocene mariners to complex hunter-gatherers: The archaeology of the California Channel Islands. *Journal of World Prehistory* 19: 169–228.

Stanford, Dennis, and Bruce Bradley, 2000. The Solutrean solution. *Discovering Archaeology* 21: 54–55.

Surovell, Todd A., and Nicole M. Waguespack, 2009. Human prey choice in the Late Pleistocene and its relation to megafaunal extinctions. In *American Megafaunal Extinctions at the End of the Pleistocene*, edited by Gary Haynes, pp. 77–105. Springer, New York.

Taylor, R. Ervin, C. Vance Haynes, and Minze Stuiver, 1996. Clovis and Folsom age estimates: Stratigraphic context and radiocarbon calibration. *Antiquity* 70: 515–525.

Tieszen, Larry L., and Thomas W. Boutton, 1989. Stable carbon isotopes in terrestrial ecosystem research. In *Stable Isotopes in Ecological Research*, edited by Philip W. Rundel, James R. Ehleringer, and Kenneth A. Nagy, pp. 167–195. Springer, New York.

Twiss, Page C. 2001. A curmudgeon's view of grass phytolithology. *Phytoliths: Applications in Earth Sciences and Human History*, edited by Jean Dominique Meunier and Fabrice Colin, pp. 7–25. A.A. Balkema Publishers, Exton, Pennsylvania.

Waters, Michael R., Thomas W. Stafford, and David L. Carlson, 2020. The age of Clovis—13,050 to 12,750 cal yr BP. *Science Advances* 6(43). https://doi.org/10.112 6/sciadv.aaz0455

White, James M., Rolf W. Mathewes, and William H. Mathews, 1985. Late Pleistocene chronology and environment of the "Ice-Free Corridor" of northwestern Alberta. *Quaternary Research* 24: 173–186.

Wilson, Michael C., 1996. Late quaternary vertebrates and the opening of the ice-free corridor, with special reference to the genus Bison. *Quaternary International* 32: 97–105.

Wolfe, Stephen, David Huntley, and Jeff Ollerhead, 2004. Relict late Wisconsinan dune fields of the northern Great Plains, Canada. *Géographie Physique et Quaternaire* 58: 323–336.

4 Pathways along the Pacific

Using early stone tools to reconstruct coastal migration between Japan and the Americas

Ian Buvit, Karisa Terry, and Masami Izuho

Arguably, first-American studies have Pleistocene migration and Palaeolithic land use at their core, since, for a century, theories explaining initial human entry into the Americas described a trek through ice age Siberia and across now-submerged land between Chukotka and Alaska. Once in the Western Hemisphere, so it went, big-game hunters were to have replaced wedge-shaped microblade cores with fluted projectile points, possibly while passing through an ice-free corridor created when North America's continental glaciers separated. In recent years, it has also been widely accepted that humans were in the Americas before Clovis, that ancestral Native Americans experienced a period of genetic isolation, probably in Beringia, prior to moving past North America's continental ice sheets, and that initial entry at least partly included a coastal corridor. These models, however, continue to leave many questions unanswered.

Despite advancements in our understanding of the ancestral Native American genome, current anthropological models fall short in explaining the timing, routes taken, and lithic technological organisation of the first Palaeolithic Asians in the Americas. One of the most problematic is the Beringian Standstill Hypothesis (BSH) that posits a period of genetic isolation somewhere in the Arctic prior to initial human migration past the ice sheets (Tamm et al. 2007; Hoffecker et al. 2014; Hoffecker et al. 2016; Hoffecker et al. 2020), not whether it happened genetically, but, rather, how it happened geographically. One issue, for example, is that without direct genetic links between first Americans and inhabitants of the 32,000-year-old Yana RHS Site (all dates here are calendar years ago) (Sikora et al. 2019) (Figures 4.1 and 4.2), there is no explanation from where any standstill population entered Beringia; a trail of archaeological evidence linking southern Siberia and Beringia does not exist. Interior archaeological sites have not produced stone tool technologies that tie Palaeolithic northeast Asian microblades to the earliest projectile points of the Americas as would be expected if ancestral Native American populations travelled traditionally explained routes. Another difficulty is that after Yana, current arguments for humans in Beringia during the Last Glacial Maximum (LGM) lack clear artefactual evidence (e.g., Burgeon et al. 2017; Rae et al. 2020; Vachula

DOI: 10.4324/9781003139553-4

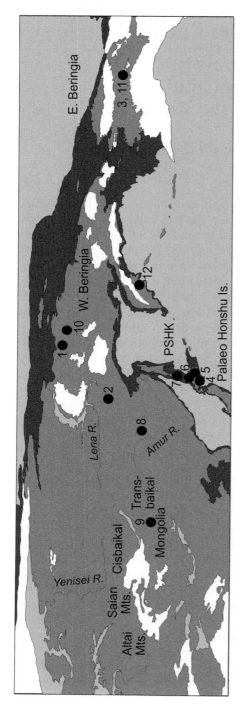

Figure 4.1 Map of the Pacific Rim showing locations mentioned in the text. Archaeological sites: 1-Yana RHS; 2-Diuktai Cave; 3-Swan Point; 4-Kashiwadai 1; 5-Kawanishi C; 6-Shimaki; 7-Ogonki 5; 8-Ust'-Ulma; 9-Studenoe; 10-Berelekh; 11-Holzman; 12-Ushki Lake.

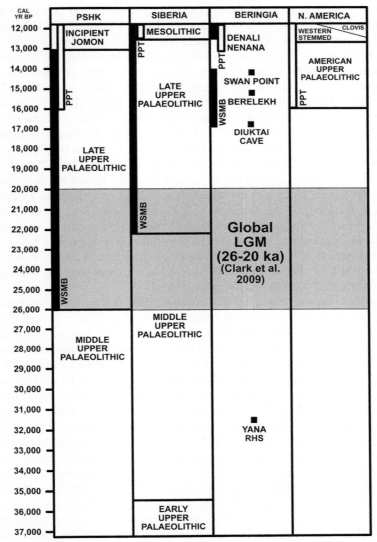

Figure 4.2 Palaeolithic cultural chronologies of northeast Asia and the Americas. Note the appearance of wedge-shaped microblade cores (WSMB) that separates middle from late Upper Palaeolithic sites across northeast Asia as black bars (>30,000–12,000 years ago). Note also the appearance of projectile points (PPT) on the Pacific Rim and North America as white bars (16,000–14,000 years ago). Beringia includes a typical middle Upper Palaeolithic assemblage at Yana RHS (32,000 years ago), then WSMB at Diuktai Cave (17,000 years ago) and Swan Point (14,500 years ago). This was followed in Alaska and Arctic Siberia by Nenana Complex sites with intriguing Chindadn points but no WSMB (13,500–13,000 years ago). Next are termed Denali assemblages with both WSMB and PPT (13,000–10,000 years ago). In North America, there is what has been termed the American Upper Palaeolithic (16,000–15,000 years ago) that predates both Western Stemmed Points (14,000–8000 years ago) and Clovis (13,000–12,500 years ago), all lacking WSMB.

2020). An undisputed presence appeared only after 17,000 years ago when sites like Diuktai Cave were occupied (see Figures 4.1 and 4.2) (Mochanov and Fedoseeva 1996). Along the same lines, it is not understood why humans remained in Beringia around the LGM when widespread human population contraction seems to have occurred elsewhere in northeast Asia (Tseitlin 1979, 129; Goebel 1999; 2002; Graf 2005; Buvit et al. 2015; Terry et al. 2016; Rybin et al. 2016; Graf and Buvit 2017; Guan et al. 2020), leaving one to ask why would the most arctic location be an exception? Most troublesome for the BSH is the lack of unquestionable material evidence where the focus for a standstill has been, in Alaska, before 14,500 years ago (Holmes 2001; Graf and Bigelow 2011; Graf and Buvit 2017; Wygal et al. 2018; Waters 2019). A generation ago, it took an incredibly high bar to break the Clovis First barrier, which was at least as important as current issues in peopling of the Americas studies, and, while Beringia is certainly one location that fits for where a standstill might have occurred, it is not the only place. Regardless, most publications continue to support a late glacial or LGM period of isolation in Alaska prior to ancestral Native Americans moving south of the ice sheets. We argue that archaeologically the idea is better supported elsewhere in northeast Asia.

As is now demonstrated by numerous studies, the ice-free corridor does not appear to have been a viable route southward until after people were in the Americas (Pedersen et al. 2016; Davis et al. 2019; but see Potter et al. 2017). Likewise, it is not understood how a single, late glacial Beringian lithic assemblage like from Diuktai Cave or Swan Point (see Figures 4.1 and 4.2), with wedge-shaped microblade cores (Figure 4.3), but lacking projectile points, gave rise to the toolkit variability found at other early sites in Beringia and the Americas, when the entire range of artefact types, subsuming almost all of the Western Hemisphere's key elements (e.g., a variety of wedge-shaped microblade core forms, large bifaces, large blades from prismatic cores, endscrapers, stemmed/lanceolate projectile points, etc.), exists on older, single occupation surfaces on Hokkaido and Sakhalin islands (Figure 4.4) (Kato and Kuwabara 1969; OBE [Obihiro Board of Education], 1992, 1993, 1995; HAOC [Hokkaido Archaeological Operation Center], 2000a, 2000b, 2000c; MBE [Makubetsu Board of Education], 2000; HAOC [Hokkaido Archaeological Operation Center], 2003a, 2003b, 2003c; Nakazawa et al. 2005; Vasilevskii 2008: 108–109; Izuho et al. 2012; Izuho 2013). Instead, we offer an alternative explanation, called PSHK Origins, where most of the earliest lithic variability in Beringia and pre-Clovis North America has its roots on the Palaeo-Sakhalin-Hokkaido-Kuril (PSHK) Peninsula, the ice age maritime extension of southern Siberia formed by islands of Japan and the Russian Federation.

Until around 12,000 years ago, Hokkaido was part of a peninsula with a narrow isthmus at the mouth of the Amur River (see Figure 4.1) (Kuzmin 1997, 133; Ono 1990). Geographically, the Amur River and its tributaries drained a large part of southern Siberia almost directly to this landform,

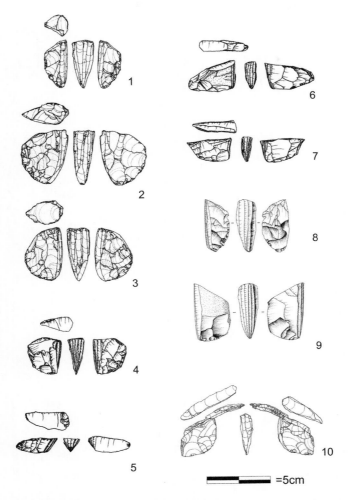

Figure 4.3 Examples of Yubetsu-type wedge-shaped microblade cores: 1–3-Kashiwadai 1, Hokkaido (26,000–25,000 years ago; HAOC [Hokkaido Archaeological Operation Center], 1999): 4–7-Studenoe 1 & 2, southern Siberia (22,000–13,000 years ago; Buvit 2008; Terry et al. 2016): 8–9-Diuktai Cave, western Beringia (17,000 years ago; Gómez Coutouly 2011): 10-Swan Point, eastern Beringia (14,500 years ago; Hirasawa and Holmes 2017).

where high-quality obsidian was accessible (Furguson et al. 2014; Yakushige and Sato 2014), as was a mosaic of other resources unavailable in Siberia. For example, while there is evidence for a group of large Pleistocene mammals associated with steppe environments (e.g., woolly mammoth [*Mammuthus primigenius*] and bison [*Bison priscus*]) in some colder, drier

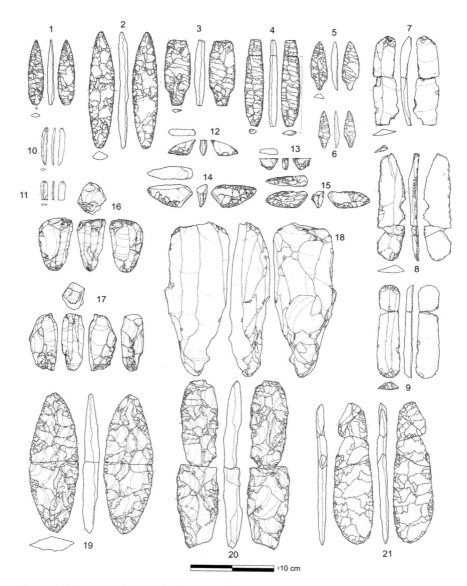

Figure 4.4 Artefacts from a single, late glacial occupation surface at the Kamishirataki 2 Site in Hokkaido, Japan (22,490–16,630 years ago) ([HAOC Hokkaido Archaeological Operation Center], 2000b) that might be analogues to examples in early Beringia and the Americas. 1–6-stemmed and lanceolate projectile points, 7–9-large blades, 10–11-microblades, 12–15-microblade cores; 16–18-blade cores, and 19–21-bifaces.

areas in the north and west of PSHK, there was also a group of animals associated with more temperate woodland environments (e.g., Naumann's elephant [*Palaeoloxodon naumanni*] and giant deer [*Sinomegaceros yabei*]) (Iwase et al. 2011) in the south and east of the peninsula (Table 4.1). Furthermore, as an on-ramp to the kelp highway, there would have been resources common to the entire Pacific Rim (Erlandson et al. 2015), especially species like salmon, which could have drawn people up unglaciated rivers from the coast to the interior. Finally, as a geographic *cul-de-sac* in proximity to small, isolated islands (Ono 1990), PSHK would have offered a perfect location for genetic isolation requisite for the formation of the ancestral Native American genome (Davis and Madsen 2020).

Past environmental data reveal a pattern of conditions on PSHK since the LGM was out of synch with interior areas of northeast Asia that conceivably pushed and pulled human groups between the interior and coast. Cold environmental conditions of the LGM began, for instance, near Lake Baikal in southern Siberia around 30,000 years ago (Bezrukova et al. 2010), but PSHK apparently was milder than the interior then, and cold conditions peaked along the coast between 24,000 and 19,000 years ago (Igarashi 2016). Another example occurred around 13,500 years ago when temperatures may have been warming elsewhere in northeast Asia and Alaska (Bezrukova et al. 2008; Graf and Bigelow 2011), and it was the coldest on PSHK as it had been for over 37,000 years (Igarashi 2016). These and other factors have led us to question longstanding ideas about the initial humans in the Western Hemisphere and to develop an alternative hypothesis couched in updated coastal migration theories to explain the northeast Asian ancestry of first Americans (e.g., Erlandson et al. 2015; Braje et al. 2020; Davis and Madsen 2020).

In short, PSHK Origins can be described as follows (see Table 4.1).

- Just before the LGM prior to 26,000 years ago, Ancient Siberians may have sought refuge from deteriorating environmental conditions of the interior by travelling south and east to the Pacific Coast, some settling on PSHK.
- By about 23,500 years ago, a group of Palaeo Siberians seemingly split from the original PSHK population, possibly making their way back into highly depopulated northeast Asia.
- After acquiring projectile point technologies that could have brought an influx of East Asian genes from Palaeo Honshu Island, those remaining, who were ancestors to Native Americans and Palaeo Beringians, would have been essentially isolated on the peninsula and nearby islands.
- Starting around 17,000 years ago, some appear to have left PSHK, travelling north along the Pacific Rim, descendants of whom reached Washington State's Columbia River within around 1000 years.
- At some point, probably between 17,000 and 14,500 years ago, Palaeo Siberians with wedge-shaped microblade cores may have reached Alaska on foot in very low numbers demonstrated by Swan Point and a few other

Table 4.1 Timeline of events.

Event, location, & age	Climate	Flora	Fauna	Archaeology
1. Interior northeast Asia is depopulated 30-25 ka.	Onset of Last Glacial Maximum (LGM) conditions in Southern Siberia[1]. Milder on PSHK until around 24 ka[2, 3].	Southern Siberia[1]: Steppe with little forest cover. Highest percentages of *Artemesia* and Poaceae. PSHK[2]: Boreal deciduous conifer forest (*Abies, Picea, Pinus*) or boreal forest (*Abies, Betula, Pinus*) interspersed with grasslands (*Gramineae* and *Cyperaceae*)	Southern Siberia[4, 5]: Tundra-Steppe species (*Bison priscus, Equus caballus, E. Hemionus hem, Ovis ammon, Coelodonta antiquitatis, Megaloceros gigantheus, Rangifer tarandus, Panthera leo*) and Mongolian Steppe species antelope (*spirocerus kiakhensis*), yak (*Poephagus baikalensis*), Mongolian gazelle (*Procapra gutturosa*), and camel (*Camelus sp.*). PSHK[6, 7, 8]*Palaeoloxodon-Sinomegaceroides* complex of Naumann's elephant (*Palaeoloxodon naumanni*), Yabe's giant deer (*Sinomegaceros yabei*)	Gaps and spikes in numbers of ^{14}C-dated archaeological sites. Microliths but no true wedge-shaped microblade cores (WSMBs) in Siberia. Appearance of true on WSMBs on PSHK ~26-25 ka.
2. Ancient Siberians originating from PSHK make their way back to depopulated northeast Asia 24-23.5 ka.	Continued cold and dry, but improving, conditions in the interior. Beginning of the LGM on PSHK.	Southern Siberia[9, 10, 11, 14]: Cold-steppe grasses (*Gramineae*) and *Artemisia* in valleys with stands of larch (*Larix*), pine (*Pinus*), and birch (*Betula*) in northerly areas, mountain-tundra plant	Southern Siberia[12, 15]: forest-steppe and tundra-steppe species including mammoth (*Mammuthus primigenius*), woolly rhinoceros (*Coelodeonta antiquitatis*), red deer (*Cervus elaphus*), antelope (*Saiga tatarica*), moose (*Alces alsces*), roe	Declining archaeological ^{14}C dates on PSHK until a 500-year gap around 23 ka. First appearance of WSMB from east to west in southern Siberia 23-22 ka.

species in alpine zones, boreal vegetation including birch (*Betula*), pine (*Pinus*), larch (*Larix*), spruce (*Picea*), and fir (*Abies*) in the mid-zone, and grass (*Gramineae*) and herb (*Artemisia*) steppe in the southern portions. PSHK[2, 3, 7, 13]. *Larix* expanded into Sakhalin and Hokkaido, becoming the main component of the open boreal deciduous coniferous forests covering the area with open forest taiga (*Larix gmelinii, Picea pumila, Picea jezonensis*) and grassy plains (*Gramineae*) and mire wetlands (spikemoss) during the LGM. Broad-leafed deciduous forests were present in lowlands and western Hokkaido, while taiga forests and grasslands

deer (*Capreolus caprolus*), and Asian wild ass (*Equus hemionus*), as well as the addition of Mongolian-steppe species in the southern region including steppe bison (*Bison priscus*), Baikal Yak (*Bos baikalensis*), ostrich (*Struthio asiaticus*), gazelle (*Procapra gutterosa*), argali sheep (*Ovis ammon*), wolf (*Canis lupus*), and bear (*Ursus spelaeus*) PSHK[6, 7, 18]. *Palaeoloxodon-Sinomegaceroides* complex of Naumann's elephant (*Palaeoloxodon naumanni*), Yabe's giant deer (*Sinomegaceros yabei*) until 23 ka and Mammoth Fauna woolly mammoth (*Mammuthus primigenius*), bison (*Bison priscus*), and moose (*Alces alces*) after 25 ka

(Continued)

Table 4.1 (Continued)

Event, location, & age	Climate	Flora	Fauna	Archaeology
3. East Asians from Paleo Honshu arrive on PSHK 23-17 ka.	PSHK experiences interstadial conditions around 19 ka.	were found throughout the rest of Hokkaido. Southern Siberia [5, 11]: Tundra steppe transitioning to forest-steppe including Pine (*Pinus cembra, Pinus sibirica*) and birch (*Betula sp.*). PSHK [2, 7, 13]: *Larix gmelinii* expanded into Sakhalin and Hokkaido, becoming the main component of the open boreal deciduous coniferous forests covering the area with open forest taiga (*Larix gmelinii, Picea pumila, Picea jezonensis*) and grassy plains (*Graminaea* and *Cyperaceae*) and mire wetlands (spikemoss) during the LGM after 23 ka broad-leaf deciduous forests disappeared from western Hokkaido.	Southern Siberia [5, 12, 16, 17]: During LGM forest-steppe and tundra-steppe species including mammoth (*Mammuthus primigenius*), woolly rhinoceros (*Coelodonta antiquitatis*), red deer (*Cervus elaphus*), antelope (*Saiga tatarica*), moose (*Alces alsces*), roe deer (*Capreolus capreolus*), and Asian wild ass (*Equus hemionus*), as well as the addition of Mongolian-steppe species in the southern region including steppe bison (*Bison priscus*), Baikal Yak (*Bos baikalensis*), ostrich (*Struthio asiaticus*), gazelle (*Procapra gutterosa*), argali sheep (*Ovis ammon*), wolf (*Canis lupus*), and bear (*Ursus spelaeus*). Post-LGM steppe and taiga species moose (*Alces alces*), reindeer (*Rangifer tarandus*), red deer (*Cervus elaphus*),	Large spike in ^{14}C-dated archaeological sites. Occupation surfaces from the Shirataki sites produced aggregations of key lithic artefacts (bifaces, large blades, WSMB, and projectile points).

bison (*Bison priscus*), horse (*Equus caballus*), roe deer (*Capreolus capreolus*), Mongolian gazelle (*Procapra guttourosa*), Asian wild ass (*Equus hemionus*), saiga (*Saiga tartarica*), wolf (*Canis lupus*), and brown bear (*Ursus arctos*). PSHK[6,7,18]: Mammoth Fauna woolly mammoth (*Mammuthus primigenius*), bison (*Bison priscus*), and moose (*Alces alces*) until 18 ka.

4. Population isolation on PSHK. 19-17 ka.	Temperature and precipitation decline on PSHK.	Southern Siberia[5,11]: Tundra steppe transitioning to forest-steppe including Pine (*Pinus cembra, Pinus sibirica*) and birch (*Betula sp.*).

Southern Siberia[5,17]: After LGM steppe and taiga species moose (*Alces alces*), reindeer (*Rangifer tarandus*), red deer (*Cervus elaphus*), bison (*Bison priscus*), horse (*Equus caballus*), roe deer (*Capreolus capreolus*), Mongolian gazelle (*Procapra guttourosa*), Asian wild ass (*Equus hemionus*), saiga (*Saiga tartarica*), wolf (*Canis*

(*Continued*)

Table 4.1 (Continued)

Event, location, & age	Climate	Flora	Fauna	Archaeology
		PSHK[2,7]: Boreal deciduous coniferous forests covering the area with open forest taiga (*Larix gmelinii, Picea pumila, Picea jezonensis*) and grassy plains (*Graminaea* and *Cyperaceae*) and mire wetlands (spikemoss).	*lupus*), and brown bear (*Ursus arctos*). PSHK[6,7,18]: Mammoth Fauna woolly mammoth (*Mammuthus primigenius*), bison (*Bison priscus*), and moose (*Alces alces*) until 18 ka.	
5. Initial human populations expanded into North America ~17 ka.	Beginning of the Kenbuchi Stadial 17 ka on PSHK[2,22]. [23,24]	Southern Siberia[5,11,19]: Tundra steppe transitioning to forest-steppe including Pine (*Pinus cembra, Pinus sibirica*), larch (*Larix*), spruce (*Picea*) and birch (*Betula sp.*). PSHK[2,7,13]: Boreal deciduous coniferous forests covering the area with open forest taiga (*Larix gmelinii, Picea pumila, Picea jezonensis*) and grassy plains (*Graminaea* and	Southern Siberia[5,17]: Steppe and taiga species moose (*Alces alces*), reindeer (*Rangifer tarandus*), red deer (*Cervus elaphus*), bison (*Bison priscus*), horse (*Equus caballus*), roe deer (*Capreolus capreolus*), Mongolian gazelle (*Procapra guttourosa*), Asian wild ass (*Equus hemionus*), saiga (*Saiga tartarica*), wolf (*Canis lupus*), and brown bear (*Ursus arctos*). PSHK[6,7,8,20,21]: Possibly Mammoth Fauna of woolly mammoth	Projectile points in the Americas ~16-15.5 ka.

		Cyperaceae) and mire wetlands (spikemoss) until 16 ka. After 16 ka, cold deciduous forests expand southward, dominated by *Larix* and *Pinus* and grasslands.	(*Mammuthus primigenius*), bison (*Bison priscus*), and moose (*Alces alces*) but moving northward, and replaced by smaller mammals such as sika deer or boar.
6. Humans completely depopulate PSHK. ~13.5 ka.	Coldest and driest on PSHK ~13.5 ka.	Southern Siberia[19, 25, 26, 5, 15]: forest-steppe transitioning to meadow-steppe 16-14.7 ka with decrease in forests, then increase forests after 14.7 ka including Pine (*Pinus cembra, Pinus sibirica*) and birch (*Betula sp.*) PSHK[2, 7, 13]: Boreal deciduous coniferous forests covering the area with open forest taiga (*Larix gmelinii, Picea pumila, Picea jezonensis*) and grassy plains (*Graminaea* and *Cyperaceae*) and mire wetlands (spikemoss) until 16 ka. After 16 ka, cold deciduous forests expand	Southern Siberia[17, 27]: Steppe and taiga species moose (*Alces alces*), reindeer (*Rangifer tarandus*), red deer (*Cervus elaphus*), bison (*Bison priscus*), horse (*Equus caballus*), roe deer (*Capreolus capreolus*), Mongolian gazelle (*Procapra guttourosa*), Asian wild ass (*Equus hemionus*), saiga (*Saiga tartarica*), wolf (*Canis lupus*), and brown bear (*Ursus arctos*). PSHK[20, 27]: Possibly smaller mammals such as sika deer or boar.

(Continued)

Table 4.1 (Continued)

Event, location, & age	Climate	Flora	Fauna	Archaeology
		southward, dominated by *Larix* and *Pinus* with grasslands.		

Notes

1 Bezrukova et al. 2010
2 Igarashi 2016
3 Igarashi and Zharov 2011
4 Baryshnikov and Markova 1992
5 Chlachula 2001b
6 Iwase et al. 2015
7 Iwase et al. 2011
8 Kawamura 1991
9 Frenzel et al. 1992
10 Grichuk 1992
11 Lbova 2000
12 Chlachula 2001a
13 Igarashi et al. 2011
14 Maloletko 1998
15 Chlachula 2001b
16 Konstantinov 1994
17 Vereshchagin and Kuz'mina 1984
18 Kawamura 1991
19 Bezrukova et al. 2009
20 Nakazawa et al. 2011
21 Takakura 2020
22 Morisaki et al. 2019
23 Wang et al. 2001
24 Yuan et al. 2004
25 Chlachula 2017
26 Lbova 2000
27 Takakura 2020

undiscovered sites, and then feasibly disappeared. Meanwhile, it is conceivable that the exodus from the peninsula climaxed 13,500 years ago when Palaeolithic populations look to have abandoned PSHK because of deteriorating environmental conditions there.

- Technological elements of the last people from the peninsula may have then penetrated interior Alaska, eventually comprising what is called the Denali Complex.

Throughout this chapter, we stress the following points.

- First, this is a working hypothesis, and we expect there to be revisions and modifications yet to come, but we feel it best explains current anthropological evidence.
- Second, PSHK is the only location in northeast Asia with an archaeological record spanning the global LGM (26,000–20,000 years ago; Clark et al. 2009) with requisite technological antecedents of early North and South American lithic assemblages, and is, therefore, a better candidate than anywhere in Siberia or Beringia for a genetic standstill.
- Third, DNA evidence from Yana RHS suggests human populations contracted from Arctic Siberia at the LGM and did not remain in place or sought nearby refuge.
- Fourth, gaps in other interior northeast Asian archaeological ^{14}C records coincide with increases in the number of dates and new sites on PSHK, a relationship that could be interpreted as population expansion to the coast and explain Ancient Siberian DNA there. According to genomic evidence, Palaeo Siberians then split from ancestral Native Americans between 24,000 and 21,000 years ago, possibly relating to the spread of wedge-shaped microblade core technology throughout late glacial Siberia and into Beringia.
- Fifth, admixture with East Asians may be seen in the archaeological record as the arrival of projectile point technologies on PSHK leading to a spike in human population levels around 19,000 years during relatively warm interstadial conditions.
- Sixth, a genetic standstill could then have occurred, perhaps 19,000 to 17,000 years ago, when PSHK was isolated from an already heavily depopulated northeast Asia: There were likely not many people around the isthmus who could have comingled with PSHK residents, and diminished populations on the peninsula would have facilitated rapid diversification of the ancestral Native American genome.
- Seventh, by around 17,000 years ago, bands of ancestral people, who utilised projectile point technologies instead of microblades, left PSHK to Beringia and the Americas, travelling a coastal corridor.
- Finally, during an extremely cold and dry period around 13,500 years ago, human populations vanished from PSHK, seen as a gap in its archaeological record. Jomon cultures from Palaeo Honshu Island then spread across Hokkaido following this apparent Palaeolithic abandonment,

events that may be related to the appearance of Denali Complex sites across Beringia.

The remainder of this chapter will detail the various events of PSHK Origins (but see Table 4.1 for an overview).

Event 1: Ancient Siberians enter PSHK around the LGM (32,000–26,000 years ago)

The model begins just prior to the LGM when middle Upper Palaeolithic humans outfitted with microliths, but no true microblades, withdrew from most of interior northeast Asia, including Siberia and Beringia, to more hospitable locations along the Pacific Coast, developing pressure-flaked, wedge-shaped microblade cores while *en route* or soon after arriving. The LGM human abandonment of Siberia has been the topic of debate for several decades (Tseitlin 1979; Goebel 1999; Kuzmin 2008), and if it indeed occurred, it would have been a catalyst of sorts that set off a chain of events leading to the peopling of the Americas. It would also explain how Ancient Siberians arrived at a geographically isolated location other than Beringia. Very little is known about northeast Asia's Palaeolithic human genomes, however, but mtDNA from the 32,000-year-old Yana RHS site suggests humans fled Arctic Siberia at the LGM. Human teeth from a pair of unrelated juveniles there prove only distantly related to later Palaeo Beringians and Indigenous Americans (Sikora et al. 2019). Stronger support might be found in the archaeological record where ^{14}C and stone tool evidence exists for declining human populations in Siberia, western Beringia, and other parts of continental Northeast Asia, along with expansion in northern Japan.

Archaeological radiocarbon dates are often used as a proxy for prehistoric Eurasian human population levels with two sides forming – one sees clear gaps in the record, indicating a hiatus in human occupation at or around the LGM (Tseitlin 1979: 129; Goebel 1999, 2002; Dolukhanov 2004; Graf 2005; Buvit et al. 2015; Buvit and Terry 2016; Graf and Buvit 2017), and the other sees continuous population levels up to, and through, the same period (Kuzmin and Keates 2005; Fiedel and Kuzmin 2007; Kuzmin 2008; Kuzmin et al. 2011). In Figure 4.5, however, when median calibrated ^{14}C dates from archaeological sites are plotted per century, the widest gaps in the records of Beringia, Mongolia, and all of southern Siberia occur around the global LGM (26,000–20,000 years ago; Clark et al. 2009). One interpretation is that humans depopulated the areas shown as a reduction in the number of ^{14}C-dated archaeological sites. This is apparently what happened when Palaeolithic groups retreated from northerly regions of Western Europe during the LGM to more temperate areas at, and near, the Atlantic Coast (Straus 2005, 2006). From what is known, it seems clear that the LGM's impact on humans and other species was widespread across Eurasia, and it is conceivable that Siberia and Beringia were equally affected as other places.

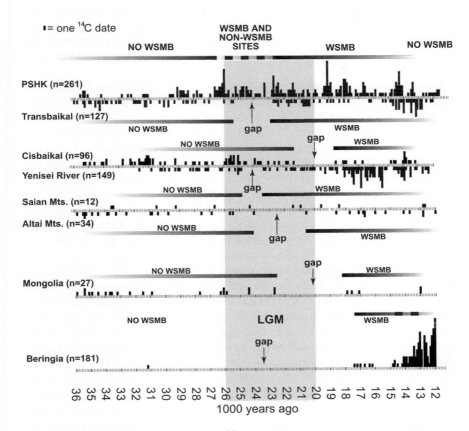

Figure 4.5 Palaeolithic archaeological ¹⁴C records of Siberia, Mongolia, Beringia, and PSHK. Each black bar represents a single date. Geographic locations with more than 25 exhibit a gap at or around the LGM separating sites with and without wedge-shaped microblade cores (WSMB) (modified from Buvit and Terry 2016: figure 4.2).

What is more, human migration out of Arctic Asia is further supported by changes in lithic technological organisation.

In every location in Figure 4.5, wedge-shaped microblade cores appeared after the gaps, where the artefacts were absent earlier, a phenomenon too widespread to be a coincidence. In other words, combining technological and chronological data, humans seemingly left most of interior northeast Asia around the LGM, only to return a few thousand years later with wedge-shaped microblade cores. Whether or not areas were completely abandoned everywhere, or repopulation happened only after the harshest conditions waned, is difficult to say. Either way, the appearance of formal wedge-shaped microblade core technology, along with concomitant increases in archaeological sites and ^{14}C dates, seem very indicative of human population expansion from other locations near the end of the LGM. A look at the PSHK ^{14}C record reveals other things were happening along the coast and the area may have been the source of post-last-glacial-maximum and late-glacial Siberian microblade core technologies.

On the peninsula, we see a gradual increase in ^{14}C dates from 30,000 years ago until a noticeably sharp rise beginning around 27,000 years ago (Figure 4.6). Then, the most dates of any century appear at 26,000 years ago, sufficiently greater than previous centuries to rule out random fluctuation (Buvit et al. 2016). In Figure 4.6, it appears that, compared to the other periods of Hokkaido prehistory, few archaeological sites existed prior to the LGM. For a while, several distinct lithic industries existed on PSHK, with some sites like Shimaki producing small-flake assemblages with abundant scrapers (Buvit et al. 2014). Little if any blades were manufactured, and microliths, but no true microblades, were produced. Other sites, like Kawanishi C, produced large blades in abundance and no microblades (Nakazawa et al. 2019). Then, the appearance of wedge-shaped microblade cores at places like Kashiwadai 1 coincided with statistically significant increases in numbers of ^{14}C dates and new archaeological sites between 27,000 and 25,000 years ago (HAOC Hokkaido Archaeological Operation Center, 1999; Izuho et al. 2012; Buvit et al. 2016). For a short time, this seemingly marked the existence of three exclusive forms of lithic technological organisation on PSHK, but two soon vanished, leaving microblades nearly ubiquitous at late-glacial archaeological sites (Ono et al. 2002; Nakazawa et al. 2005). Microblades were an innovation where very small (often <1 cm wide and a few cm long), consistently blade shaped artefacts were detached from several distinct core forms by pressure flaking, or sometimes by percussion, to serve as extremely sharp insets in composite tools like knives and projectile points. This increased group mobility and lithic toolkit sustainability because hundreds of meters of cutting edge could be manufactured from light weight, portable stone cores, innovations that conceivably revolutionized how northeast Asians confronted palaeolandscapes.

Event 2: Groups from PSHK with wedge-shaped microblade cores returned to continental northeast Asia (23,500 years ago)

From the standpoint of DNA, Event 2 is better supported than Event 1. Both mtDNA (Sikora et al. 2019) and Y-chromosomes (Pinotti et al. 2019) show that Palaeo Siberians split from Indigenous Americans and Palaeo Beringians (Moreno-Mayar et al. 2018) sometime between 24,000 and 21,000 years ago. Additionally, Sikora et al. (2019) link the subsequent expansion of Palaeo Siberians with the spread of wedge-shaped microblade core technology throughout northeast Asia.

To add, Graf (2008: 407–408, 2009) noticed a post-LGM, east-to-west temporal cline of Yubetsu wedge-shaped microblade core technology across

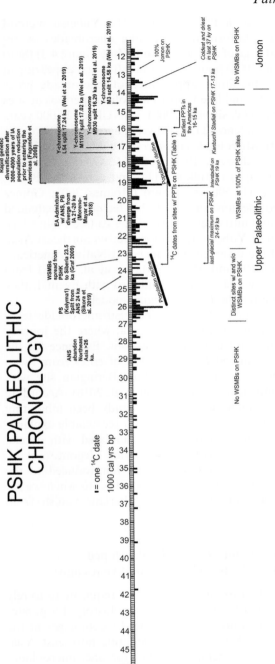

Figure 4.6 Major events in the PSHK Origins model in relation to the peninsula's late-glacial archaeological ^{14}C chronology. These include a significant increase in the number of dates and archaeological sites starting around 26,500 years ago, possibly corresponding to an influx of Ancient Siberians from interior northeast Asia around the LGM. This is followed by a noticeable drop in the number of dates until around 23,500 years ago during the coldest conditions of the LGM on PSHK, and close to when Paleo Siberians, equipped with Yubetsu microblade technology, made their way back into Siberia and surrounding areas. By 19,000 years ago, interstadial conditions returned to PSHK (Igarashi 2016), along with increases in numbers of radiocarbon dates and archaeological sites (Buvit et al. 2016), which could indicate human populations increased. Igarashi (2016) describes a period of deteriorating environmental conditions on PSHK starting around 17,000 years ago until around 13,500 years ago, when it was the coldest and driest since humans arrived on the peninsula. From 19,000 until 17,000, there are declining numbers of ^{14}C dates, again possibly indicating drops in population sizes. Around 17,000 years ago is also when we see the formation of several ancestral Native American paternal lineages, perhaps when paternal bands, equipped with distinct lithic assemblages, migrated into Beringia and the Americas along a coastal corridor. Abbreviations: PS = Palaeo Siberian, ANS = Ancient Northern Siberian, WSMB = wedge-shaped microblade cores, PSHK = Palaeo-Sakhalin-Hokkaido-Kuril Peninsula, EA = East Asian, PB = Palaeo Beringian, IA = Indigenous American, PPT = projectile point.

southern Siberia and argued for a PSHK origin. The Yubetsu method "involved forming a striking platform for detaching microblades by removing spalls from the lateral edge of a bifacial blank" (see Figure 4.3) (Sato and Tsutsumi, 2007: 57). Graf (2008) pointed out that after first appearing in Hokkaido, the directional, ubiquitous appearance of microblade cores across Siberia divides the middle and late Upper Palaeolithic periods at sites like Ogonki 5 (23,000 years ago) (Vasilevskii 2008), Ust'-Ulma (23,000 years ago) (Derev'anko 1998), and Studenoe-2 (22,000 years ago) (see Figures 4.1 and 4.2) (Konstantinov 1994). Microblade cores from Diuktai Cave (17,000 years ago) (Mochanov and Fedoseeva 1996, 165–166; Gómez Coutouly 2016), Berelekh (15,000 years ago) (Pitulko 2011), and across the Bering Strait at Swan Point, Alaska (14,500 years ago) (Holmes 2001; Gómez Coutouly and Holmes 2018), demonstrate that this technology extended across the mammoth-steppe and into Arctic North America (see Figure 4.1). Therefore, if the transmission of Yubetsu technology across northeast Asia, gaps in the ^{14}C records, and the divergence of ancestral Native Americans described by Pinotti et al. (2019) and Moreno-Mayar et al. (2018) from Siberians (Sikora et al. 2019) are related, then combined with Event 1, they would comprise a cornerstone for the PSHK Origins model.

As far as it spread geographically, however, Yubetsu microblade technology seems to have petered out in eastern Beringia, where the journey might have been incredibly difficult to continue on foot. Evidence for the initial technology in Alaska is limited to the handful of cores from Swan Point, nothing more. Holzman, the only other East Beringian site with comparable dates, lacks diagnostic artefacts (Wygal et al. 2018). According to PSHK Origins, however, humans would have already been inhabiting coastal areas when the two sites were occupied, and a reasonable explanation could be that these few interior inhabitants assimilated with others. Some molecular evidence might support this scenario. Flegontov et al. (2019), for example, assert that Siberian DNA was reintroduced to the Native American gene pool around 14,000 years ago, close to when we see the earliest microblade cores, and thus, Palaeo Siberians vanish from Alaska.

Event 3: Lithic artifact assemblages on PSHK developed variability comparable to pre-Clovis (26,000–17,000 years ago)

PSHK sites have produced no Pleistocene human fossils requiring us to rely entirely on its archaeological record to understand its prehistory. Large sites existed near lithic raw material sources that produced assemblages at the Shirataki group of sites like nowhere else in Palaeolithic northeast Asia. Shirataki will be discussed in more detail later, but for one, microblades continued to be an important part of the stone-tool repertoire, developing into a variety of different styles (Nakazawa et al. 2005). At some time, a Palaeo Honshu-type hunting strategy that included stemmed and lanceolate projectile points arrived on PSHK, perhaps in several waves, and integrated with existing tool kits collectively called Shirataki assemblages (Aikens and Higuchi 1982, 79; Ono et al. 2002; Yakushige and Sato 2014). Although the Shirataki sites have produced many ^{14}C dates (Table 4.2) (Buvit et al. 2016), they range from about 28,00 to 13,000 years ago, and precise chronological control is difficult. One observation in Table 2, however, is that over 90% are between 23,000 and 16,000 years ago, without a single millennium dominating, and the 7000-year period might best represent the timeframe for projectile points antecedent to early North and South American artefacts on PSHK. In contrast, Japanese archaeologists have mostly agreed the technology arrived on Hokkaido only 16,000 years ago (Ono et al. 2002; Yakushige and Sato 2014), but this would require extraordinarily rapid transmission to the Americas (Davis et al. 2019; Pratt et al. 2020). Instead, they may have started to arrive earlier on PSHK and peaked around 20,000–19,000 years ago when we see a spike in archaeological ^{14}C dates there in Figure 4.6. This would allow time for a genetic standstill and for transmission of point forms to the Americas several thousand years later.

With projectile points, Hokkaido assemblages still lacked ceramics (Iizuka 2018) but included large blades, bifaces, and wedge-shaped microblade cores comparable to early Beringia and the Americas (Adovasio et al. 1977; Powers and Hoffecker 1989; Jenkins et al. 2012; Halligan et al. 2016; Williams et al. 2018; Waters et al. 2018; Davis et al. 2019; Dillehay et al. 2019). Figure 4.7a is a map showing the distribution of sites along the Pacific Rim with projectile points but no microblades and vice versa. Figure 4.7b compares projectile points from Palaeolithic PSHK to early sites in the Americas. A key similarity is the variability both Hokkaido, and pre-Clovis sites produce, variability not found in interior, late-glacial microblade assemblages, or anywhere else along the continental routes proposed by other models until after humans with projectile point technology unquestionably inhabited PSHK and Americas.

Event 4: Rapid genetic diversification accompanied human population contraction (19,000–17,000 years ago)

As stated previously, a possible population increase might be seen as a spike in ^{14}C dates on PSHK around 19,000 years ago (see Figure 4.6), coinciding with what Igarashi (2016) describes as interstadial conditions. The cluster of ^{14}C dates around that time might represent an influx of East Asians, possibly with projectile point technologies, from Palaeo Honshu Island, and explain their admixture with isolated ancestors of Indigenous Americans originally from interior northeast Asia. It was not long, however, that the situation

Table 4.2 Palaeolithic ¹⁴C chronology of the Shirataki Group of Sites.

Laboratory #	Method	Site	$\delta^{13}C$	¹⁴C Age BP	cal BP 1-σ	Median age cal BP
Beta-112907	AMS	Kamishirataki 8	−23.7	23,640 ± 310	28,010–27,500	27,780
Beta-112908	AMS	Kamishirataki 8	−24	22,230 ± 110	26,600–26,250	26,450
Beta-126157	AMS	Okushirataki 1	−24.3	18,890 ± 140	22,910–22,560	22,750
Beta-112906	AMS	Kamishirataki 8	−23.3	18,870 ± 160	22,910–22,530	22,740
Beta-112900	AMS	Kamishirataki 8	−24.2	18,770 ± 170	22,820–22,460	22,650
Beta-101793	AMS	Kamishirataki 8	−22.4	18,510 ± 270	22,640–22,010	22,630
Beta-112889	AMS	Kamishirataki 2	−25.2	18,620 ± 160	22,680–22,330	22,490
Beta-101788	AMS	Kamishirataki 8	−24.3	18,580 ± 60	22,510–22,380	22,450
Beta-126156	AMS	Okushirataki 1	−24.2	18,360 ± 140	22,400–22,060	22,220
Beta-112878	AMS	Okushirataki 1	−23.5	18,250 ± 190	22,330–21,900	22,100
Beta-112888	AMS	Kamishirataki 2	−24.2	18,050 ± 190	22,140–21,620	21,870
PLD-3315	Decay Counting	Hattoridai 2	−22.8	17,900 ± 60	21,810–21,590	21,700
Beta-150432	AMS	Kamishirataki 2	−25.3	17,740 ± 110	21,660–21,300	21,480
Beta-112885	AMS	Kamishirataki 2	−24.4	17,670 ± 180	21,630–21,100	21,370
Beta-112898	AMS	Kamishirataki 8	−25.6	16,410 ± 150	19,980–19,610	19,800
Beta-112877	AMS	Okushirataki 1	−25.6	16,030 ± 130	19,520–19,170	19,340
Beta-186201	AMS	Kamishirataki 8	−23.6	16,000 ± 90	19,460–19,180	19,310
Beta-112882	AMS	Okushirataki 1	−23.6	15,850 ± 200	19,370–18,900	19,140
Beta-112879	AMS	Okushirataki 1	−25	15,850 ± 150	19,300–18,920	19,130
Beta-112880	AMS	Okushirataki 1	−25	15,850 ± 150	19,300–18,920	19,130
Beta-112881	AMS	Okushirataki 1	−27	15,850 ± 130	19,270–18,940	19,130
Beta-126163	AMS	Kamishirataki 8	−25	15,790 ± 110	19,180–18,910	19,050
Beta-136459	AMS	Hattoridai 2	−23.7	15,730 ± 70	19,050–18,880	18,970
Beta-112897	AMS	Kamishirataki 8	−25.7	15,660 ± 240	19,200–18,670	18,940
Beta-150445	AMS	Hattoridai 2	−24.9	15,610 ± 50	18,910–18,800	18,850
Beta-136460	AMS	Hattoridai 2	−23.2	15,700 ± 70	19,020–18,860	18,940
Beta-136461	AMS	Hattoridai 2	−23.1	15,660 ± 70	18,970–18,820	18,900
Beta-150445	AMS	Hattoridai 2	−24.9	15,610 ± 50	18,910–18,800	18,850

Beta-112873	AMS	Okushirataki 1	-25.1	15,570 ± 130	18,950–18,690	18,830
Beta-112874	AMS	Okushirataki 1	-25.6	15,570 ± 190	19,020–18,620	18,830
Beta-112875	AMS	Okushirataki 1	-24.1	15,270 ± 150	18,700–18,370	18,530
Beta-112883	AMS	Okushirataki 1	-24.3	15,110 ± 130	18,530–18,200	18,360
Beta-112876	AMS	Okushirataki 1	-23.6	15,030 ± 120	18,420–18,100	18,260
Beta-112909	AMS	Kamishirataki 8	-25.3	14,140 ± 110	17,390–17,050	17,210
Beta-18200	AMS	Kamishirataki 8	-24	13,930 ± 80	17,050–16,740	16,890
PLD-3321	Decay Counting	Shirataki 3	-24.4	13,865 ± 50	16,930–16,670	16,790
IAAA-51633	AMS	Shirataki 3	-22.6	13,810 ± 70	16,860–16,560	16,710
PLD-3312	Decay Counting	Hattoridai 2	-25.2	13,800 ± 50	16,830–16,560	16,690
Beta-126153	AMS	Hattoridai 2	-25.3	13,790 ± 50	16,810–16,540	16,680
Beta-112884	AMS	Kamishirataki 2	-25.4	13,750 ± 170	16,880–16,350	16,630
IAAA-51634	AMS	Shirataki 3	-21.2	13,740 ± 70	16,750–16,440	16,600
PLD-3313	Decay Counting	Hattoridai 2	-22.5	13,725 ± 45	16,690–16,430	16,570
Beta-126152	AMS	Hattoridai 2	-24.8	13,680 ± 50	16,600–16,360	16,500
Beta-150440	AMS	Hattoridai 2	-24.1	13,430 ± 80	16,280–16,040	16,160
PLD-3322	Decay Counting	Shirataki 3	-24.8	12,940 ± 45	15,570–15,340	15,460

seems to have changed relatively rapidly. Again, based on the number of ^{14}C dates in Figure 4.6, human populations on PSHK appear to have steadily declined until around 17,000 years ago, the beginning of the Kenbuchi Stadial (Igarashi 2016), when, like around 23,500 years ago, just after the LGM peaked, PSHK human population levels may have risen enough to prompt exodus.

Event 5: Human populations expanded into North America (17,000–16,000 years ago)

PSHK origins would explain differences between early microblade cores at Swan Point and later forms in Denali assemblages (13,000–10,000 years ago). With older Yubetsu cores associated with Palaeo Siberians (Sikora et al. 2019) and Campus cores, such as at the Upward Sun River site, with Palaeo Beringians (13,000 years ago) (Moreno-Mayar et al. 2018), at least two distinct waves of people reached Alaska. Furthermore, while the genetic structure of those who produced Nenana assemblages with their Chindadn points in Beringia is not known, artefacts in the same Hokkaido collections (see Figure 4.7b, artefacts in section 1 [at least 16,000 years ago]) are strikingly similar to younger examples from Nikita Lake, Chukotka (see Figure 4.7b, artefacts in section 12 [13,700 years ago]) (Pitulko et al. 2016) and Walker Road, Alaska (see Figure 4.7b, artefacts in section 11 [13,200 years ago]) (Hoffecker et al. 1993). Furthermore, although it is too young to be part of the earliest group of ancestral sites (Goebel et al. 2010), Ushki Lake is a stratified example with a younger Yubetsu microblade core complex in Layer VI (Gómez Coutouly and Ponkratova 2016) overlying a stemmed point complex in Cultural Layer VII (Dikov 1977: 50–51), demonstrating their relative temporal relationship in an area nearer the Beringian coast than Diuktai Cave or Swan Point (see Figures 4.1 and 4.7).

Event 6: Humans completely depopulate PSHK (~13,500 years ago)

Archaeological evidence indicates major drops in human populations on the peninsula, followed by clear shifts in the technological organisation on the landscape, at the height of the Kenbuchi Stadial 13,500–13,000 years ago. Reporting on the disappearance of microblades on Hokkaido, Takukara (2020) concludes that two distinct technological complexes straddle either side of a gap in the archaeological record, this time near the end of the Pleistocene. He contends that at first, before the gap, a few Jomon sites appeared on PSHK concurrent with, but mutually exclusive of, Palaeolithic localities, but later, after the gap, only Jomon existed. Around the same time, the number of ^{14}C-dated Denali sites substantially increased in Alaska (Graf and Bigelow 2011). If a relationship exists, then we would expect Denali sites to reflect the lithic variability most similar to Palaeolithic PSHK

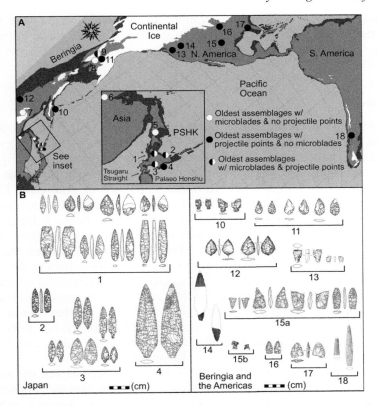

Figure 4.7 (A) Map showing locations of Hokkaido sites with Palaeolithic artefacts: 1-Hattoridai, Kamishirataki 2, Okushirataki 1, Shirataki 3, Shirataki 18, & Kamishirataki 8 (24-16,000 years old) (HAOC [Hokkaido Archaeological Operation Center, 2000a; 2003a); 2-Nakamoto (Kato and Kuwabara 1969); 3-Ozora (OBE [Obihiro Board of Education], 1993), Satsunai-N (MBE [Makubetsu Board of Education], 2000), Minamimachi (OBE [Obihiro Board of Education], 1995), and 4-Kushiro (found in a fishing net ~25 km off the coast of Hokkaido) (Nishi 1991). The map also shows northeast Asian sites with Yubetsu/Diuktai microblade cores potentially associated with Paleo Siberians (see Figure 3 for examples): 5-Ogonki-5 (23,000 years ago) (Vasilevskii 2008); 6-Ust'-Ulma (23,000 years ago) (Derev'anko 1998); 7-Diuktai Cave (17,500 years ago) (Dikov 1977), and 8-Swan Point (14,500 years ago) (Holmes 2001). A Denali Complex site associated with Paleo Beringians: 9-Upward Sun River (12,000 years ago) (Moreno-Mayar et al. 2018). 10-Ushki Lake, where stemmed projectile points stratigraphically underlie wedge-shaped microblade cores. Nenana Complex sites: 11-Moose Creek/Walker Road (13,500 years ago) (Hoffecker, Powers, and Goebel 1993) & 12-Nikita Lake (13,500 years ago) (Pitulko, Pavlova, and Basilyan 2016). And, early sites in the Americas with projectile points, but no microblades: 13-Paisley Cave (14,500 years ago) (Jenkins et al. 2012), 14-Coopers Ferry (13,000 years ago) (Davis et al. 2019); 15a-Friedkin (15,500 years ago) (Waters et al. 2018), 15b-Gault (15,500 years ago) (Collins et al. 2013; Williams et al. 2018), 16-Meadowcroft Rockshelter (15,500 years ago) (Adovasio et al. 1977), 17-Page-Ladson (14,500 years ago) (Halligan et al. 2016), and 18-Monte Verde (14,500 years ago) (Dillehay et al. 2019). (B) late Paleolithic projectile point variability of PSHK (Left) and pre-Clovis Americas (Right). Site and artefact numbers refer to the list in 4.7A, except note that examples labelled 1 are from a single late-glacial occupation surface at Kamishirataki 2.

assemblages. An initial comparison indicates this might be the case: Both comprise several microblade core forms, stemmed and lanceolate projectile points, large bifaces, and blades (Nakazawa et al. 2005; Goebel and Buvit 2011).

Discussion

Ideas linking Palaeolithic Japan and the first Americans are not new. Since at least the 1960s, researchers on both sides of the Pacific have proposed connections between late-glacial lithic assemblages, focusing primarily on the transmission of wedge-shaped microblade cores between Hokkaido and Alaska (Morlan 1967: 208–209; Kobayashi 1970: 42–43; Andrefsky 1987). These ideas, however, did not gain traction, but more recently, emphasis on common marine resources and projectile point forms between Japan and western North America has gained more favour (Erlandson and Braje 2011; Erlandson et al. 2015; Braje et al. 2017; Davis et al. 2019; Davis and Madsen 2020; Braje et al. 2020: 9). The new models serve as alternatives to the long-held paradigm that Palaeolithic humans undertook a terrestrial migration across Siberia and the Bering land bridge where the focus was on the trail of microblades, not projectile points. Arguably, this journey across the mammoth-steppe took place with microblade technology but halted in Alaska and did not likely lead directly to pre-Clovis North and South America.

Because no Pleistocene human fossils exist on PSHK, critics of this model have relied on Holocene-age Jomon biology to reject the hypothesis presented here (Scott et al. 2016; Hoffecker et al. 2020). There is, however, archaeological evidence against population continuity between the Palaeolithic and Jomon on PSHK. Takakura (2020) argues that the disappearance of microblades around the Younger Dryas coincides with the appearance of new cultural groups, albeit in small numbers, on the peninsula likely from Palaeo Honshu Island. Accordingly, if Palaeolithic populations vanished from PSHK near the Younger Dryas, and if no biological link exists between them and later peoples, then one cannot rule out the possibility that nothing is known about the Pleistocene human genetics of Hokkaido. Without a material link between the Palaeolithic and Jomon, it is possible that there is no blood relationship between them, and the latter's genome is irrelevant to PSHK Origins. Genetics aside, there are also characteristics the Pleistocene migrations described here share with other prehistoric examples.

Before Holocene expansion in the southern Pacific Ocean, human populations typically exceeded an island's carrying capacity, contracted, and then some inhabitants sailed to new locations. Further, it seems that socio-cultural intensification, seen, for example, as higher settlement densities, fish ponds, and food-producing fields, occurred concurrently with overall population declines (Carson and Hung 2014). This type of cycle (i.e., population rise, followed by contraction and intensification, and finally migration) also occurred with ancestral Pueblo populations in the Mesa Verde region starting around AD 600. Around AD 1275, it is well known that people completely abandoned North America's Colorado Plateau (Kohler et al. 2008). There are clear similarities in all the examples, including PSHK Origins – population increases, followed by population declines with cultural intensification, and finally, expansions into new areas. On PSHK, intensification, as well as what may have led to much of the earliest North and South American lithic artefact variability, may be demonstrated at the Shirataki sites.

Shirataki comprises 100 prehistoric sites found near the Akaishiyama obsidian source in the upper reaches of the Yubetsu River, Hokkaido, Japan. The sites are extremely important in Japanese Palaeolithic prehistory. For more than 50 years, understanding general lithic technological assemblage patterns, their temporal and spatial distributions, how they formed in terms of human behaviour, and their relationship with natural environments during the Late Pleistocene and early Holocene, have been discussed by local Japanese archaeologists in light of evidence from the Shirataki sites (Yoshizaki 1961; Shirataki Research Group 1965; Kimura 1995; Hall and Kimura 2002; Kimura and Girya 2015; Naoe 2014; Naoe, Suzuki, and Sakamoto 2016). Furthermore, Shirataki microblade assemblages are used to address several key issues in greater northeast Asia, such as the reconstruction of Yubetsu core reduction and long-distance transport of obsidian (Yoshizaki 1961; Morlan 1967; Aikens and Higuchi 1982). It is possible that before now, Shirataki assemblages have been overlooked in Palaeo American studies because in past decades when the Clovis-First hypothesis dominated most thinking, Palaeolithic archaeology in Japan was negatively affected by a scandal involving a prominent researcher salting ice-age sites (Normile 2001). As we move past that event, a new understanding of possible relationships begins to unfold, and as such, a closer look at the Shirataki sites is warranted in light of first American studies.

Between 1995 and 2008, rescue campaigns of the Shirataki sites preceded highway construction (HAOC Hokkaido Archaeological Operation Center, 2000d, 2001, 2002, 2004a, 2004b, 2006, 2007a, 2007b, 2008, 2009, 2010, 2011, 2012, 2013, 2015) amounting to an excavation area of nearly 1,230,000 m^2 and more than 6,700,000 lithic artefacts (weight = 13.6 t), the vast amount of which can be refit almost in their entirety allowing archaeologists to investigate reduction sequences in extraordinary detail. The Shirataki assemblages are divided into 16 stone industries that involve, in part, tools made from trapezoids, flakes, blades, and microblades, as well as bifacial, stemmed and lanceolate points (see Figure 4.4) Naoe, Suzuki, and Sakamoto 2016). Moreover, Shirataki sites are chronologically consistent with the general temporal framework of Hokkaido prehistory, where various Upper Paleolithic assemblages fall between 30,000 and 12,000 years old (Izuho et al. 2012; Izuho et al. 2018), likely suggesting that diversity in lithic industries reflects changes in human adaptation to late Pleistocene ecosystems.

Contextually, almost all Shirataki assemblages are buried in loamy deposits of sub-aeolian to colluvial facies shallower than 30 cm from the surface. According to geomorphological research conducted in the upper Yubetsu River Valley, Shirataki sites are stratigraphically above the 30,000-year-old Ds-Oh tephra (Nakamura, Hirakawa, and Naganuma 1999; Nakamura and Hirakawa 2000). The spatial distributions of artefacts, however, show mild post-depositional disturbance due to solifluction and limited movement on gentle slopes, enough to make the age of the lithic scatters difficult to interpret (Izuho and Hayashi 2015). Many of the AMS ^{14}C dates on charcoal described as associated with stone artefacts sometimes contradict with archaeologists' techno-typological expectations. Notwithstanding, some past attempts at gaining chronological control of the Shirataki assemblages are consistent with the understanding of Japan's Palaeolithic prehistory (Naoe 2014; Naoe et al. 2016).

Even though his assessment is far from sufficient, and thus the chronology requires a critical re-examination, Naoe's (2014, 2016) framework is largely consistent with what is understood about the Upper Palaeolithic of Hokkaido (Izuho et al. 2018). For example, at the Kyu-Shirataki 3 site, studies of formation processes and post-depositional disturbance suggest that the microblade component was buried in thick, sandy mud of aeolian/slope-wash facies, while the site's trapezoids were from sandy silt of flood plain facies in the lower gully fill deposits, suggesting that the microblades are younger than the trapezoids. In general, however, many of the artefacts are found together in seemingly the same cultural layers and occupation surfaces (e.g., Kamishirataki 2; HAOC Hokkaido Archaeological Operation Center 2000a), indicating they coexisted after coming together at different times in the past. As such, numerical ages of the Shirataki sites should be studied in much more detail but approached with some caution for now (Izuho et al. 2015).

The microblade components of the Shirataki assemblages notwithstanding, however, their stemmed and lanceolate projectile points are especially important to PSHK Origins. Naoe et al. (2016) classified Shirataki bifacial stemmed points into four groups labelled IV through I (Figure 4.8), and to avoid confusion in terminology, we refer to their Groups I, II, and IV "tanged" points instead as "stemmed" points. The group III points are wider and have clear, contracting basal stems with distinct shoulders like those found at the Kamishirataki 8 and Shirataki 18 sites estimated at older than 16,000 years (Naoe et al. 2016). Bifacial points from Groups II and I are generally referred to as Tachikawa points (Yoshizaki 1961; Pratt et al. 2020) as at Kamishirataki 2 and Shirataki 3. These points were made by intensive bifacial reduction creating slender bodies with parallel sides, as well as short, narrow, contracting stems with distinct shoulders. Naoe et al (2016) chronologically placed them between 16,000 and 13,000 years old. Group IV includes specimens with denticulated edges such as from Kyushirataki 5, resembling stemmed points mainly found at Incipient Jomon sites on Honshu estimated at younger than 13,000 years old. While better chronological

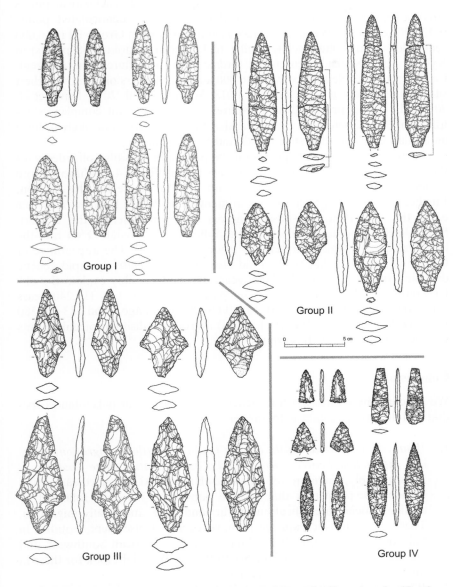

Figure 4.8 Examples of Naoe et al.'s (2016) PSHK projectile point classification. Note that if fractured correctly at the shoulders, all Group III points (>16,000 years old) bear a resemblance to the examples in Figure 4.7B (on the right of 14) from Cooper's Ferry, Idaho, and the Friedkin, Texas (at the far left of 15a).

control is desired, it seems that groups I and II Tachikawa points (16,000–13,000 years old) are older than Group IV denticulated points (<13,000 years old). While the bifacial stemmed points in Group III (>16,000 years old) are morphologically similar to Hakuhen Sentoki (H-S points) in Kyushu and Korea, their blank preparation procedures are different. Considering possible ages between 16,000 and 15,500 years old for the earliest sites in the Americas (Waters et al. 2018; Williams et al. 2018; Davis et al. 2019), the model presented here requires a projectile point complex established on PSHK before this, meaning that Group III are most likely to have an affiliation to the earliest in North and South America.

The arrival of Group III points on the peninsula possibly coincided with the spike in ^{14}C dates around 19,000 years ago in Figure 4.6. Without more evidence to the contrary, there is no reason to reject this notion. Furthermore, if point technologies originated on Palaeo Honshu, as seems to be the case, their transmission across the open water of the Tsugaru Strait, likely in boats, could explain an influx of East Asian DNA into the ancestral Native American genome prior to isolation. The appearance of Group III forms at that time would also allow for relatively rapid genetic diversification until around 17,000 years ago and possibly explain the differentiation of Indigenous American paternal lineages L54 (17,240 years ago), M1107 (17,020 years ago), M930 (16,290 years ago), and M3 (14,580 years ago) described by Wei et al. (2018) if groups utilising projectile points were leaving PSHK at those times.

Conclusions

We conclude by addressing four general questions about palaeolandscapes that PSHK Origins might help answer.

1. *What can we learn about the ancient landscape settings, migration routes, and interaction between people and these migration-route landscapes?* Most notably, there are recognisable characteristics in northeast Asian Palaeolithic prehistory that involve population increases followed by declines, social intensification on the landscape, and finally migration. These characteristics appear to have occurred elsewhere, notably in Palaeolithic Europe, with the Anasazi of the American Southwest, and among ancient mariners of the South Pacific. We might look for these in the archaeological record elsewhere.

2. *What are the implications for understanding how people use, create, and respond to migration routes in different parts of the world and different time periods?* Predictable events may occur. Based on PHSK Origins described here, we assess that when times were good, people would not migrate. It appears that only when population rise was not able to keep pace with environmental, and likely other, stressors, we see the manifestation of events at the core of the PSHK Origins model.

3. *What are the implications in modern and future contexts of people coping with rising sea levels around coastlines?* We would expect to see intensification on the landscape reflected in changes in the social and technological organisation (e.g., humanity will likely invent temporary solutions pushing engineering to new levels like what prehistoric northeast Asians did with microblade technology) before migration away from coastal areas. Then, we would expect a sharp population rise in interior areas, perhaps eventually followed by the complete abandonment of drowning regions nearest the oceans.

4. *What more research still will be recommended?* Clearly, better chronological control of the Shirataki sites is in order, but this will require hands-on work with existing collections and research at sites that may be difficult or impossible to revisit. Efforts along these lines, however, must occur. Another important path moving forward will be the mapping of now-submerged ancient shorelines to find potential sites. These hold the archaeological keys to refuting or supporting PSHK origins. We must also continue sourcing obsidian from the oldest collections along the Pacific Rim to trace migration routes and understand how stone tool technology was used on palaeolandscapes. Likewise, it might be worth examining if the transmission of projectile point forms continued along the Pacific Rim into the Holocene, perhaps as trade goods, for example. Some of these challenges might be overcome with high-coverage cladistic analyses, or 3D geometric morphometrics, comparing stone tool assemblages from Japan to early sites in the Americas. Finally, a single shred of Palaeolithic DNA evidence from PSHK would solve many questions raised here, and fieldwork should continue until some are found.

Acknowledgements

We wish to thank Mike Perrin and anonymous reviewers whose comments helped make PSHK Origins a stronger argument. No grants or other funding was used to support writing this chapter.

References

Adovasio, J. M., J. D. Gunn, J. Donahue, and R. Stuckenrath, 1977. Meadowcroft Rockshelter, 1977: An overview. *American Antiquity* 43: 632–651. https://doi.org/10.2307/279496

Aikens, C. Melvin, and Takayasu Higuchi, 1982. *Prehistory of Japan*. Academic Press, New York.

Andrefsky, William, Jr., 1987. Diffusion and Innovation from the Perspective of wedge shaped cores in Alaska and Japan. In *The Organization of Core Technology*, edited by Jay K. Johnson and Carol A. Morrow, pp. 13–43. Westview Press, Boulder, Colorado.

Barshinikov, G. F., and A. K. Markova, 1992. Main mammal assemblages between 24,000 and 12,000 yr BP. In *Atlas of Paleoclimates and Paleoenvironments of the Northern Hemisphere: Late Pleistocene-Holocene*, edited by B. Frenzel, M. Pecsi, and A. Velichko, pp. 61. Hungarian Academy of Sciences, Budapest.

Bezrukova, E. V., S. K. Krivonogov, H. Takahara, P. P. Letunova, K. Schichi, A. A. Abzaeva, N. V. Kulagina, and Yu. S. Zabelina, 2008. Lake Kotokel as a stratotype for the late glacial to Holocene in Southeastern Siberia. *Doklady Earth Sciences* 420: 658–663. https://doi.org/10.1134/S1028334X08040296.

Bezurovka, E., A. Abzaeva, P. Letunova, N. Kulagina, and L. Orlova, 2009. Evidence of environmental instability of the Lake Baikal Area after the last glaciation (Based on Pollen Records from Peatlands). *Archaeology, Ethnology, and Anthropology of Eurasia* 37: 17–25. https://doi.org/10.1016/j.aeae.2009.11.002.

Bezrukova, Elena V., Pavel E. Tarasov, Nadia Solovieva, Sergey K. Krivonogov, and Frank Riedel, 2010. Last glacial-interglacial vegetation and environmental dynamics in Southern Siberia: Chronology, forcing and feedbacks. *Palaeogeography, Palaeoclimatology, Palaeoecology* 296: 185–198. https://doi.org/10.1134/S1875372 813020108.

Braje Todd J., Tom D. Dillehay, Jon M. Erlandson, Richard G. Klein, and Torben C. Rick, 2017. Finding the First Americans. *Science* 358: 8–10. https://doi.org/1 0.1126/science.aao5473.

Braje, Todd J., Jon M. Erlandson, Torben C. Rick, Loren Davis, Tom Dillehay, Daryl W. Fedje, Duane Foese, et al., 2020. Fladmark +40: What have we learned about a potential Pacific Coast Peopling of the Americas? *American Antiquity* 85: 1–21. https://doi.org/10.1017/aaq.2019.80.

Burgeon, Lauriane, Ariane Burke, and Thomas Higham, 2017. Earliest human presence in North America dated to the Last Glacial Maximum: New radiocarbon dates from Bluefish Caves, Canada. *Plos One.* https://doi.org/10.1371/journal.pone.0169486.

Buvit, Ian, 2008. Geoarchaeological Investigations in the Southwestern Transbaikal Region, Russia. Unpublished doctoral dissertation, Washington State University, Pullman.

Buvit, Ian, Masami Izuho, Yorinao Shitaoka, Tsutomu Soda, and Dai Kunikita, 2014. Late pleistocene geology and paleolithic archaeology of the Shimaki Site, Hokkaido, Japan. *Geoarchaeology* 29: 221–237. https://doi.org/10.1002/gea.21474.

Buvit, Ian, Karisa Terry, Masami Izuho, Mikhail V. Konstantinov, and Aleksander V. Konstantinov, 2015. Last Glacial Maximum human occupation of the Transbaikal, Siberia. *PaleoAmerica* 1: 374–376. https://doi.org/10.1179/205555 7115Y.0000000007.

Buvit, Ian, and Karisa Terry, 2016. Outside Beringia: Why the Northeast Asian Archaeological record does not support a long Beringian standstill. *PaleoAmerica* 2: 281–285. https://doi.org/10.1080/20555563.2016.1238277.

Buvit, Ian, Masami Izuho, Karisa Terry, Mikhail V. Konstantinov, and Aleksander V. Konstantinov, 2016. Radiocarbon dates, microblades and late pleistocene human migrations in the Transbaikal, Russian and the Paleo-Sakhalin-Hokkaido-Kuril Peninsula. *Quaternary International* 425: 120–139. https://doi.org/10.1016/ j.quaint.2016.02.05.

Carson, Mike T., and Hsiao-chun Hung, 2014. Semiconductor theory in migration: Population receivers, homelands and gateways in Taiwan and Island SE Asia. *World Archaeology* 46: 502–515. https://doi.org/10.1080/00438243.2014.931819.

Chlachula, Jiri, 2001a. Pleistocene Climate change, natural environments and palaeolithic occupation of the Upper Yenisei Area, South-Central Siberia. *Quaternary International* 80–81: 101–130. https://doi.org/10.1016/S1040-6182(01)00022-2.

Chlachula, Jiri, 2001b. Pleistocene climate change, natural environments and palaeolithic occupation of the Angara-Baikal Area, East Central Siberia. *Quaternary International* 80–81: 69–92. https://doi.org/10.1016/S1040-6182(01)00020-9.

Chlachula, Jiri, 2017. Chronology and environments of the Pleistocene peopling of North Asia. *Archaeological Research in Asia* 12: 33–53. https://doi.org/10.1016/j.ara.2017.07.006.

Clark, Peter U., Arthur S. Dyke, Jeremy D. Shakun, Anders E. Carlson, Jorie Clark, Barbara Wohlfarth, Jerry X. Mitrovica, Steven W. Hostetler, and A. Marshall McCabe, 2009. The Last Glacial Maximum. *Science* 325: 710–714. https://doi.org/10.1016/S1040-6182(01)00020-9.

Collins, Michael B., Dennis J. Stanford, Darrin L. Lowery, and Bruce A. Bradley, 2013. North America before Clovis: Variance in temporal/spatial cultural patterns 27,000-13,000 cal yr BP. In *Paleoamerican Odyssey*, edited by Kelly E. Graf, Caroline V. Ketron, and Michael R. Waters, pp. 521–539. Texas A&M University Press: College Station.

Davis, Loren G., and David B. Madsen, 2020. The coastal migration theory: Formulation and testable hypotheses. *Quaternary Science Reviews* 249. https://doi.org/10.1016/j.quascirev.2020.106605.

Davis, Loren G., David B. Madsen, Lorena B. Valdivia, Thomas Higham, David A. Sisson, Sarah M. Skinner, Daniel Stueber, et al., 2019. Late Upper Paleolithic occupation at Cooper's Ferry, Idaho, USA shows Americas Settled before ~16,000 years ago. *Science* 365: 891–897. https://doi.org/10.1126/science.aax9830.

Derev'anko, A., 1998. Sites of the Selemdzha Type. In *The Paleolithic of Siberia: New Discoveries and Interpretations*, edited by Anatoly P. Derev'anko, Demitri B. Shimkin, and W. Roger Powers, 281–283. University of Illinois Press, Urbana.

Dikov, N. N., 1977. *Arkheologicheskie Pamyatniki Kamchatki, Chukotki i Verkhnei Kolimi* [Archaeological Sites of Kamchatka, Chukotka, and the Upper Kolyma]. Nauka, Moscow.

Dillehay, Thomas D., et al. 2019. A late ice-age settlement in Southern Chile. *Scientific American* 251: 106–117. https://www.jstor.org/stable/24969460.

Dillehay, T. D., C. Ocampo, J. Saavedra, M. Pino, L. Scott-Cummings, P. Kovacik, C. Silva, and R. Alvar, 2019. New excavations at the late Pleistocene Site of Chinchihuapi I, Chile. *Quaternary Research* 92: 70–80. https://doi.org/10.1017/qua.2018.145.

Dolukhanov, P. M., 2004. Prehistoric environment, human migrations and origin of pastoralism in Northern Eurasia. In *Impact of the Environment on Human Migration in Eurasia*, edited by E. M. Scott, A. Y. Alekseev, and G. I. Zaitseva, pp. 225–242. Kluwer Academic Publishers, Dordrecht, Netherlands.

Erlandson, Jon M., and J. Todd Braje, 2011. From Asia to the Americas by boat? Paleogeography, paleoecology, and stemmed points of the Northwest Pacific. *Quaternary International* 1–2: 28–37. https://doi.org/10.1016/j.quaint.2011.02.030.

Erlandson, Jon M., Todd J. Braje, Kristina M. Gill, and Michael H. Graham, 2015. Ecology of the Kelp Highway: Did marine resources facilitate human dispersal from Northeast Asia to the Americas? *The Journal of Island & Coastal Archaeology* 10: 392–411. https://doi.org/10.1080/15564894.2014.1001923.

Fagundes, Nelson J. R., Ricardo Kanitz, Roberta Eckert, Ana C. S. Valls, Mauricio R. Bogo, Francisco M. Salzano, David Glenn Smith, et al., 2008. Mitochondrial population genomics supports a single Pre-Clovis Origin with a coastal route for peopling of the Americas. *The American Journal of Human Genetics* 82: 583–592. https://doi.org/10.1016/j.ajhg.2007.11.013.

Fiedel, Stuart J., and Yaroslav V. Kuzmin, 2007. Radiocarbon date frequency as an index of intensity of paleolithic occupation of Siberia: Did humans React Predictably to Climate Oscillations? *Radiocarbon* 49: 741–756. https://doi.org/10.1017/s0033522200042624.

Flegontov, Pavel, N. Ezgi Altınışık, Piya Changmai, Nadin Rohland, Swapan Mallick, Nicole Adamski, Deborah A. Bolnick, et al., 2019. Palaeo-Eskimo genetic ancestry and the peopling of Chukotka and North America. *Nature* 570: 236–240. https://doi.org/10.1038/s41586-019-1251-y.

Furguson, Jeffrey R., Michael D. Glascock, Masami Izuho, Msayuki Mukai, Keiji Wada, and Hiroyuki Sato, 2014. Multi-method characteristics of obsidian source compositional groups in Hokkaido Island (Japan). In *Methodological Issues for Characterisation and Provenance Studies of Obsidian in Northeast Asia*, edited by Akira Ono, Michael D. Glascock, Yaroslav V. Kuzmin, and Yoshimitsu Suda, 13–32. BAR International Series 2620 Archaeopress, Oxford.

Frenzel, B., H. Beug, K. Brunnacker, D. Busche, P. Frankenberg, P. Fritz, M. Geyh, H. Hagedorn, J. Hovermann, A. Kessler, W. Konigswald, K. Krumsiek, W. Lauer, H. Mensching, H. Moser, K. Munnich, Chr. Sonntag, and R. Vinken, 1992. Climates during the last interglacial. In *Atlas of Paleoclimates and Paleoenvironments of the Northern Hemisphere: Late Pleistocene-Holocene*, edited by B. Frenzel, M. Pecsi, and A. Velichko, pp. 90–92. Hungarian Academy of Sciences, Budapest.

Goebel, Ted, 1999. Pleistocene human colonization of Siberia and peopling of the Americas: An ecological approach. *Evolutionary Anthropology* 8: 208–227. https://doi.org/10.1002/(SICI)1520-6505(1999)8:6<208::AID-EVAN2>3.0.CO;2-M.

Goebel, Ted, 2002. The 'Microblade Adaptation' and recolonization of Siberia during the late Upper Pleistocene. In *Thinking Small: Global Perspectives on Microlithization*, edited by R. G. Elston and S. L. Kuhn, pp. 117–132. Archaeological Papers of the American Anthropological Association Number 12. Wiley, Hoboken, New Jersey.

Goebel, Ted, and Ian Buvit, 2011. Introducing the archaeological record of Beringia. In *From the Yenisei to the Yukon: Explaining the Lithic Assemblage Variability in Late Pleistocene/Early Holocene Beringia*, edited by Ted Goebel and Ian Buvit, 1–32. Texas A&M University Press, College Station.

Goebel, Ted, Sergei B. Slobodin, and Michael R. Waters, 2010. New dates from Ushki-1, Kamchatka, Confirm 13,000 cal BP age for earliest paleolithic occupation. *Journal of Archaeological Science* 37: 2640–2649. https://doi.org/10.1016/j.jas.2010.05.02.

Gómez Coutouly, Yan A., 2011. Identifying pressure flaking modes at Diuktai Cave: A case study of the Siberian Upper Paleolithic Microblade tradition. In *From the Yenisei to the Yukon: Interpreting Lithic Assemblage Variability in Late Pleistocene/Early Holocene Beringia*, edited by Ted Goebel and Ian Buvit, pp. 75–90. Texas A&M University Press, College Station.

Gómez Coutouly, Yan A., 2016. Migrations and interactions in prehistoric Beringia: The Evolution of Yakutian Lithic Technology. *Antiquity* 90: 9–31. https://doi.org/10.15184/aqy.2015.176.

Gómez Coutouly, Yan Axel, and Irina Y. Ponkratova, 2016. The late pleistocene microblade component of Ushki Lake (Kamchatka, Russian Far East). *PaleoAmerica* 2: 303–331. https://doi.org/10.1080/20555563.2016.1202722.

Gómez Coutouly, Yan A., and Charles E. Holmes, 2018. The microblade industry from Swan Point Cultural Zone 4b: Technological and cultural implications from the earliest human occupation in Alaska. *American Antiquity* 83: 735–752. https://doi.org/10.1017/aaq.2018.38.

Graf, Kelly E., 2005. Abandonment of the Siberian Mammoth-Steppe during the LGM: Evidence from the calibration of ^{14}C-dated archaeological occupations. *Current Research in the Pleistocene* 22: 2–5.

Graf, Kelly E., 2008. Uncharted Territory: Late Pleistocene Hunter-Gatherer Dispersals in the Siberian Mammoth-Steppe. Unpublished doctoral dissertation, University of Nevada, Reno.

Graf, Kelly E., 2009. The good, the bad, and the ugly: Evaluating the radiocarbon chronology of the middle and late Upper Paleolithic in the Enisei River Valley, South-Central Siberia. *Journal of Archaeological Science* 36: 694–707. https://doi.org/10.1016/j.jas.2008.10.014.

Graf, Kelly E., and Nancy H. Bigelow, 2011. Human response to climate during the younger Dryas Chronozone in Central Alaska. *Quaternary International* 242: 434–451. https://doi.org/10.1016/j.quaint.2011.04.030.

Graf, Kelly E., and Ian Buvit, 2017. Human dispersal from Siberia to Beringia: Assessing a Beringian standstill in light of the archaeological evidence. *Current Anthropology* 58(17): S583–S603. https://doi.org/10.1016/j.quascirev.

Grichuk, V., 1992. Late Pleistocene vegetation history. In *Late Quaternary Environments of the Soviet Union*, edited by A. Velichko, 155–178. University of Minnesota Press, Minneapolis.

Guan, Ying, Xiaomin Wang, Fagang Wang, John Olsen, Shuwen Pei, Zhenyu Zhou, and Xing Gao, 2020. Microblade Remains from the Xishahe site, North China and their implications for the origin of Microblade Technology in Northeast Asia. *Quaternary International* 535: 38–47. https://doi.org/10.1016/j.quaint.2019.03.029.

Hall, M., and H. Kimura, 2002. Quantitative EDXRF Studies of Obsidian Sources in Northern Hokkaido. *Journal of Archeological Science* 29: 3259–3266. https://doi.org/10.1006/jasc.2002.0706.

Halligan, J., M. R. Waters, A. Perrotti, I. J. Owens, J. M. Feinberg, M. D. Bourne, B. Fenerty, B. Winsborough, D. Carlson, D. C. Fisher, T. W. Stafford, Jr., and J. S. Dunbar, 2016. Pre-Clovis Occupation 14,550 years ago at the Page-Ladson site, Florida, and the Peopling of the Americas. *Science Advances* 2: e1600375. https://doi.org/10.1126/sciadv.1600375.

HAOC (Hokkaido Archaeological Operation Center), 1999. *Chitose-shi Kashiwadai 1 Iseki: Ippan Kokudo 337-go Shin-Chitose Kuko Kanren Koji Yochinai Maizo Bunkazai hakkutsu Chosa Hokokusho* [The Chitose City Kashiwadai 1 Site: Archaeological Reports on the Excavation of National Route 337 during Roadwork within the Shin-Chitose Airport]. Hokkaido Archaeological Operation Center Foundation, Ebetsu.

HAOC (Hokkaido Archaeological Operation Center), 2000a. *Shirataki Iseki Gun II: Dai 1 Bunsatsu (Honbun Hen); Shirataki-mura, Kami-Shirataki 2 Iseki, Kami-Shirataki 6 Iseki, Kitashi Yubetsu 4 Iseki – Ippan Kokudo 450-go Shirataki-mura*

Shirataki Maruseppu Doro Koji Yochinai Maizo Bunkazai Hakkutsu Chosa Hokokusho [The Shirataki Sites II: Volume 1 (Main Text); Shirataki Village, Kami-Shirataki 2, Kami-Shirataki 6, Kitashi Yubetsu 4–Archaeological Reports on the Excavation of National Route 450 during Roadwork within the Shirataki Road Improvement Worksite]. Hokkaido Archaeological Operation Center Foundation, Ebetsu.

HAOC (Hokkaido Archaeological Operation Center), 2000b. *Shirataki Iseki Gun II: Dai 2 Bunsastsu (Sekki Jissoku Bunpuzu Hen); Shirataki-mura, Kami-Shirataki 2 Iseki, Kami-Shirataki 6 Iseki, Kitashi Yubetsu 4 Iseki – Ippan Kokudo 450-go Shirataki-mura Shirataki Doro Kairyo Koji Yochinai Maizo Bunkazai Hakkutsu Chosa Hokokusho* [The Shirataki Sites II: Volume 2 (Stone Tool Measurements and Distribution Maps); Shirataki Village, Kami-Shirataki 2, Kami-Shirataki 6, Kitashi Yubetsu 4–Archaeological Reports on the Excavation of National Route 450 during Roadwork within the Shirataki Road Improvement Worksite]. Hokkaido Archaeological Operation Center Foundation, Ebetsu.

HAOC (Hokkaido Archaeological Operation Center), 2000c. *Shirataki Iseki Gun II: Dai 3 Bunsatsu (Shashin Zuhan Hen); Shirataki-mura, Kami-Shirataki 2 Iseki, Kami-Shirataki 6 Iseki, Kitashi Yubetsu 4 Iseki – Ippan Kokudo 450-go Shirataki-mura Shirataki Doro Kairyo Koji Yochinai Maizo Bunkazai Hakkutsu Chosa Hokokusho* [The Shirataki Sites II: Volume 3 (Photographs); Shirataki Village, Kami-Shirataki 2, Kami-Shirataki 6, Kitashi Yubetsu 4–Archaeological Reports on the Excavation of National Route 450 during Roadwork within the Shirataki Road Improvement Worksite]. Hokkaido Archaeological Operation Center Foundation, Ebetsu.

HAOC (Hokkaido Archaeological Operation Center), 2000d. *Shirataki Isekigun I* [The Shirataki Group of Sites I]. Hokkaido Archaeological Operation Center Foundation, Ebetsu.

HAOC (Hokkaido Archaeological Operation Center), 2001. *Shirataki Isekigun II* [The Shirataki Group of Sites II]. Hokkaido Archaeological Operation Center Foundation, Ebetsu.

HAOC (Hokkaido Archaeological Operation Center), 2002. *Shirataki Isekigun III* [The Shirataki Group of Sites III]. Hokkaido Archaeological Operation Center Foundation, Ebetsu.

HAOC (Hokkaido Archaeological Operation Center), 2003a. *Shirataki Iseki Gun IV: Dai 3 Bunsatsu (Shashin Zuhan Hen); Shirataki-mura, Oku Shirataki 11 Iseki, Kami-Shirataki 8 Iseki, Kami-Shirataki 6 Iseki (2), Shirataki Dai 30 Jiten Iseki – Ippan Kokudo 450-go Shirataki-mura Shirataki Doro Kairyo Koji Yochinai Maizo Bunkazai Hakkutsu Chosa Hokokusho* [The Shirataki Sites II: Volume 3 (Photographs); Shirataki Village, Oku-Shirataki 11, Kami-Shirataki 8, Kami-Shirataki 6 (Part 2), Shirataki Waypoint 30–Archaeological Reports on the Excavation of National Route 450 during Roadwork within the Shirataki Maruseppu Road Worksite]. Hokkaido Archaeological Operation Center Foundation, Ebetsu.

HAOC (Hokkaido Archaeological Operation Center), 2003b. *Shirataki Iseki Gun IV: Dai 2 Bunsatsu (Sekki Jissoku Bunpuzu Hen); Shirataki-mura, Oku Shirataki 11 Iseki, Kami-Shirataki 8 Iseki, Kami-Shirataki 6 Iseki (2), Shirataki Dai 30 Jiten Iseki – Ippan Kokudo 450-go Shirataki-mura Shirataki Doro Kairyo Koji Yochinai Maizo Bunkazai Hakkutsu Chosa Hokokusho* [The Shirataki Sites II: Volume 2 (Stone Tool Measurements and Distribution Maps); Shirataki Village, Oku-Shirataki 11,

Kami-Shirataki 8, Kami-Shirataki 6 (Part 2), Shirataki Waypoint 30–Archaeological Reports on the Excavation of National Route 450 during Roadwork within the Shirataki Maruseppu Road Worksite]. Hokkaido Archaeological Operation Center Foundation, Ebetsu.

HAOC (Hokkaido Archaeological Operation Center), 2003c. *Shirataki Iseki Gun IV: Dai 1 Bunsatsu (Honbun Hen); Shirataki-mura, Oku Shirataki 11 Iseki, Kami-Shirataki 8 Iseki, Kami-Shirataki 6 Iseki (2), Shirataki Dai 30 Jiten Iseki – Ippan Kokudo 450-go Shirataki-mura Shirataki Doro Kairyo Koji Yochinai Maizo Bunkazai Hakkutsu Chosa Hokokusho* [The Shirataki Sites II: Volume 1 (Main Text); Shirataki Village, Oku-Shirataki 11, Kami-Shirataki 8, Kami-Shirataki 6 (Part 2), Shirataki Waypoint 30–Archaeological Reports on the Excavation of National Route 450 during Roadwork within the Shirataki Maruseppu Road Worksite]. Hokkaido Archaeological Operation Center Foundation, Ebetsu.

HAOC (Hokkaido Archaeological Operation Center), 2004a. *Shirataki Isekigun IV* [The Shirataki Group of Sites IV]. Hokkaido Archaeological Operation Center Foundation, Ebetsu.

HAOC (Hokkaido Archaeological Operation Center), 2004b. *Shirataki Isekigun V* [The Shirataki Group of Sites V]. Hokkaido Archaeological Operation Center Foundation, Ebetsu.

HAOC (Hokkaido Archaeological Operation Center), 2006. *Shirataki Isekigun VI* [The Shirataki Group of Sites VI]. Hokkaido Archaeological Operation Center Foundation Ebetsu: Ebetsu.

HAOC (Hokkaido Archaeological Operation Center), 2007a. *Shirataki Isekigun VII* [The Shirataki Group of Sites VII]. Hokkaido Archaeological Operation Center Foundation, Ebetsu.

HAOC (Hokkaido Archaeological Operation Center), 2007b. *Shirataki Isekigun VIII* [The Shirataki Group of Sites VIII]. Hokkaido Archaeological Operation Center Foundation, Ebetsu.

HAOC (Hokkaido Archaeological Operation Center), 2008. *Shirataki Isekigun IX* [The Shirataki Group of Sites IX]. Hokkaido Archaeological Operation Center Foundation, Ebetsu.

HAOC (Hokkaido Archaeological Operation Center), 2010. *Shirataki Isekigun X* [The Shirataki Group of Sites X]. Hokkaido Archaeological Operation Center Foundation, Ebetsu.

HAOC (Hokkaido Archaeological Operation Center), 2011. *Shirataki Isekigun XI* [Shirataki Group of Sites XI]. Hokkaido Archaeological Operation Center Foundation, Ebetsu.

HAOC (Hokkaido Archaeological Operation Center), 2012. *Shirataki Isekigun XII* [The Shirataki Group of Sites XII]. Hokkaido Archaeological Operation Center Foundation, Ebetsu.

HAOC (Hokkaido Archaeological Operation Center), 2013. *Shirataki Isekigun XIII* [The Shirataki Group of Sites XIII]. Hokkaido Archaeological Operation Center Foundation, Ebetsu.

HAOC (Hokkaido Archaeological Operation Center), 2015. *Shirataki Isekigun XIV* [The Shirataki Group of Sites XIV]. Hokkaido Archaeological Operation Center Foundation, Ebetsu.

Hirasawa, Yu, and Charles E. Holmes, 2017. The relationship between microblade morphology and production technology in Alaska from the perspective of the

Swan Point Site. *Quaternary International* 442: 104–117. https://doi.org/10.1016/j.quaint.2016.07.021.

Hoffecker, J. F., W. R. Powers, and T. Goebel, 1993. The Colonization of Beringia and the peopling of the New World. *Science* 259: 46–53. https://doi.org/10.1126/science.259.5091.46.

Hoffecker, John F., Scott A. Elias, and Dennis H. O'Rourke, 2014. Out of Beringia? *Science* 343: 979–980. https://doi.org/10.1126/science.1250768.

Hoffecker, John F., Scott A. Elias, Dennis H. O'Rourke, G. Richard Scott, and Nancy H. Bigelow, 2016. Beringia and the global dispersal of modern humans. *Evolutionary Anthropology* 25: 64–78. https://doi.org/10.1002/evan.21478.

Hoffecker, John F., Scott A. Elias, and Olga Potapova, 2020. Arctic Beringia and Native American Origins. *PaleoAmerica* 6: 158–168. https://doi.org/10.1080/20555563.2020.1725380.

Holmes, Charles E., 2001. Tanana River Valley Archaeology circa 14,000 to 9000 B.P. *Arctic Anthropology* 38: 154–170. https://www.jstor.org/stable/40316728.

Igarashi, Yaeko, 2016. Vegetation and climate during the LGM and the last deglaciation on Hokkaido and Sakhalin Islands in the Northwest Pacific. *Quaternary International* 425: 28–37. https://doi.org/10.1016/j.quaint.2016.05.018.

Igarashi, Yaeko, and Alexander Zharov, 2011. Climate and vegetation change during the late Pleistocene and Early Holocene in Sakhalin and Hokkaido, Northeast Asia. *Quaternary International* 237: 24–31. https://doi.org/10.1016/j.quaint.2011.01.005.

Igarashi, Yaeko, Masanobu Yamamoto, and Ken Ikehara, 2011. Climate and Vegetation in Hokkaido, Northern Japan, since the LGM: Pollen records from Core GH02-1030 off Tokachi in the Northwestern Pacific. *Journal of Asian Earth Sciences* 40: 102–1110. https://doi.org/10.1016/j.jseaes.2010.08.001.

Iizuka, Fumie, 2018. The timing and behavioral context of the Late-Pleistocene adoption of ceramics in Greater East and Northeast Asia and the First People (Without Pottery) in the Americas. *PaleoAmerica* 4: 276–324. https://doi.org/10.1080/20555563.2018.1563406.

Iwase, Akira, Jun Hashizume, Masami Izuho, Keiichi Takahashi, and Hiroyuki Sato, 2011. Timing of Megafaunal Extinction in the Late Pleistocene on the Japanese Archipelago. *Quaternary International* 255: 114–124. https://doi.org/10.1016/j.quaint.2011.03.029.

Iwase, Akira, Keiichi Takahashi, and Masami Izuho, 2015 Further study on the Late Pleistocene megafaunal extinction in the Japanese Archipelago. In *Emergence and Diversity of Modern Human Behavior in Paleolithic Asia*, edited by Yousuke Kaifu, Masami Izuho, Ted Goebel, Hiroyuki Sato, and Akira Ono, pp. 325–344. Texas A&M University Press, College Station.

Izuho, M., and K. Hayashi, 2015. Kyushirataki 3 Iseki no Taisekibutsu Ryudo Bunseki oyobi Dojyo Kagakusei Bunseki Kekka Hokoku [Report of the Analyses of Sediments and Soil Chemistry from the Kyushirataki-3 Site. In *Shirataki Isekigun XIV* [The Shirataki Group of Sites XIV], edited by the Hokkaido Archaeological Operation Center Foundation, pp. 106–115. Hokkaido Archaeological Operation Center Foundation, Ebetsu.

Izuho, Masami, 2013. Human technological and behavioral adaptation to landscape changes around the last glacial maximum in Japan: A focus on Hokkaido. In

Paleoamerican Odyssey, edited by Kelly E. Graf, Caroline V. Ketron, and Michael R. Waters, 45–64. Texas A&M University Press, College Station.

Izuho, Masami, Fumito Akai, Yuichi Nakazawa, and Akira Iwase, 2012. The Upper Paleolithic of Hokkaido: Current Evidence and its Geoarchaeological Framework. In *Environmental Changes and Human Occupation in East Asia during OIS3 and OIS 2*, edited by Akira Ono and Masami Izuho, 109–128. BAR International Series 2352 Archaeopress, Oxford.

Izuho, Masami, Dai Kunikita, Yuichi Nakazawa, Noriyoshi Oda, Koichi Hiromatsu, and Osamu Takahashi, 2018. New AMS dates from the Shukubai-Kaso Site (Loc. Sankakuyama), Hokkaido (Japan): Refining the chronology of small flake-based assemblages during the Early Upper Paleolithic in the Paleo-Sakhalin-Hokkaido-Kuril Peninsula. *PaleoAmerica* 4: 134–150.

Jenkins, Dennis L., Loren G. Davis, Thomas W. Stafford, Jr., Paula F. Campos, Bryan Hockett, George T. Jones, Linda S. Cummings, et al., 2012. Clovis Age Western stemmed projectile points and Human Coprolites at the Paisley Caves. *Science* 337: 223–228. https://doi.org/10.1126/science.1218443.

Kato, Shimpei, and Mamoru Kuwabara, 1969. *Nakamoto Iseki: Hokkaido Sendoki Iseki no Hakkutsu Hokoku* [The Nakamoto Site: A Hokkaido Pre-Ceramic Site Report]. Eiritsu Shuppan, Tokyo.

Kawamura, Yoshinari, 1991. Quaternary mammalian faunas in the Japanese Islands. *The Quaternary Research* 30: 213–220. https://doi.org/10.4116/jaqua.30.213.

Kimura, H., 1995. Obsidian-Humans-Technology. In *Paleoekologiya i Kultury Kamennogo Veka Severnoi Azii i Sopredelnykh Territorrii, Tom 2* [Palaeoecology and Stone-Age Cultures of Northern Asia and Adjacent Territories, Volume 2], edited by A. Derev'yanko, pp. 302–314. Russian Federation Institute of Archaeology and Ethnography, Novosibirsk.

Kimura, H., and E. Girya, 2015. Human activity patterns at the Horokazawa Toma Upper Paleolithic Stone Tool manufacturing site in the Shirataki Obsidian Source Area: Combining excavation with experimentation. *Quaternary International*. https://doi.org/10.1016/j.quaint.2015.04.015.

Kobayashi, Tatsuo, 1970. Microblade Industries in the Japanese Archipelago. *Arctic Anthropology* 7: 38–58. https://www.jstor.org/stable/40315740.

Kohler, Timothy A., Mark D. Varien, Aaron M. Wright, and Kristin A. Kuckelman, 2008. Mesa Verde Migrations. *American Scientist* 96: 146–153. https://doi.org/1 0.1511/2008.70.146.

Konstantinov, Mikhail V., 1994. *Kameni Vek Vostochovo Regiona Baikal'skoi Azii* [The Stone Age of the Eastern Region of Baikal Asia]. Russian Federation: Chita State Pedagogical Institute, Chita.

Kuzmin, Yaroslav V., 1997. Pleistocene geoarchaeology of the Russian Far East: Updated results. *Anthropologie* 32: 131–136. https://www.jstor.org/stable/26295787.

Kuzmin, Yaroslav V., 2008. Siberia at the Last Glacial Maximum: environment and archaeology. *Journal of Archaeological Research* 16:163–221. https://doi.org/10.1 007/s10814-007-9019-6.

Kuzmin, Yaroslav V., and Susan G. Keates, 2005. Dates are not just data: Paleolithic Settlement patterns in Siberia derived from radiocarbon records. *American Antiquity* 70: 889–892. https://doi.org/10.2307/40035874.

Kuzmin, Ya. V., L. A. Orlova, V. N. Zenin, L. V. Lbova, and V. N. Dementev, 2011. Radiouglerodnoe Datirovanie Paleolita Sibiri i Dal'nevo Vostoka Rossii:

Materiali k Katalagu ^{14}C Dat (po Sostoyaniyu na Konets 2010 g.) [Radiocarbon Dates of Paleolithic Siberian and Far Eastern Russia: Material from the Catalogue of ^{14}C Dates (at the end of 2010)]. *Stratum plus 1*: 171–200.

Lbova, Ludmila, 2000. *Palaeolit Severnoi Zoni Zapanogo Zabaikal'ia* [The Palaeolithic of the Northern Zone of the Western Transbaikal]. Russian Federation: Buryat Science Center, Ulan-Ude.

Maloletka, A., 1998. The Quaternary Paleogeography of North Asia. In *The Paleolithic of Siberia: New Discoveries and Interpretations*, edited by Anatoly P. Derevianko, Dimitri B. Shimkin, and W. Roger Powers, 14–22. University of Illinois Press, Urbana.

MBE (Makubetsu Board of Education), 2000. *Satsunai Yon Iseki: Nochi Kairyo ni Tomonau Hakkutsu Chosa Hokokusho* [The Satsunai 4 Site: Archaeolological Survey for Agricultural Improvement]. Makubetsu Town Board of Education, Makubetsu.

Mochanov, Yuri A, and Svetlana A. Fedoseeva, 1996. Dyuktai Cave. In *American Beginnings*, edited by Fredrick Hadleigh West, 164–174. University of Chicago Press, Chicago.

Moreno-Mayar, J. V., B. A. Potter, L. Vinner, M. Stenrucken, S. Rasmussen, J. Terhorst, J. A. Kamm, et al., 2018. Terminal pleistocene Alaskan genome reveals first founding population of Native Americans. *Nature* 553: 203–207. https://doi.org/10.1038/nature25173.

Morisaki, Kazuki, Noriyoshi Oda, Dai Kunikita, Yuka, Sasaki, Yasuko Kuronuma, Akira Iwase, Takeshi Yamazaki, Naoichiro Ichida, and Hiroyuki Sato, 2019. Sedentism, pottery and inland fishing in Late Glacial Japan: A reassessment of the Maedakochi Site. *Antiquity* 93: 1442–1459. https://doi.org/10.15184/aqy.2019.170.

Morlan, Richard E., 1967. The preceramic of Hokkaido. *Arctic Anthropology* IV(1): 164–220. https://www.jstor.org/stable/i40013028.

Nakamura, Yugo, Kazuomi Hirakawa, and Takashi Naganuma 1999. Hokkaido Shirataki Iseki to Shuhen Chiiki no Tephra [Tephras at and around a Paleolithic Site in the Shirataki Basin, Eastern Hokkaido, Japan]. *Chigaku Zasshi* 108: 616–628.

Nakamura, Yugo, and Kazuomi Hirakawa, 2000. Daisetsu Ohachidaira Tephra no Ganseki Kisaigakuteki Tokucho [Petrographic Properties of Daisetsu Ohachidaira Tephra, Hokkaido, Japan]. *Kazan* 45: 281–288.

Nakazawa, Yuichi, Masami Izuho, Jun Takakura, and Satoru Yamada, 2005. Toward an understanding of technological variability in microblade assemblages in Hokkaido, Japan. *Asian Perspectives* 44(2): 276–292. https://doi.org/10.1353/asi.2005.0027.

Nakazawa, Yuichi, Akira Iwase, and Masami Izuho, 2011. Human responses to the Younger Dryas in Japan. *Quaternary International* 242: 1–18. https://doi.org/10.1016/j.quaint.2010.12.026.

Nakazawa, Yuichi, Akira Iwase, Toshiro Yamahara, and Minoru Kitazawa, 2019. A Functional Approach to the use of the Earliest Blade Technology in Upper Paleolithic Hokkaido, Northern Japan. *Quaternary International* 515: 53–65. https://doi.org/10.1016/j.quaint.2017.10.049.

Naoe, Y., 2014. Hokkaido niokeru Kyusekki Jidai kara Jyomon Jidai Sosoki ni Soto suru Sekkigun no Nendai to Hen'nen [Chronology and radiocarbon dates from the Early Upper Paleolithic to the Incipient Jomon in Hokkaido]. *Palaeolithic Research* 10: 23–39.

Naoe, Yasuo, Hiroyuki Suzuki, and Naofumi Sakamoto, 2016. Dai 11 Sho Shirataki Isekigun no Sekijin Giho [Chapter 11: The Blade Reduction Methods in the Shirataki Group of the Sites]. In *Human Society in the Pleni-Glacial*, edited by Hiroyuki Sato, Masami Izuho, and Satoru Yamada, pp. 209–234. Rokuichi Shobo Publishing, Tokyo.

Nishi, Yukitaka, 1991. Kushiro-oki Hakken no Yuzetsu Sentoki ni Tsuite [Regarding the tanged point discovered off of Kushiro, Hokkaido Coast]. *Memoirs of the Kushiro City Museum* 16: 21–24.

Normile, Dennis, 2001. Japanese fraud highlights media-driven research ethic. *Science* 291: 34–35. https://doi.org/10.1126/science.291.5501.34.

OBE (Obihiro Board of Education), 1992. *Obihiro-shi Maizo Bunkazai Chosa Hokoku Dai 11 Satsu: Obihiro Ochiai Iseki* [Obihiro City Archaeological Reports, Volume 11: The Obihiro Ochiai Site]. Obihiro City Board of Education, Obihiro, Japan.

OBE (Obihiro Board of Education), 1993. *Obihiro-shi Maizo Bunkazai Chosa Hokoku Dai 12 Satsu: Obihiro Ozora Iseki* [Obihiro City Archaeological Reports, Volume 12: The Obihiro Ozora Site]. Obihiro City Board of Education, Obihiro, Japan.

OBE (Obihiro Board of Education), 1995. *Obihiro-shi Maizo Bunkazai Chosa Hokoku Dai 14 Satsu: Obihiro Minami-cho Iseki* [Obihiro City Archaeological Reports, Volume 14: The Obihiro Minami Machi Site]. Obihiro City Board of Education, Obihiro, Japan.

Ono, Akira, Hiroyuki Sato, Takashi Tsutsumi, and Yuichiro Kudo, 2002. Radiocarbon Dates and archaeology of the Late Pleistocene in the Japanese Islands. *Radiocarbon* 44: 477–494. https://doi.org/10.1017/S0033822200031854.

Ono, Y., 1990. Kita No Rikkyo [The Northern Land Bridge]. *The Quaternary Research* 29(3): 183–192.

Pedersen, Mikkel W., Anthony Ruter, Charles Schweger, Harvey Friebe, Richard A. Staff, Kristian K. Kjeldsen, Marie L. Z. Mendoza, et al., 2016. Postglacial viability and colonization in North America's Ice-Free Corridor. *Nature* 537: 45–49. https://doi.org/10.1038/nature19085.

Pinotti, Thomaz, Anders Bergstrom, Maria Geppert, Matt Bawn, Dominique Ohasi, Wenao Shi, Daniela R. Lacerda, et al., 2019. Y Chromosome sequences reveal a short beringian standstill, rapid expansion, and early population structure of native American Founders. *Current Biology* 29: 149–157. https://doi.org/10.1016/j.cub.2018.11.029.

Pitulko, Vladimir V., 2011. The Berelekh Quest: A review of forty years of research in the Mammoth Graveyard in Northeast Siberia. *Geoarchaeology* 26: 5–32. https://doi.org/10.1002/gea.20342.

Pitulko, Vladimir V., E. Y. Pavlova, and A. E. Basilyan, 2016. Mass Accumulations of mammoth (mammoth 'graveyards') with indications of past human activity in the Northern Yana-Indighirka Lowland, Arctic Siberia. *Quaternary International* 406: 202–217. https://doi.org/10.1016/j.quaint.2015.12.039.

Potter, Ben A., Joshua D. Reuther, Vance T. Holliday, Charles E. Holmes, D. Shame Miller, and Nicholas Schmuck, 2017. Early colonization of Beringia and Northern North America: Chronology, routes, and adaptive strategies. *Quaternary International* 444: 36–55. https://doi.org/10.1016/j.quaint.2015.12.039.

Powers, William R., and John F. Hoffecker, 1989. Late Pleistocene settlement in the Nenana Valley, Central Alaska. *American Antiquity* 54: 263–287. https://doi.org/1 0.2307/281707.

Pratt, Jordan, Ted Goebel, Kelly Graf, and Masami Izuho, 2020. A Circum-Pacific perspective on the origin of stemmed points in North America. *PaleoAmerica* 6: 64–108. https://doi.org/10.1080/20555563.2019.1695500.

Rae, J. W. B., W. R. Gray, R. C. J. Willis, I. Esenman, B. Fitzhugh, M. Fotheringham, E. F. M. Littley, et al., 2020. Overturning circulation, nutrient limitation, and warming in the glacial North Pacific. *Science Advances* 6: eabd1654. https://doi.org/10.1126/sciadv.abd1654.

Rybin, Evgeny P., Arina M. Khatsenovich, Byambaa Gunchinsuren, John W. Olsen, and Nicolas Zwyns, 2016. The impact of the LGM on the development of the Upper Paleolithic in Mongolia. *Quaternary International* 425: 69–87. https://doi.org/10.1016/j.quaint.2016.05.001.

Sato, Hiroyuki, and Takashi Tsutsumi, 2007. The Japanese Microblade Industries: Technology, raw material procurement, and adaptations. In *Origin and Spread of Microblade Technology in Northern Asia and North America*, edited By Yaroslav V. Kuzmin, Susan G. Keates, and Chen Shen, pp. 53–78. Simon Fraser University: Burnaby, B. C.

Scott, G. Richard, Kirk Schmitz, Kelly N. Heim, Kathleen S. Paul, Roman Schomberg, and Marin A. Pilloud, 2016. Sinodonty, Sundadonty, and the Beringian Standstill Model: Issues of timing and migrations to the New World. *Quaternary International* 466: 233–246. https://doi.org/10.1016/j.quaint.2016.04.027.

Shirataki Research Group, 1965. *Shirataki Iseki no Kenkyu* [Investigations of the Shirataki Site]. Chigaku Dantai Kenkyukai, Tokyo.

Sikora, Martin, Vladimir V. Pitulko, Vitor C. Sousa, Morten E. Allentoft, L. Vinner, Simon Rasmussen, Ashot Margaryan, et al., 2019. The population history of Northeastern Siberia since the Pleistocene. *Nature* 570: 182–188. https://doi.org/10.1038/s41586-019-1279-z.

Straus, Lawrence Guy, 2005. Solutrean settlement of North America? A review of reality. *American Antiquity* 65: 219–226. https://doi.org/10.2307/2694056.

Straus, Lawrence Guy, 2006. Of stones and bones: Interpreting site function in the Upper Paleolithic and Mesolithic of Western Europe. *Journal of Anthropological Archaeology* 25: 500–509. https://doi.org/10.1016/j.jaa.2006.03.004.

Takakura, Jun, 2020. Rethinking the disappearance of Microblade Technology in the Terminal Pleistocene of Hokkaido, Northern Japan: Looking at the archaeological and paleoenvironmental evidence. *Quaternary* 3. https://doi.org/10.3390/quat3030021.

Tamm, Erika, Toomas Kivisild, Maere Reidla, Mait Metspalu, David G. Smith, Connie J. Mulligan, Claidio M. Bravi, et al., 2007. Beringian standstill and spread of Native American Founders. *PLoS ONE* 2: e829. https://doi.org/10.1371/journal.pone.0000829.

Terry, Karisa, Ian Buvit, and Mikhail V. Konstantinov, 2016. Emergence of a Microlithic Complex in the Transbaikal Region of Southern Siberia. *Quaternary International* 425: 88–99. https://doi.org/10.1016/j.quaint.2016.03.012.

Tseitlin, S. M., 1979. *Geologiia Paleolita Severnoi Azii* [Geology of the Paleolithic of Northern Asia]. Nauka, Moscow.

Vasilevskii, A. A., 2008. *Kamennii Vek Ostrova Sakhalin* [The Stone Age of Sakhalin Island]. Sakhalin Book Publishers, Iuzhno-Sakhalin, Russian Federation.

Vershchagin, N. and I. Kuz'mina, 1984. Late Pleistocene Mammal Fauna of Siberia. In *Late Quaternary Environments of the Soviet Union*, edited by A. A. Velichko, 219–226. University of Minnesota Press, Minneapolis.

Vachula, Richard, 2020. Alaskan Lake Sediment Records and their implications for the Beringian Standstill Hypothesis. *PaleoAmerica* 6: 303–307. https://doi.org/10.1 080/20555563.2020.1818171.

Wang, Y. J., H. Cheng, R. L. Edwards, Z. S. An, J. Y. Wu, C. -C. Shen, and J. A. Dorale, 2001. A High-resolution Absolute-dated Late Pleistocene Monsoon Record from Hulu Cave, China. *Science* 294: 2345–2348. https://doi.org/10.1126/science.1064618.

Waters, Michael R., 2019. Late Pleistocene exploration and settlement of the Americas by Modern Humans. *Science* 365: eaat5447. https://doi.org/10.1126/science.aat5447.

Waters, M. R., J. L. Keene, S. L. Forman, E. R. Prewitt, D. L. Carlson, J. E. Wiederhold, 2018. Pre-Clovis projectile points at the Debra L. Friedkin site, Texas–Implications for the Late Pleistocene peopling of the Americas. *Science Advances* 4: eaat4505. https://doi.org/10.1126/sciadv.aat4505.

Wei, Lan-Hai, Ling-Xiang Wang, Shao-Qing Wen, Shi Yan, Rebekah Canada, Vladimir Gurianov, Yun-Zhi Huang, et al., 2018. Paternal origin of Paleo-Indians in Siberia: Insights from Y-Chromosome sequences. *European Journal of Human Genetics* 26: 1687–1696. https://doi.org/10.1038/s41431-018-0211-6.

Williams, T. J., M. B. Collins, K. Rodrigues, W. J. Rink, N. Velchoff, A. Keen-Zebert, A. Gilmer, C. D. Fredrick, S. J. Ayala, and E. R. Prewitt, 2018. Evidence of an early projectile point technology in North America at the Gault Site, Texas, USA. *Science Advances* 4: eaar5954. https://doi.org/10.1126/sciadv.aar5954.

Wygal, Brian T., Kathryn E. Krasinski, Charles E. Holmes, and Barbara A. Crass, 2018. Holzman South: A Late Pleistocene archaeological site along the Shaw Creek, Tanana Valley, Interior Alaska. *PaleoAmerica* 4(1): 90–93. https://doi.org/10.1080/20555563.2017.1408358.

Yakushige, M., and H. Sato, 2014. Shirataki Obsidian exploitation and circulation in Prehistoric Japan. *Journal of Lithic Studies* 1: 319–342. https://doi.org/10.2218/jls.v1i1.769.

Yoshizaki, M., 1961. Shirataki Iseki to Hokkaido no Mudoki Bunka [The Shirataki Site and Nonceramic Cultures in Hokkaido]. *Ethnological Research* 26 1: 13–23.

Yuan, D., H. Cheng, R. L. Edwards, C. A. Dykoski, M. J. Kelly, M. Zhang, and J. Qing, 2004. Timing, duration, and transitions of the Last Interglacial Asian Monsoon. *Science* 304: 575–578. https://doi.org/10.1126/science.1091220.

5 From wetlands to deserts

The role of water in the prehistoric occupation of eastern Jordan

Lisa A. Maher, A. J. White, Jordan Brown,
Felicia De Peña, and Christopher J. H. Ames

In many parts of the Middle East, both in the past and present, sustainable occupation is dependent on highly variable water resources within sensitive local ecosystems. A recent study of long-term adaptations of human societies to changing environments in the Middle East, and the vulnerability of these societies to rapid changes, especially in water availability, highlights that our relationship(s) to the environment can be characterised by both precariousness and resiliency (Jones et al. 2019). However, it also reminds us of the importance of examining local palaeolandscapes and site-specific conditions that provide nuance to our understandings of particular decisions, especially as humans become increasingly impactful in transforming environments and shaping the direction of climate change. Gleick (2014) and Kelley et al. (2015), for example, both draw attention to the critical role of drought and the power struggles created by uneven access to diminishing water supplies, as one of many issues that gave rise to the current wave of conflict in Syria. Indeed, it seems that hydroscapes are becoming increasingly important to the sustainable economies and geo-politics of this region; Gleick (2014) suggests that mitigation against future conflict should thus include equitable water management strategies. As a locale of early and sustained agriculture and urbanism over the last c. 20,000 years, it seems that there are lessons to be learned from the archaeological past (Jones et al. 2019) on the topics of sustainability and resilience in the face of environmental change.

In the Azraq Basin, an extensive inland drainage basin in eastern Jordan (described in detail below), dramatic landscape changes from wetlands to desert resulted in major shifts in settlement and landscape use over time (Figure 5.1). The Azraq Basin has always been a delicate ecosystem. Existing aquifers accumulated water gradually over thousands of years and, until recently, natural spring discharge was the only major outflow of this hydrological system (Al-Kharabsheh 2000; El-Naqa et al. 2007; Noble 1998). Today, the basin's aquifers are one of Jordan's largest freshwater sources and, thus, have been deeply drilled and heavily pumped to supply the rapidly growing populations of Jordan's urban centres with fresh water. Since the 1990s, several springs in the central Azraq Oasis have ceased to flow, and the

DOI: 10.4324/9781003139553-5

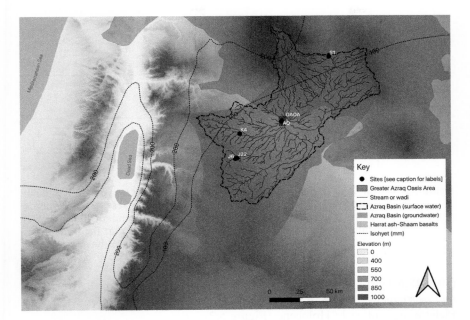

Figure 5.1 Overview of Azraq Basin in regional context. The extent of the Basin, as defined by surface drainage (shown), is delineated by a thick dashed line (from HydroSHEDS; Lehner et al. 2008). The approximate extent of the groundwater basin drained by the springs of the Azraq Oasis is shown as shaded centre-right of the image (after EXACT 1998). The Harrat ash-Shaam basalt field is shown in grey (from Pollastro et al. 1999). Selected isohyets (100 and 200 mm) are indicated by dashed lines (approximate, after EXACT 1998 and Steinel et al. 2016). Key archaeological sites are shown as dots: Ayn Qasiyya (AQ), Jilat 6 (J6) and 22 (J22), Kharaneh IV (K4), and Shubayqa 1 (S1). The Greater Azraq Oasis Area (GAOA), including Ayn Qasiyya, Ayn Sawda, C Spring, the Druze Marsh, the Lion Spring, and Shishan Marsh, is indicated by a shaded rectangle (after Ames and Cordova 2015). Topography derived from GMTED2010 (Danielson and Gesch, 2011).

once-rich marshland has largely dried up, with devastating effects for local communities. The dwindling Azraq Oasis is the last remaining system of water reserves and refuge for the wildlife that it sustains (Nelson 1973, 1985). Groundwater here has been radiometrically dated to between 25 and 12 kya (Noble 1998). This means that the water that supplies the town of Azraq with household water today had fallen as rain around 25 to 12 kya, making a direct link between the impacts of water sustainability for people of the past and those living here today. Below, we explore this link further with a case study examining the evidence for, and role of, local environmental change at the Early and Middle Epipalaeolithic site of Kharaneh IV

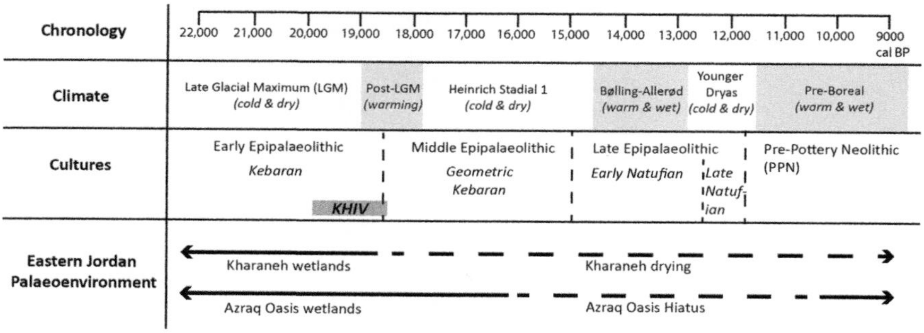

Figure 5.2 Chronology of the major phases of climate change over the Terminal Pleistocene and Early Holocene detected in Southwest Asia, alongside contemporary archaeological entities and broad trends in palaeoenvironmental interpretations for the Azraq Basin. Note that the boundaries between both archaeological entities and climate phases have associated errors and are not necessarily as abrupt as shown here.

(19.8–18.6 kya) and situate it within the context of changing hydroscapes and settlement in the Azraq Basin, and beyond, throughout the late Pleistocene and early Holocene (Figure 5.2).

Reconstructing palaeoenvironments of eastern Jordan

Palaeoenvironmental research in the Azraq Basin suggests that balancing the effects of climate change and human demands on the landscape have been a long-standing concern as conditions have dramatically changed throughout the Pleistocene, with significant impacts on human population movements and landscape use over time. At the end of the Last Glacial Maximum (LGM; 26.5–19 kya; Table 5.1), it seems the balance tipped toward increasing desertification (Jones, Maher, Macdonald, et al. 2016; Jones, Maher, Richter, et al. 2016). Recent work at the Epipalaeolithic site of Kharaneh IV (and others nearby) indicates settlement of this intensively used aggregation site around this time, followed by abandonment at the start of a second drying period, Heinrich Stadial (HS1; 18–15.5 kya, see Figure 5.2 and Table 5.1). In contrast to narratives that associate cultural change in Southwest Asia with climatic amelioration, intensively occupied sites in eastern Jordan during this time are associated with increased rates of evaporation, shrinking wetlands and disappearing resources. We discuss here reasons why the occupants of this area chose to aggregate and why they may have been buffered against aridity noted elsewhere, where persistent, concentrated resources in a region becoming more desert-like may have helped make seasonal aggregation an attractive strategy (Jones, Maher, Richter, et al. 2016; Maher 2017)

Table 5.1 Regional changes in climatic conditions of the Southern Levant during the Pleistocene-Holocene transition are by major climatic events. The effects of global changes on regional hydrology and climate during each period were collected from geoarchaeological, geomorphological, marine core, palynological, and/or speleothem research conducted at one or more sites from each region.

Southern Levantine palaeoclimate reconstructions

	Sinai Peninsula	Negev Desert	Mediterranean Coast	Western Highlands	Rift Valley	Eastern Desert	Jordan Plateau	Southern Syria
Location & Type	Gebel Maghara, Egypt Wadi	Nahal Sekher, Israel Wadi	Carmel Cave, Israel Cave	Soreq Cave, Israel Cave	Lake Lisan, Jordan Palaeolake basin	Azraq Basin & Druze Marsh, Jordan Endorheic basin	Qa'el-Jafr, Jordan Endorheic basin	Dederiyeh Cave & Bouqaia Basin, Syria Cave and alluvial fan
Type of Analysis	Geomorphology and marine cores	Geomorphology	Speleothems, geoarchaeology, and palynology	Speleothems	Geomorphology	Geomorphology, geoarchaeology, sedimentology, faunal and palaeobotanical analysis	Sedimentology and micropalaeontology	Geomorphology and geoarchaeology
Scale of Analysis	Regional	Regional	Localised to catchment	Localised to catchment	Regional	Regional	Regional	Localised to catchment & Regional
Last Glacial Maximum	Cold and hyper-arid. Palaeolakes	Cold and dry. Sand dune	Cold and dry. Paleosol formation	Cold and moist. Year-round supply	Cold with West-East seasonal	Cold and wet. Marshy but began to dry	Cool and wet. Wetlands present on	Cold and dry. Surface water present due to

(Continued)

Table 5.1 (Continued)

Southern Levantine palaeoclimate reconstructions

	Sinai Peninsula	Negev Desert	Mediterranean Coast	Western Highlands	Rift Valley	Eastern Desert	Jordan Plateau	Southern Syria
	begin to recede.	incursion begins.	during drier conditions.	of drip water available in topsoil.	cyclones (±50 mm of precipitation). Large lake present.	up around 25,000 cal BP.	landscape begin to recede by 25,000 cal BP.	reduced evaporation.
Heinrich Stadial 1	Cold and dry. Pronounced aridity, sand dune incursion.	Cold and arid. Peak of sand dune incursion.	Cold and wet. Clay deposits suggest wetland environment.	Cold and moist. Spring meltwater contributes to water table, water available in topsoil and in cave.	Cool and arid. Increased wind and hypersalinity of lake.	Cold and wet. Marshy but progressively drier, springs provided localised water sources.	Cool and dry. Water tables dropped, and brief wet/dry cycles occurred.	Cool and wet. Wet conditions and fluvial activity present.
Bølling-Allerød	Warmer and moister. Increased precipitation creates small lakes. Wadis blocked by sand dunes, playas and small lakes form.	Warmer and moister. Wadis blocked by dunes, playas and small lakes form.	Warmer and wet. Clay deposits suggest wetland environment.	Drier. Decrease in seasonality (wet/dry) and increase in aridity.	Warmer and moister. Lake levels rise dramatically.	Cool and dry. Possible damming of wadi induces localised marshes.	Cool and dry. Water tables dropped and brief wet/dry cycles.	Warmer and wet. Wet conditions and fluvial activity present.

Younger Dryas	Cool and dry. Arid phase.	Cool and drier. Seasonal and/or episodic fill of low-velocity water in dammed wadies.	Dry. Unconformity in stratigraphy, possible dry period.	Drier. Decrease in seasonality (wet/dry) and increase in aridity.	Cool and dry. Lake levels decrease.	Cool and dry. Dry lakebed with eolian activity.	Cool and dry. Water tables dropped and brief wet/dry cycles.	Cool and dry. Arid phase.
Pre-Boreal	Warmer and moister. Increased precipitation creates small lakes.	Warmer and dry. Arid conditions return.	Warmer and wet. Newly formed marshland.	Warmer and drier. Decrease in seasonality (wet/dry) and increase in aridity.	Warmer and dry. Lake levels continue to decrease.	Warmer and dry. Dry lakebed with eolian activity.	Warm and arid. Ephemeral and sporadic playas	Warmer and wet. Wet conditions and fluvial activity present.
Citations	Hamdan and Brook, 2015; Robinson et al. 2006; Roskin et al. 2014	Goring-Morris and Goldberg, 1990; Roskin et al. 2014	Kadosh et al. 2004	Bar-Mathews et al. 1999; Bar-Matthews 2011, Abed 2000; Orland et al. 2012	Ghazleh & Kempe, 2009; Rohling et al. 2013	Cordova et al. 2013; Ames and Cordova, 2015; Jones and Richter, 2011; Jones et al. 2016; Jones et al. 2016	Davies, 2005; Enzel 2008; Huckriede and Wiesemann, 1968; Rech et al. 2017; Mischke et al. 2015	Iriarte et al. 2011; Oguchi et al. 2008

The Azraq Basin as a Pleistocene refugium?

Archaeological research in the Azraq Basin focuses on reconstructing long-term changes in the availability of water and how human societies in the region adjusted their ways of life to continually changing habitats. Several recent interdisciplinary projects in the Levant have provided us with fairly detailed palaeoenvironmental datasets of climate, moisture, precipitation, and temperature changes over the last several hundreds of thousands of years (e.g., Bar-Matthews and Ayalon 2003; Bar-Matthews et al. 1997; Cordova 2007; Cordova et al. 2013; Davies 2005; Enzel et al. 2008; Enzel and Bar-Yosef 2017; Rambeau 2010; Rech 2011; Robinson et al. 2006). These data have been used to interpret palaeoclimatic changes (and their impacts) over the larger region (see Table 5.1). Yet, as we know today, the region is highly varied and represents a mosaic of microenvironments (Belfer-Cohen and Bar-Yosef 2000; Goring-Morris et al. 2009) with steep temperature and precipitation gradients. Simply put, our current regional reconstructions do not always do justice to the nuanced and different ways that a larger climate regime would have been expressed environmentally across space and over time within Southwest Asia (Jones, Maher, Richter, et al. 2016; Jones et al. 2019). The work synthesised here for the Azraq Basin is proving particularly enlightening on this front; changing habitats and land-use sustainability were likely key factors in the choices of settlement location during the last 20,000 years and beyond throughout the basin.

A general mismatch between larger regional trends in climate change and the particularities noted for the Azraq Basin has led several researchers to suggest that the Azraq Basin or, at least, parts of it, may have served as a refugium for human populations at various times throughout the Pleistocene. Indeed, the idea of the Azraq area serving as a prehistoric refugium has been suggested by archaeologists since the 1980s (Ames et al. 2014; Ames et al. In press; Copeland and Hours 1989; Copeland and Hours 1989; Cordova et al. 2013; Garrard et al. 1994; Garrard and Byrd 2013; Maher 2017; Rollefson et al. 1997). The Azraq Basin as a refugium may have been important to human groups in the Epipalaeolithic, especially in relation to sustaining populations at aggregation sites during the LGM. Noble (1998) noted that water can reside in the Azraq aquifers for thousands to tens of thousands of years. As suggested by Jones and Richter (2011) and also Cordova et al. (2013), the Azraq Oasis hydrology may not have been in sync with regional climatic changes, and notably dry or harsh periods elsewhere may have seen the persistence of lake and marsh habitats here. Thus, the normally dry and cold LGM conditions noted elsewhere in the Levant would have been buffered in Azraq by the aquifer-fed springs in the oasis, maintaining wetlands that attracted substantial faunal and human populations. A similar refugium situation, with an abundance of springs and wetlands during the LGM associated with Early Epipalaeolithic occupations, has been suggested for the Wadi Jilat to the southwest of the Azraq

Oasis (Garrard and Byrd 1992, 2013), the Azraq Basin, and the Wadi al-Hasa > 100 km to the southwest beyond the Azraq catchment (Clark et al. 2001; Coinman 1998; Hill 2001).

The proximity of most of these sites to robust springs is a double-edged sword: On the one hand, it provides the likelihood of well-preserved palaeoecological archives; on the other, it risks the possibility that these sites are not representative of the basin as a whole. As a step toward balancing this concern – guided by recent work in the Hasa and Jafr Basins (Catlett et al. 2017; Ginat et al. 2018; Mischke et al. 2015; Rech et al. 2017) – here we present site-scale palaeoenvironmental data from Kharaneh IV, a non-central locale in the Azraq Basin.

A case study: Kharaneh IV

Kharaneh IV is situated approximately 1 km southwest of Qasr Kharaneh, a large Islamic caravanserai, at an elevation of c. 640 m asl (Figure 5.3). It is located in the western end of the Azraq Basin (see Figure 5.1), within a wider landscape that proved an attractive location for human settlement throughout the Epipalaeolithic (Ames et al. 2014; Betts 1998; Copeland and Hours 1989; Cordova et al. 2013; Garrard and Byrd 2013; Jones and Richter 2011).

Previous work at the site

In 1959, G. L. Harding first mentioned Kharaneh IV as seen from Qasr Kharaneh, where he describes it as "a magnificent Upper Palaeolithic-Mesolithic site, where the ground is covered with thousands of flint implements and flakes" (1959: 146). Harding (Harding 1959: 45) posed a resounding question when he wondered how a "small white dog" he observed on multiple occasions at Qasr Kharaneh could survive in an area that is seemingly devoid of water. The site was later mentioned in a survey of sites in the Azraq Basin conducted in the 1970s by A. Garrard and N. Stanley Price (Garrard 1975–1977; Abed 2000). During the 1980s, Mujahed Muheisen conducted small test excavations at Kharaneh IV in 1981 and 1983 and wrote two dissertations (Muheisen 1983, 1988c) and published two preliminary reports on his work at the site (Muheisen 1988a, 1988b).

Muheisen's work demonstrates the importance of Kharaneh IV as one of the largest and most archaeologically dense Palaeolithic sites in the region; his small soundings quickly revealed a complex suite of archaeological remains, including architecture and human burials. Sites containing such a wide array of features were extremely rare from this period in Jordan; indeed, Kharaneh IV remains one of only two known "mega-sites" (as Muheisen referred to them) and is the only one being excavated at this time. In addition, the location of such a large occupation site within the Azraq Basin, proposed as a well-watered wetland environment during the Late

Figure 5.3 Aerial image of Kharaneh IV showing the extent of the site and with the locations of major excavation areas and geoarchaeological trenches highlighted and labelled. In Areas A and E, the deep soundings (AS42 and BS58, respectively) are marked in the larger trench by a small darkened square. A portion of the 2019 pedestrian survey discussed in the text is marked by the shaded area along the northern perimeter of the site. In the foreground of the site is the Wadi Kharaneh, flowing west to east following the local topography and draining towards the central Azraq Oasis.

Pleistocene, suggests that Kharaneh IV has the potential to provide us with key data on palaeoclimate, prehistoric technology, mortuary practices, sedentism, architecture, and plant use prior to the Neolithic period.

Recent excavations

Renewed work at the site commenced in 2006 when L. Maher and T. Richter conducted a small field season at Kharaneh IV to create a detailed map of the site's surface. Excavations at the site began in 2008 as part of the Epipalaeolithic Foragers in Azraq Project (EFAP), and to date, EFAP has completed eight excavation seasons (2008–2010, 2013, 2015–2016, 2018–2019) and five study seasons (2006, 2007, 2011 and 2017). Recent GPS and aerial drone mapping of the site confirms that Kharaneh IV covers more than 21,000 m^2, making it the largest known hunter-gatherer site in the region. The repeated occupation led to the formation of a complicated, high-resolution stratigraphic record containing evidence for hut structures, hearths, postholes, symbolic and mundane caches, flint-knapping activities, food processing, consumption and disposal areas, and human burials (Maher 2016; Maher et al. 2012). The immense size of the site, as well as its richness in stone tools, fauna, worked bone objects, ochre, marine shell beads, and archaeobotanical remains, provides excellent evidence for technological innovation, food surpluses (involving storage and feasting), and caching of utilitarian and symbolic objects by hunter-gatherer groups at Kharaneh IV (Maher et al. 2016; Spyrou 2019).

During the 2008–2013 field seasons, we were able to document the site's complete vertical stratigraphy and now better understand the excellent preservation conditions at the site, particularly in regards to charcoal samples that have provided a sequence of dates for occupation at Kharaneh IV (Macdonald et al. 2018; Maher et al. 2012; Richter et al. 2013). Our most recent work (2015–2018) focused on the excavation of at least four hut structures within the Early Epipalaeolithic occupations. These features are exceptionally rare for the Early Epipalaeolithic period and represent some of the only known habitation structures from the Early Epipalaeolithic in the Levant (Maher 2019; Maher and Conkey 2019; Maher et al. 2012). Within the deposits of one of these structures, we discovered a human burial placed inside the hut immediately prior to the intentional burning of the structure as the remains are similarly burnt (Maher et al. 2021). While hut structures and human burials are known from other Early and Middle Epipalaeolithic sites (Maher et al. 2012; Nadel 2003), this is the first time that human remains have been found in clear association with structures prior to the Natufian (Maher et al. 2021). In 2019 we focused on excavating the spaces in between hut structures in order to articulate and reconstruct the structured use of space here, as well as opening up a new area in the northern portion of the site (see Figure 5.3) and conducting an off-site geoarchaeological survey.

Extensive dating of the site's well-preserved deposits indicates that the site was occupied between 19,800 and 18,600 years ago (Richter et al. 2013), spanning the Early and Middle Epipalaeolithic and, in this 1200-year span, multi-season, prolonged and repeated habitation of the site created one of the largest Palaeolithic sites in the region. Kharaneh IV is thus ideally suited to address questions of changing hunter-gatherer lifeways and novel human-environment interactions *prior to* the origins of villages and agriculture due to its extraordinarily well preserved archaeological features, as well as the evidence for intensive occupation. The enormity of the site and an unusually high density of archaeological remains suggests that the site functioned as an aggregation locale, a focal point on the landscape where people congregated to participate in diverse economic, social, technological, and symbolic or ideological activities. Aggregation sites function as community building places, seasonally occupied by numerous hunter-gatherer groups at the same time, as part of an aggregation/dispersal mobility pattern (Maher and

Conkey 2019). Previous work at the site on the movement of material objects and technological knowledge to and from Kharaneh IV suggests these hunter-gatherer groups were involved in long-distance exchange networks enacted in an intensively used regional landscape (Maher et al. 2016; Maher 2019). The study of Kharaneh IV's various occupants can illuminate the multi-faceted, nuanced and complex relationships among and between these groups and the culturally constituted environment in which they lived, providing a useful context for understanding the ever-changing dynamics of hunter-gatherer lifeways and landscape construction.

Summary of early palaeoenvironmental investigations at Kharaneh IV

Garrard et al. (1985) characterised the generalised stratigraphy around Kharaneh IV and in the surrounding valley as lower gravel deposits containing Upper Palaeolithic artefacts, overlain by clays of a possible lacustrine origin that form the terrace that contains Kharaneh IV. The authors noted that there are no obvious signs of a basin or bedrock barrier to have allowed the formation of a lake but posited that calcreted wadi gravels may have acted as a dam. Garrard et al. (1985) reported a homogenous sandy loam above the clays that is possibly a loess similar to deposits they identified in nearby Wadi Jilat (Figure 5.4). They interpreted this sequence as representing an initial wet period with greater vegetation than at present that occurred during the deposition of clays, followed by a dry period marked by the deposition of loess that coincided with the occupation of Kharaneh IV. At some point after the deposition of cultural deposits at Kharaneh IV, the wadi became an erosional environment characterised by increased incision and terrace formation.

Besançon et al. (1989) describe the geomorphology of the Wadi Kharaneh in some detail. They noted that although the wadi has likely existed for some time, the wadis have only incised the area approximately 20 m below the bedrock platform on which Qasr Kharaneh is located. They attribute this minimal downcutting to both low annual rainfall in the region throughout the Holocene and a low gradient in this part of the Azraq Basin. Besançon et al. (1989) identified distinct terraces in Wadi Kharaneh: An upper terrace of unknown age, a middle terrace with Late Acheulian bifaces, and a lower terrace that includes Kharaneh IV. They describe the stratigraphy of the lower terrace as 30 cm of fine, light beige silt containing Epipalaeolithic artefacts above a clay horizon containing white-to-greenish particles of calcareous material, traces of roots and rare, possibly Late Palaeolithic artefacts. Below the clay is a basal pebble layer of large, unpatinated flints, including some Late Palaeolithic artefacts. Besançon et al.'s (1989) study highlights the complexity of Wadi Kharaneh's history; although the pebble layer is likely an old and intact basal depositional layer (perhaps the remains of a former channel), the wadi has experienced multiple depositional and erosional phases, filling in and being exhumed with changes in regional climate.

Figure 5.4 Views from Kharaneh IV of the surrounding landscape to the southwest (A) and south (B). To the southwest (A), the camels are grazing on the lowest terraces adjacent to the wadis that coalesce at the site. To the south (B), the middle terrace is visible on the left side of the photograph across the wadi.

On-site work

Since the first full field season at Kharaneh IV in 2008, EFAP has paired excavations with geoarchaeological investigations on- and off-site. A more thorough description of this work has been published elsewhere (Jones, Maher, Richter, et al. 2016; Jones, Maher, Macdonald, et al. 2016) and is thus only summarised here. In the main excavation areas, Area A and Area B, deep soundings were excavated at the edges of large horizontal exposures with the goal of documenting the site's full stratigraphic record and, in particular, correlating the deeply buried on-site deposits with off-site contexts examined in surrounding terraces (see Figure 5.3). The first of these deep soundings was excavated in Area B and represented our attempts in 2008 to remove the backfill from Muhseisen's deepest 2 × 2 m excavation area R/S2/60 that also served as our starting point for more extensive excavations (Figure 5.5; see also Figure 5.3). We relocated his old area and then extended and cut back the west section to fully document the site's stratigraphy here, including excavating a 1 × 1 m section of this area down to a depth of 1.9 m below the surface. After 1.7 m of artifactually dense and well-preserved occupational levels, we reached sterile deposits – a very hard, brown clay – and continued to excavate this clay for another 20–30 cm, followed by augering another c. 50 cm deep. Indeed, this extremely compact brown clay is the one exception to the carbonate-rich marls that mark the basal deposits at the site elsewhere (see below). Micromorphological and bulk sediment samples from this clay are currently being analysed and should offer insight into the significant lateral variation documented between Area B and other deep excavation areas and trenches across the site.

In Area A, excavations focused primarily on exposing contiguous areas of relatively shallow Middle Epipalaeolithic occupation deposits. However, a deep sounding in AS42 was excavated to explore the depth and character of cultural deposits in this western side of the site and articulate these with stratigraphy noted from deep soundings elsewhere (Area B, Area C, Area E and Geotrench 2009) (Macdonald et al. 2018). Here, a 1 × 1 m square was excavated to a depth of 2.3 m below the surface (637.29 m asl); 'sterile' deposits were never reached. At a depth of c. 70 cm below surface (638.97 m asl), the loose silty artefact-dense occupational deposits abruptly end, where we encountered a series of pale green and grey wetland mud and marl deposits. At the boundary between the artefact-dense occupational deposits and underlying wetland deposits is an erosional unconformity, below which there is a thin grey clay-rich deposit dense in large-mammal fauna. Under this clay was a greenish marl deposit that is extremely well-sorted and compact, with some rust-coloured iron staining; it is similar to compact deposits noted elsewhere on-site (in AZ51) and off-site in the surrounding terrace trenches (Jones et al. 2016). Below the marl is a compact brown clay-rich sediment and below this is another chalky marl lens. The deepest excavated deposit was loose brown sediment with a very low artefact density and rust-coloured iron stains throughout. While artefact densities dropped significantly after 70 cm below the surface, both Early Epipalaeolithic (Early Kebaran) microliths and faunal remains continued to be found.

Square AZ51 (in Area C) is a 1 × 1 m square excavated over the course of a week in 2009 to a depth of 1.2 m below the surface (637.90 m asl) (see Figure 5.3; see also Figure 5.5). It was placed in a small depression between Areas A and B with the aim of linking stratigraphy between these two main areas, including determining whether any Early Epipalaeolithic deposits may be present. The bottom of the excavated deposits here are 50 cm deeper than

Figure 5.5 Section photographs of the deep trenches discussed in the text; in each photograph, the boundary between dense cultural deposits and underlying palaeo-wetland deposits is marked with a dashed line in the section. (A) Muheisen's deep sounding R/S2/60 in Area B showing the cutback and extended western section. (B) AS42 in Area A showing alternations in carbonate-rich marls and brown clays discussed in the text (Macdonald et al. 2018). (C) AZ51 in Area E showing the entire section (top) and a close-up view of the greenish, carbonate-rich marls (bottom). (D) Geotrench 2009 showing the northern section (top) and a close-up view of the east section.

R/S2/60 in Area B (see below) but still about 1 m above the level of the modern terrace upon which the site sits. The uppermost deposits here are clearly Middle Epipalaeolithic in age; we documented a compact earthen surface cut by a pit. The lithic assemblage is dominated by wide trapeze/rectangles, typical of the later Middle Epipalaeolithic deposits elsewhere on site (Area A), and the fauna is dominated heavily by gazelle. After approximately 60 cm of dense occupational deposits consisting primarily of silty clays, a greenish/brown clay is encountered, and both lithic and faunal frequencies drop off dramatically. During flotation, we recovered high densities of freshwater ostracod shells (M. Jones, personnel communication 2010), suggesting that the clay was deposited within shallow water that existed in the vicinity of the site prior to Epipalaeolithic occupation. At the very bottom of this square, we encountered an archaeologically sterile, very hard white clay that suggests a period of severe aridity. It seems that occupation of this portion of the site occurred after the drying up of this wetland. It also suggests that earlier occupations of the site in the Early Epipalaeolithic may have occurred adjacent to a small freshwater source. This excavation area helped to answer if the Middle and Early Epipalaeolithic occupations overlap in this portion of the site (they do) and also contributed to documenting the complex palaeoenvironmental history of the site.

In order to link together the geomorphological work done in the surrounding landscape by M. Jones (Jones et al. 2016; Jones et al. 2016) with our excavations and on-site stratigraphy, we opened a long trench (9 × 1 m) running northward and upslope from the site boundary (at the wadi) up toward Area A. This trench, designated Geotrench 2009, allowed us to define the edges of the subsurface occupation of the site and to assess the natural and cultural processes that created and modified the site in relation to the formation of the landscape in general (see Figure 5.3; see also Figure 5.5). The southernmost 5 m of the trench, leading from the wadi upslope into the site, contained no subsurface archaeology. However, at 6 m northward along the trench, we found the beginnings of subsurface archaeology with dense deposits of *in situ* lithics and fauna, including the discovery of a small knapping area containing cores, debris, and retouched tools that are currently undergoing refit analysis. Although erosional tailings and surface deflation might slightly exaggerate the size of the site, these findings suggest that the southern boundaries of Kharaneh IV's surface and subsurface deposits coincide (see also description of the 2019 pedestrian survey below, in relation to the site's northern boundary).

The trench also allowed us to document Early Epipalaeolithic occupations underlying the Middle Epipalaeolithic horizons that dominate this portion of the site. This overlap indicates that much of the extent of the site area was in use throughout the Early and Middle Epipalaeolithic periods; the Early and Middle occupations were not merely adjacent. Lastly, the deposits in this trench document several episodes of inundation (marked by the deposition of greenish-grey marls) and drying (marked by three separate horizons of carbonate deposition) prior to occupation of the site area.

Off-site work

In 2008 and 2009, M. Jones conducted a landscape survey of the area around Kharaneh IV, including upstream and downstream of the Wadi Kharaneh and its tributaries, to document the terraces and local hydrology, past and present. Jones also excavated several small geological trenches to document and correlate their stratigraphy and collected bulk sediment samples for particle size and geochemical analyses and samples for radiocarbon and optically stimulated luminescence (OSL) dating (Jones, Maher, Richter, et al. 2016; Jones, Maher, Macdonald, et al. 2016). Through these five sedimentological sections, we were able to document two main depositional units – Holocene wadi fills and Pleistocene wetland deposits – both with lateral variation, as well as the relative extent of these palaeo-wetlands that appear to dry up between 23,000 and 19,000 cal BP (Jones, Maher, Macdonald, et al. 2016). In 2009, Jones mapped the site and immediate surroundings with a differential GPS to provide a detailed topographic map of the local landscape. This was further augmented by aerial drone photogrammetry conducted over the site in 2015 to produce a high-resolution image of the site and immediate surroundings (see Figure 5.3).

Jones et al. (2016) analysed grain size and sediment chemistry from bulk sediment samples taken on-site at Geotrench 2009 and off-site from five locations in the lower wadi terraces several hundred meters northwest and southeast of the site (Figure 5.6). Grain size is similar in on-site and off-site samples and is predominantly silts and clays, although on-site samples are slightly coarser. On-site samples display much higher magnetic susceptibility, which Jones et al. (2016) attributed to high levels of cultural burning in virtually all archaeological contexts. Off-site samples have lower percentages of SiO_2 and MgO but higher percentages of CaO and carbonate content, which Jones et al. (2016) used to interpret off-site terrace deposits as carbonate-enriched wetland sediments. Jones et al. (2016) dated off-site terrace deposits by OSL to approximately 23,000–19,000 BP, suggesting that wetland conditions existed at Kharaneh IV over that range and likely overlapped with the beginning of the site's occupation. Jones et al. (2016) addressed Garrard et al.'s (1985) concern for how a body of water could have collected in Wadi Kharaneh when there are no signs of a basin or bedrock barrier to have allowed the formation of a small lake by suggesting that wetland marls formed as groundwater discharge deposits.

Site-formation processes and site conservation efforts in defining site boundaries

While an extensive desert pavement is present on the platform containing Qasr Kharaneh to the north (Besançon et al. 1989), Wadi Kharaneh does not contain well-established pavements, most likely due to alluvial activity in the wash and few clasts within the white lower terrace marls. The exception

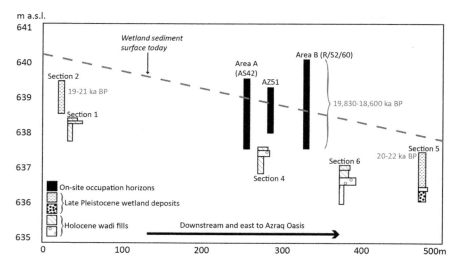

Figure 5.6 Five off-site sections and three on-site excavation areas document relative topography of the site and its surrounding environs (redrawn and modified from Maher 2017). An east-west transect through the Wadi Kharaneh shows the relative heights of the occupational horizons at Kharaneh IV and the off-site sections discussed in the text.

is at Kharaneh IV, where large flint artefacts acted in the same role as natural clasts in pavements, either accumulating through the deflation of fine particles, or the upward migration of flints through fine particles, or aeolian infilling beneath the flints (Higgitt and Allison 1998; McFadden et al. 1987; Wells et al. 1985). Although deflation during the formation of the flint pavement destroyed what were once the uppermost deposits of Kharaneh IV, the pavement protected underlying deposits as a stable ground surface and is the primary reason the site exists in an otherwise highly erosive aeolian environment.

Thanks to the conservation efforts of Muheisen, the site of Kharaneh IV is owned and maintained by the Department of Antiquities of Jordan, affording protection to the site from local development. EFAP constructed and maintains a low mudbrick wall around the site to protect it from vehicular and pedestrian traffic. Muheisen's original excavation areas have been heavily eroded and today are visible as areas slumped into the site's surface. In addition, archaeological material (flint and bone) has been washed off-site from heavy winter rains along its northern and eastern boundaries. Through the aerial survey conducted during the 2015 field season, we are monitoring the extent of this erosion. Updated photogrammetry in 2019 (and beyond) allows us to track this surface erosion along the northern extent of the site (see below). The most immediate causes of destruction now, however, are looter pits dug into the fragile and near-surface deposits and four-wheel-drive traffic over the site, which gouges into the mound, churns up and breaks artefacts and destroys fragile features like hearths. Ongoing monitoring and site conservation efforts thus remain a top priority.

Recent palaeoenvironmental work at Kharaneh IV

2019 pedestrian survey

The narrative of interwoven social and ecological shifts provided by EFAP's geoarchaeological work at Kharaneh IV to date highlights a complex set of questions regarding demography, resource management, and the structuring of social and natural space. Through reconstructing the formation and evolution of Kharaneh IV and its environs (Jones, Maher, Richter, et al. 2016; Jones, Maher, Macdonald, et al. 2016), we stand to gain insight into the long-term dynamics of hunter-gatherer lifeways in eastern Jordan. Excavations on-site at Kharaneh IV have already provided intriguing answers to some of these questions. Yet, due to the spatially extensive nature of hunter-gatherer lifeways – as well as evidence for broad interaction-spheres around Kharaneh IV, in particular – these lines of inquiry are also pursued beyond the site itself and into the surrounding area. In particular, how do we define the boundaries of Kharaneh IV? Do contemporaneous sites exist in the immediate surroundings of the site? Are they part of the occupation of the larger site?

Previous survey work in the area found surface evidence of Middle Palaeolithic, Upper Palaeolithic, Natufian, and Neolithic use of the landscape around Kharaneh IV (Garrard and Stanley Price 1975–1977) but did not find substantial evidence of other Epipalaeolithic occupations in the landscape surrounding Kharaneh IV (Garrard and Byrd, 2013). To further investigate the site boundaries, off-site activities, and near-site landscape use during the Epipalaeolithic, EFAP conducted a 45,000 m^2 pedestrian survey of the area directly north of Kharaneh IV in 2019, which consisted of six, 450-m-long and 20-m-wide transects oriented from south to north (Figure 5.7). We chose this region to determine the surface and subsurface extent of Kharaneh IV-related material beyond its existing site boundary, a feature currently defined by the extent of a low mudbrick wall constructed around the site that delineates the site's boundaries on the basis of a drop in (not a cessation of) surface artefact density. Rather than identifying every artefact encountered along the survey, we devised a systematic sampling strategy to record artefact density with distance from Kharaneh IV by counting Epipaleolithic diagnostic artefacts (over a 4 m^2 area) and all artefacts (over a 0.25 m^2 area) at 30 m intervals along the transects (see Figure 5.7). Diagnostics include geometric microliths, non-geometric microliths, crested blades, core-face rejuvenation pieces, partial-ridged blades, narrow-faced cores, and broad-faced cores (see Maher et al. 2012 for definitions). We used a handheld GPS to maintain transect bearings

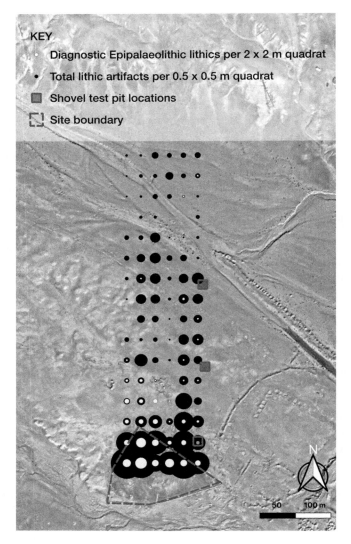

Figure 5.7 Quadrat survey of lithic artefact density (white and black circles) northward from Kharaneh IV. Circle area is proportional to artefact count (with preliminary correction for interobserver bias). Circles are centred at location of SW corner of quadrat. Counts of diagnostic Epipalaeolithic artefacts are shown as white circles. Total lithic artefact counts are shown in black. Shaded squares indicate locations of shovel test pits (see text). Approximate site boundaries are shown by the dashed line at the bottom of the image. Note the wadi running south of site boundary and calcareous terraces north of survey area.

and to locate artefact count positions. Following completion of the pedestrian survey, we dug three shovel test pits (STP), one close to the site (30 m north of boundary), one at a moderate distance (140 m north of boundary), and one far from the site (260 m north of boundary), to determine the subsurface extent of Kharaneh IV beyond its surficial site boundary (see Figure 5.7). We used a shovel and trowel to dig approximately 0.25 m^2 holes and stopped digging when the sediment became homogenous and difficult to penetrate. We used a 3 mm sieve to screen test pit sediment for artefacts. Data shown here are preliminary and will be presented in full in a subsequent publication.

The results of the pedestrian survey show that the total number of artefacts and diagnostic artefacts decrease rapidly with distance from Kharaneh IV and stabilise at low values approximately 100 m north of the site (see Figure 5.7). We encountered artefacts at each STP; however, only STP 3 contained significant archaeological deposits. STP 1 produced several non-diagnostic flakes approximately 12 cm below the surface near the contact between red and grey silt. We found several flakes just below the surface of STP 2. We identified a combustion feature approximately 7 cm below the surface in STP 3. The feature curves downward in a convex shape and contains 10 flakes, a blade, and part of a non-initial core tablet 5–15 cm below the surface in stratigraphic association with the feature.

The rapid decrease in artefact count with distance from Kharaneh IV and low artefact density in the STPs nearest the site indicate that Kharaneh IV does not extend much farther to the north than its current boundary. This finding suggests that either (a) a majority of activities at Kharaneh IV were intensely focused to a relatively small area demarcated by its modern site boundary or (b) the periphery of the site has been eroded and transported away over thousands of years to its current extent. Because diagnostic artefacts contemporary with Kharaneh IV diminish with distance from the site, it is clear that some erosion and transportation of Kharaneh IV material has occurred. However, the transportation of such material appears to be largely confined to within 100 m of the site. These tailings of surface artefacts around the site's perimeter represent a combination of downward settling and deflation of once-intact occupational deposits around the edges of the site alongside minimal downslope erosion of flints that accumulated atop of the mound with the ongoing erosion of the site. This matches previous determinations of the site's southern boundaries determined by Geotrench 2009 that suggested that the original determinations of site size by Muheisen (1983) are valid and largely reflected by accompanying subsurface stratigraphy (Jones et al. 2016).

It is most likely that activities at Kharaneh IV were concentrated and centred on its current position. Some activities at Kharaneh IV, including hunting, gathering, playing, and camping in smaller groups, must also have occurred at some distance from the site's modern boundary, but their lack of archaeological visibility, in both this survey and others by EFAP and previously (Garrard and Byrd 2013; Garrard and Stanley Price 1975–1977;

Muheisen 1983) is conspicuous. It is possible that off-site areas were used far less intensively than on-site areas or that evidence for off-site activities exist and have not yet been identified, but it is also possible that such deposits have eroded and the cultural deposits protected by Kharaneh IV surface material are largely the only remaining sediments from the time of the site's occupation on the landscape. It is perhaps conceivable that the undated combustion features identified in STP 3 could represent off-site activities associated with Kharaneh IV beyond its modern site boundary. However, Kharaneh IV's many hearths, postholes, pits, flintknapping areas, middens, and hut structures (Maher et al. 2012; Ramsey et al. 2018), as well as isotopic evidence (from gazelle cementum) indicating multi-seasonal site occupation (Henton et al. 2017), and the site's high artefact density and large size (Macdonald et al. 2018; Maher 2019) demonstrate a strong and lasting connection of Epipalaeolithic populations specifically to the area within Kharaneh IV's modern boundary.

2018–2019 geoarchaeological survey, excavations & sampling

In recent EFAP field seasons, we excavated a shallow, 13-m-long geological exploratory trench, Geotrench 2018, and a 2.3 m-deep sounding, RH1, to better characterise the site's stratigraphy determine its horizontal and vertical subsurface extent, and acquire new data for palaeoenvironmental analysis. We began Geotrench 2018 at the site's southern boundary and continued north until we encountered dense artefact deposits (see Figure 5.3). Geotrench 2018 measured 13 m long, 0.6 m wide, and varied between 20 and 84 cm deep. In 2019, we converted a "robber hole" that had penetrated dense surficial deposits in the act of looting into a geological exploration unit, RH1, located on the south-central slope of the site (see Figure 5.3). RH1 measured 1.4 m across at the surface, 0.5 m across at its base, and was 2.3 m deep. For both trenches, we used shovels and pickaxes during excavation and used trowels and brushes to expose sidewall stratigraphy for documentation and sampling. We obtained 88 sediment samples from multiple locations in the sidewall of Geotrench 2018 and 51 sediment samples from the northeast sidewall of RH1 at 5 cm intervals for ongoing geochemical research using stable isotope and faecal stanol analysis to better understand the relationship between hydroclimate and human populations at Kharaneh IV.

We identified seven sedimentological units in Geotrench 2018 and 12 distinct units in RH1. In Geotrench 2018, Layer A is a thin (<5 cm) surface palimpsest of lithic artefacts and loose brown sediment that covers the length of the trench (Figure 5.8). Layer B is a 15 cm-thick, reddish-brown silt that underlies Layer A for most of the trench and is equivalent to the "wadi silt" described by Jones et al. (2016). Layer B's abrupt and wavy lower contact suggests it is an erosional surface and a disconformity formed at some time after Kharaneh IV's occupation. Layer C is a light grey, 25 cm-thick marl that extends through the length of the trench beneath Layer B.

Figure 5.8 Annotated stratigraphy of Geotrench 2018 from two sections showing layers A, B, C, D and E (left image) and A, C, D, F, and G (right image). See text for discussion of each layer.

The layer is significantly more compact than overlying units, breaks into large aggregates, and contains gypsum and carbonate evaporites at its top. Underlying Layer D is similar to Layer C in compaction and texture, but it is dark brown and contains root traces that indicate a higher organic content than in other layers. We stopped digging at the upper centimetres of compact white marl, Layer G, similar in appearance to Layer C.

Toward the north terminus of Geotrench 2018, we encountered two layers, E and F, between Layers B and C. Layer E is similar to Layer C in texture but is lighter in colour and very loose. Layer F is a lateral continuation of Layer E that is defined by the regular appearance of *in situ* artefacts that we began to encounter approximately 11 m north into the trench. It is the only layer in the trench observed to contain bone, in addition to lithics and charcoal. The occurrence of a higher number and larger variety of artefacts suggests that Layer F marks the boundary of objects associated with Kharaneh IV's main Area B occupation.

RH1 is defined by dense cultural deposits to a depth of 60 cm (Layers 1–5) followed by a transitional unit, Layer 6, that displays a decrease in artefact density and sediment size and an increase in sediment compaction (Figure 5.9). Artefact density is near sterile in Layer 7 and continues at low levels to depth. Layers 8 and 9 are very compact, light through dark brown marls that contain large carbonate and gypsum precipitates and oxidised patches. Layer 9's base contains freshwater snail shells (*Valvata saulcyi*, Amr and Abu Baker 2004), root traces, and pockets of gleyed soil. Layer 10 is a homogenous, sterile, white fine sand underlain by basal Layers 11 and 12, comprised of very compact brown marl with unworked flint nodules.

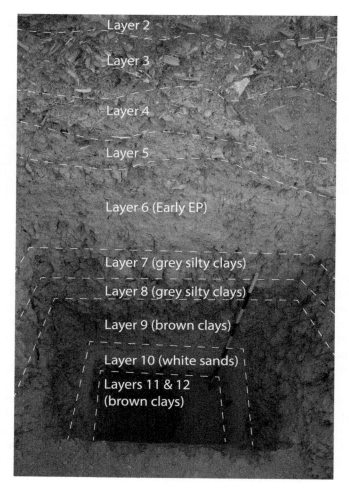

Figure 5.9 Annotated stratigraphy of the northeastern section of RH1 showing layers 1–12. See text for discussion of each layer.

Initial occupation of Kharaneh IV

Geoarchaeological excavations from the EFAP 2018 and 2019 seasons, in addition to previous investigations (see above), show a consistent stratigraphic pattern to the basal deposits throughout the site. These units, underlying the main occupational deposits in all areas, shed valuable insights into the nature of the earliest occupations at the site. The lowest units are grey-brown compact marls that have low artefact concentrations and display evidence for shallow standing water, including root traces, freshwater snail shells, and gleyed sediment patches. Above these units are marls of similar compaction and artefact density that contain large carbonate and gypsum precipitates and oxidised sediment stains that indicate an increase in evaporation and a retraction of standing water. The upper units are carbonate-compacted clays with few natural or archaeological inclusions that underlie the loose, artefact-dense cultural deposits of Kharaneh IV and likely represent further local desiccation.

Our recent palaeoenvironmental investigations support previous work (Besançon et al. 1989; Jones, Maher, Macdonald, et al. 2016; Garrard, Byrd, et al. 1985) that attributes the lower Kharaneh IV marls to formation in a wetland environment that was already retracting by the beginning of Kharaneh IV's occupation. Despite this general agreement, our recent excavations identified significant lateral variation across the site in its lower deposits and the occurrence of archaeological material within these deposits (see above). For example, in Area B, there is an abrupt boundary – an unconformity – between the dense Early Epipaleolithic occupation levels and the sterile compact brown clay with no evidence for carbonate infiltration or precipitation. In AS42, however, the basal deposits alternate between brown and greenish marls, and carbonate precipitation interfingers within these latter marls; sterile deposits were never reached despite a depth of over 2 m. Although marl units in Geotrench 2018 are similar to those recorded by Jones et al. (2016) in geoarchaeological excavations at the same elevation approximately 40 m to the west, they display different thicknesses, colours, and contacts with underlying deposits. This suggests a variation in vegetation cover and water content that created a patchwork of wet and dry areas that is inconsistent with a homogenous depositional environment like a lake. It supports the presence of groundwater discharge deposits angling "downstream" of the groundwater seep front, resulting in lateral facies variation. The Kharaneh IV wetland was a mosaic of vegetated and differentially submerged lands that provided habitat for a diverse range of species (Martin et al. 2010; Ramsey et al. 2018; Ramsey et al. 2016; Ramsey and Rosen 2016) similar to Jones' reconstruction for the Azraq Shishan during the Epipalaeolithic occupation of Ayn Qasiyya (see below) and the Azraq wetlands today.

Abandonment at Kharaneh IV

The sedimentary record leading to the beginning of occupation at Kharaneh IV is intact and allows us to form inferences on the site's initial settlement and formation of its resultant deposits. Isolated flint artefacts in lower marl deposits at Kharaneh IV indicate that the site vicinity was occupied intermittently by hunter-gatherer groups of the Azraq Basin before it was intensively settled, perhaps as a favoured locale for hunting game or gathering plant resources in the wetlands. Evidence for desiccation through evaporite precipitation, oxidation stains, and a decline in organic content show that people began to concentrate their activities at Kharaneh IV, providing evidence for sustained site use and aggregation, as the wetland retracted and exposed reliably dry land along its (shrinking) margins, suggesting a strong association between water availability and human activity at the site.

Kharaneh IV's stratigraphic record ends with uppermost deposits dated to approximately 18,600 BP (Richter et al. 2013). This site's surface is a palimpsest that prevents us from directly determining the palaeoenvironmental conditions that were present during the site's depopulation, but nearly coincident depopulations at Jilat 6 and Ayn Qasiyya (Richter et al. 2013) and a Late Pleistocene sedimentary hiatus suggesting little to no spring discharge at Ayn Qasiyya (Jones and Richter 2011) implicate regional drought in Kharaneh IV's decline. This is further supported by palaeoclimatic data from Soreq Cave, Israel (Bar-Matthews and Ayalon 2003; Grant et al. 2012) and Lake Lisan (Palaeo Dead Sea) (Bartov et al. 2003; Torfstein et al. 2013) that indicate drier conditions near the end of Kharaneh IV's occupation that coincide with HS1 (Bard et al. 2000). If regional desiccation reached the Azraq Basin, it is possible that the Kharaneh wetland, which had already experienced significant drying some thousand years previously, no longer could support large populations at Kharaneh IV. It is likely that water availability (or increased water scarcity) played an important role in the establishment of the site, but also in Epipalaeolithic people's decisions to locate elsewhere (Richter and Maher 2013) around the start of Heinrich Stadial 1.

Beyond Kharaneh IV: reconstructing prehistoric settlement in the larger Azraq Basin

The abundance of Pleistocene archaeological sites in the Azraq Basin (see Figure 5.1) has long acted as a first-order indication of considerable environmental change between that time and the present (Garrard and Byrd 2013). Today, the basin is only sparsely populated, especially with restricted access to deeply buried groundwater resources. The densest modern settlements in the basin sit close to the springs of the Azraq Oasis, situated along the northwestern margin of Qa' al-Azraq. Until recently (El-Naqa et al. 2007), these settlements subsisted on regular access to spring water and shallow groundwater aquifers that sustained settled populations in the environs of the oasis for many centuries (Nelson 1973). Indeed, older inhabitants of the town of Azraq share childhood memories of swimming in the seasonal Qa' and the perennial pools of Azraq Shishan (L. Maher, personal communication). In this regard, Pleistocene occupation of the oasis and its surrounding environs is not surprising.

However, dense archaeological sites located in parts of the basin now virtually devoid of settlement suggest strongly that the hydrological and environmental character of the basin as a whole – if not the oasis at its centre – has seen considerable change (Ames et al. 2014; Copeland and Hours 1988; Cordova et al. 2008, 2013; Garrard and Byrd 2013; Jones and Richter

2011; Maher 2017; Richter 2017). The geographic distribution of these sites varies through time, suggesting that hominins occupying the area adjusted their patterns of mobility and settlement to the changing environment (Richter and Maher 2013; Richter et al. 2013). However, the diversity of subsistence strategies, settlement intensities, cultural proclivities, and taphonomic histories that these sites represent means that we cannot directly infer the environmental history of the Azraq Basin from site distributions alone. Similarly, the heterogeneity of environmental responses to climate change in arid regions makes reconstructing palaeoenvironments on the basis of palaeoclimate archives alone a tall order.

To bridge archaeological and palaeoclimatological datasets, a number of research projects within the Azraq Basin have undertaken a joint archaeological and palaeoecological study of various localities (see below). These investigations have begun to paint a picture of how environments and settlement patterns have developed differently over time in various locales across the Azraq Basin (Byrd and Garrard 2017; Maher 2017; Richter 2017). As these local archives of environment and settlement grow more numerous, the question of regional palaeolandscape becomes newly tractable. After all, the social and ecological dynamics of human groups respond not only to the local environment but also to that of the broader region in which they live. This statement applies all the more strongly to prehistoric populations likely to have been at least seasonally mobile. The scarcity of reliable, basin-averaging palaeoenvironment archives (e.g., lakes) in eastern Jordan means that regional palaeoenvironmental reconstruction cannot be attempted from any single sedimentary record. Therefore, a detailed understanding of the hydrological and geomorphological structure of the Azraq Basin represents a crucial constraint on the possible state of this environmental system at various points in the past (Jones et al. 2016; Kushnir et al. 2017).

Physiography, geomorphology & hydrology of the basin

The Azraq Basin drains approximately 12,750 km^2 of the limestone and basalt desert plateau that stretches outward and upward in all directions from the Azraq Oasis (see Figure 5.1). The northern and eastern edges of the basin are marked by the southern flank of the Jabal al-Arab (or Jabal ad-Duruz) and the north-south ridge of the basaltic Harrat ar-Rujayla, while the remaining borders are less distinct, bisecting the Sirhan Depression in the southeast and then wandering across the limestone plateau between Azraq and Zarqa to the west (Besançon et al. 1989; Dottridge 1998; Shahbaz and Sunna 2000).

The present extent of the basin appears to have taken shape around the Oligocene, during which time the northwest-trending faults that cross it were most active and which also saw the beginning of the sustained volcanic activity that formed the Harrat ash-Shaam basalt plateau that covers approximately 11,000 km^2 of the basin's area (Allison et al. 1998; Avni 2017).

The modern surface is composed of basalt flows of various ages – Oligocene to recent (Ilani et al. 2001) – and marls, limestones, and cherts of both Cretaceous and Tertiary age, generally exposed to the southwest of Azraq to the northeast (Ibrahim et al. 2001). Characteristic landforms include chains of buttes and scoria cones developed along basaltic dykes, wadis formed by the sediments of ephemeral drainages, marabs, which are the pan-like broadenings in wadi courses that host fine-grained but generally non-saline deposits, and qa'as – the evaporative pans that form at the centre of centrally draining catchments or depressions and which generally consist of saline silt deposits (Allison et al. 2000; Allison et al. 1998). It is worth noting that marabs, by virtue of being not only well-watered but also well-drained by sporadically through-flowing wadis, often sustain seasonal grazing that qa'as do not (Dottridge 1998).

Drainage density differs between limestone and basalt substrates, as well as between basalt flows of different ages, as does the size and distribution of qa'as (Allison et al. 1998). This variation speaks to a complex interplay of surface hydrology and volcanic activity throughout much of the Pleistocene. The shared propensity for extrusive basalt lineaments and wadi courses to develop along antecedent zones of structural weakness (e.g., fault traces) makes it likely that the interplay between the two systems was important to the hydrological development of the basin (Avni 2017; Drury 1998). As indicated by the radiocarbon groundwater studies of Noble (1998) and Bajjali and Abu-Jaber (2001), most recharge to the aquifers of the Azraq Basin occurs at altitude (greater than 850 m, or so) and at temperatures characteristic of the southern slopes of the Jabal al-Arab in winter. This story is corroborated by groundwater flow modelling (Al-Kharabsheh 2000; El-Naqa et al. 2007; Noble 1998). Some recharge also occurs over the western highlands, which rise just east of the Jordan Valley, as the Azraq groundwater basin extends much further west than does its surface equivalent, which is truncated by the westward-draining basin of the Wadi al-Mujib (Dottridge 1998; Salameh et al. 2018).

The basin exhibits an arid Mediterranean-type climate, with hot, dry summers and cool, relatively wet winters, receiving roughly 200 mm of rainfall per year in the northwest, in the foothills of the Jabal al-Arab, and decreasing to less than 50 mm/yr toward the southeast (EXACT 1998). The north receives more winter rainfall, with a sharper peak in December and considerable drop-offs to either side, while rainfall in the south is spread more evenly throughout the October–May wet season. Soils in the Azraq Basin are primarily aridisols and entisols but vary considerably depending on the age of (particularly) the basalts that form their parent material. Flint and basalt "pavements" are common and may derive from aeolian or fluvial processes (Allison et al. 1998). The eastern Mediterranean cyclone is the dominant driver of precipitation in the Azraq Basin today and likely would have been important for the development of the palaeoenvironment of the Azraq Basin throughout the Pleistocene (Enzel et al. 2008; Kushnir et al.

2017). After all, even in the harsh setting of the modern Azraq Basin, a single rainstorm brings greenery and animal life to the farthest-flung desert landscapes (Nelson 1973, 1985; Tansey and Millington 2001; Wasse et al. 2020); one wonders how much more so this may have been the case prior to large-scale erosion of soils, even if the water table had not been considerably higher than today throughout the basin. Vegetation in the Azraq Basin today is largely composed of sparsely distributed, drought-tolerant woody taxa, with limited areas of denser vegetation surrounding agricultural settlements and irrigated fields.

Three main aquifers make up the Azraq groundwater basin, of which only the Upper Aquifer, composed of the Harrat ash-Shaam basalts and the Rijam limestone, is relevant for our purposes. The Middle and Lower Aquifers would generally not have contributed to water resources available to prehistoric humans since they are generally deeply buried and saline. The offset of different aquifer units along fault systems drives important groundwater exfiltration patterns in the basin; the connection between the springs of the Azraq Oasis and Fuluk Fault is particularly notable (El-Naqa et al. 2007; Ibrahim et al. 2001). The activity of fault-fed springs is linked to the level of the water table: Water table depression under intensive pumping regimes finally exhausted the Azraq springs at the end of the last century, after millennia of reliable flow (Dottridge 1998; Nelson 1973). Meanwhile, wet winter in the high-infiltration basalt highlands of the northern Azraq Basin can meaningfully raise water table elevations in that subregion. Due to highly variable groundwater flow rates, such recharge may not be felt in the central portion of the basin for decades to millennia, making possible the "fossil groundwater" phenomenon described by Lloyd and Farag (1978). Noble (1998) and Bajjali and Abu-Jaber (2001), among others, have demonstrated that such pluvial pulses are not the sole source of Azraq Basin groundwater and that active modern recharge is in fact occurring. Nevertheless, it remains important to consider the time delays inherent in groundwater dynamics as we consider palaeoenvironment scenarios, particularly in a groundwater basin as extensive and heterogeneous with respect to conductivity as the Azraq Basin.

Prehistory and history of environment and human settlement in the Azraq Basin

Several studies conducted in the environs of the Azraq Oasis (Ames and Cordova 2015; Cordova et al. 2013; Jones and Richter 2011) have yielded tantalising glimpses into parallel archaeological and ecological development of a number of locales (see Figure 5.1). In concert with summary analyses of settlement patterns and geomorphology across the basin as a whole (see below), it is becoming possible to constrain plausible palaeo-states of the Azraq Basin environment (Table 5.2).

Table 5.2 Local palaeoclimate reconstructions of the Late Pleistocene-Holocene transition within the Azraq Basin. Notable changes in climate and hydrology during each period were collected from geoarchaeological, geomorphological, sedimentological, faunal and/or palaeobotanical records available from Kharaneh IV, Shubayqa 1, Jilat 6 and 22, and Ayn Qasiyya.

Azraq Basin palaeoclimate reconstructions			
	Kharaneh IV	*Shubayqa 1*	*Jilat 6 & 22*	*Ayn Qasiyya*
Location	West Azraq Basin	Northeast Azraq Basin	Southwest Azraq Basin	Central Azraq Basin, Azraq Oasis
Type of Analysis	Sedimentology, geomorphology	Geomorphology, faunal analysis	Sedimentology	Sedimentology, geomorphology, plant ecology (phytoliths)
Scale of Analysis	Localised- Wadi Kharaneh	Localised- Qa'Shubayqa	Localised- Wadi Jilat	Localised- Ayn Qasiyya spring
Last Glacial Maximum	Cold and wet. Marshy wetlands with some areas of open water potentially existed but began to dry up around 22,000 cal BP.	(Not yet available)	Wet but a shift to more arid conditions occurred. Playa marsh conditions gave way to seasonal aridity.	Cool and wet. Increasingly dry conditions through the LGM. Stable land surfaces existed and were cut in some places by marshy deposits. Calm, open water at spring site.
Heinrich Stadial 1	Wet and cold. Marshy but progressively drier, springs provided localised water source.	(Not yet available)	Dry. Minimal pedogenesis suggests increased and persistent aridity.	Dry? No net deposition, possible dry springs. Potential for seasonal collection of surface runoff in Qas.
Bolling-Allerod	Cool and dry. Possible damming of the wadi induced localised marshes	(Not yet available)	Dry. Arid conditions with sand deposition.	Dry? No net deposition, possible dry springs. Potential for seasonal collection of surface runoff in Qas.
Younger Dryas	Dry. Dry lakebed with eolian activity.	Cool and wet. Faunal remains indicate a wetland environment with both	(Not available)	Dry? No net deposition, possible dry springs. Potential

				for seasonal collection of surface runoff in Qas.
Pre-Borial	Warmer and dry. Dry lakebed with eolian activity	shallow pools and large deep bodies of water. (Not yet available)	(Not available)	Warm and wet. High energy flooding and localised erosion.
Citations	Jones et al. 2016; Maher, 2017	Richter et al. 2017; Yeomans, 2018	Garrard and Byrd, 1992; Garrard et al. 1994; Garrard and Byrd, 2013	Cordova et al. 2013; Jones and Richter, 2011; Ramsey et al. 2015

The Druze Marsh and Azraq Shishan

Since the initial discovery in 1956 of Lower Palaeolithic remains buried beneath the springs of the central Azraq basin (Kirkbride 1989; Zeuner et al. 1957), additional research has established the region as an important early Palaeolithic landscape. Alongside many more instances of Lower and Middle Palaeolithic remains buried beneath the wetlands (Hunt and Garrard 1989; Jones and Richter 2011; Rollefson et al. 1997; Rollefson 2000), Lower and Middle Palaeolithic surface remains have been documented throughout the western and northern wadi sectors that flow into the central basin (Copeland and Hours 1989; Copeland 1989; Garrard and Byrd 2013; Rollefson 1984). Aside from the Late Acheulean knapping floor excavated at C Spring (Copeland 1989a, 1991), there has been little systematic excavation of the earlier Palaeolithic record in Azraq, and thus our understanding of this time period is riddled by a lack of radiometric ages and limited palaeoenvironmental data. However, recent investigations in the former Azraq Druze and Azraq Shishan wetlands have taken preliminary steps to address these gaps in our knowledge.

A series of deep soundings in the northern of the two wetlands, Azraq Druze or the Druze Marsh, has revealed a sedimentological sequence characterised by cycles of wetting and drying that extends back at least 300 ka. Interspersed between pulses of wetland deposits are Lower, Middle and Upper Palaeolithic occupation horizons (Ames et al. 2014). Obtaining radiometric dates has so far proven difficult for the Lower and Middle Palaeolithic occupations here, but reconstruction of the palaeolandscape indicates that as arid conditions set in the wet environments retracted back toward the spring locations, which acted as desert refugia for hominins in the area (Ames and Cordova 2015; Cordova et al. 2013).

Excavation in the southern wetlands, Azraq Shishan or the Shishan Marsh, has revealed a strikingly similar sedimentological record to the one in the Druze Marsh – again indicative of cyclical wetting and drying characterised by open water deposits separated by erosional discontinuities and carbonate precipitation (Ames et al. In press). However, the only archaeological material encountered thus far has been an expansive horizon of Late Acheulean stone tools and faunal remains near the base of the sequence that is dated to 260 ± 40 ka (Nowell et al. 2016). The deposit containing the archaeological remains is bounded by a possible lake deposit below and overlying aeolian silt. Palaeoenvironmental reconstruction that combines analysis of the sediments, plant microfossils, and faunal remains indicates that the Lower Palaeolithic inhabitants were taking advantage of a wetland margin environment that developed as the region was experiencing a millennial-scale drying trend. The preservation of blood protein residues on the tools implies considerable butchering activities were taking place (Nowell et al. 2016), perhaps as hominins took advantage of the isolated freshwater resources that would have attracted medium and large fauna –

namely, elephant, rhinoceros, gazelle, wild cattle and camel all identified in the faunal assemblage here – as the surrounding region was becoming more arid (Pokines et al. 2019).

Combined, this recent work in the former Azraq wetlands indicates that early Palaeolithic hominins occupied the Azraq basin since the Middle Pleistocene, and in the process, experienced dramatic fluctuations in the availability of freshwater resources from large, perennial wetland complexes (possibly palaeolakes) to isolated spring-fed pools. Although more work remains to be conducted, it appears that during wetter palaeoenvironmental conditions, hominins were occupying the wadi systems and perhaps the margins of the wetlands/lakes. Whereas during more arid intervals, in-habitants sought refuge in the concentrated and isolated resources of the central Azraq spring-fed wetlands (Ames and Cordova 2015). The importance and precarity of water resources for inhabitants in this region today is thus not solely a recent phenomenon but a dynamic that extends back to at least the Middle Pleistocene.

Sedimentological records from Ayn Qasiyya provide a well-dated sequence of environmental change in the oasis throughout the last glacial and glacial to interglacial transition, from roughly 65,000 to 9000 years BP (Besançon and Sanlaville 1988; Besançon et al. 1989; Copeland 1989; Cordova et al. 2013; Hunt and Garrard 1989; Jones and Richter 2011).

Prior to the LGM, the central oasis was well fed by its two main springs, Ayn Soda and Ayn Qasiyya. Work at sites of the same names, and others, indicate overall wetter periods during Marine Isotope Stages 5 and 4 (130–50 kya), with abundant standing water marked by the deposition of fine green clays and marsh deposits (Cordova et al. 2013; Jones and Richter 2011; Richter 2014). Middle Palaeolithic occupation occurred along the margins of the wetlands. During this period, some evidence of periodic high-velocity input from streams or rivers was found in the form of small and medium-sized natural flint cobbles interspersed with Levallois and Mousterian artefacts (Cordova et al. 2013; Jones and Richter 2011; Ramsey et al. 2015; Ramsey and Rosen 2016; Richter 2014; Rollefson et al. 1997).

During the LGM, water levels fell, probably due to reduced spring output. Upper Palaeolithic material is sparse and appears in reworked contexts associated with a dry marsh bed, aeolian activity and carbonate formation, and corresponds to MIS 3 (Cordova et al. 2013). During this retraction of the marshes, channels cut across the exposed landscape, which when water levels increased again in the LGM, became focal points of marsh formation, as depressions filled with standing water or wetland soils and associated with Early Epipalaeolithic material culture (Jones and Richter 2011).

Excavations at Ayn Qasiyya note a substantial Early Epipalaeolithic site, dated to between 20,900 and 19,200 cal BP (Richter 2014; Richter et al. 2014; Richter, Stock, et al. 2010). Aside from occupational debris of chipped stone and faunal remains that suggest a series of hunter-gatherer residential campsites (Richter, Alcock, et al. 2010), the remains of an adult male were found bound and entombed in the palaeomarsh (Richter, Stock, et al. 2010). Sedimentary records from several sections at Ayn Qasiyya suggest a hiatus in deposition between 16,000 and 10,500 years ago that likely relates to a period of wetland contraction. By the end of the Late Epipalaeolithic and Early Holocene, the wetland once again expanded, signifying a return of wetter conditions when the occupation of the area is documented around these springs. Early Holocene clastic sediments suggest alluviation rather than marsh sediments, while in the later Holocene, wetlands expand again where spring activity and marsh conditions seem to be re-established with warmer temperatures. Importantly, groundwater reservoirs at the oasis mean that regardless of these fluctuations, the oasis would have provided permanent water sources to sustain human populations (and the flora and fauna they would have also relied on) throughout the last glacial period and well into the Holocene (Jones, Maher, Richter, et al. 2016; Jones and Richter 2011).

Beyond the marsh: the Black Desert and Shubayqa

The northern and northeastern reaches of the Azraq Basin – stretching from the Jabal al-Arab down through eastern Jordan into Saudi Arabia – is dominated by a series of volcanoes whose Pleistocene eruptions (and subsequent erosion) have created a dense basalt rock desert overlying the limestone substrate (see Figure 5.1). This arid landscape has no nearby sources of permanent water beyond the oases at Azraq and Burqu. Instead, it is dissected by a network of seasonal wadis draining south from the Jabal al-Arab highlands and interspersed by several low-lying playas that can hold water into the summer months and sustain temporary farming and grazing lands still used by local Bedouin today. A. Betts conducted an extensive survey of the black desert area in the 1980s and 1990s in the northern and northwestern reaches of the basin, documenting and testing a number of Late Epipalaeolithic (Natufian) and Neolithic sites, including Shubayqa 1 (Betts 1998). Notably, Upper Palaeolithic and Early Epipalaeolithic sites were rare here.

Renewed work in the Qa' Shubayqa in the northern portion of the Azraq Basin include the excavation of a Late Epipalaeolithic village, as well as landscape survey and geomorphological sampling of the playa (Richter 2017). Well-fed by numerous seasonal wadis, the Qa' or playa is a 12 km² depressed basin in an otherwise basalt boulder landscape of the Jordanian *Harra* (Betts 1998). The Qa' Shubayqa is part of the larger Wadi Rajil, major drainage of the Jabal Arab/Druze transporting water south to the Qa' Azraq; it fills even in years when Qa' Azraq doesn't. Today the Qa' forms a seasonal body of standing water that supports wildlife (notably gazelle) and wheat fields still used by local inhabitants. Radiocarbon and OSL dates from a 4.5 m core of sediment from the playa suggest at least a Holocene age (M. Jones, personal communication, 2014).

During the Pleistocene, it is likely that the Qa' was wetter than today. Ongoing avifaunal research suggests that both small pools and larger, deeper bodies of water were present (Richter et al. 2017; Yeomans 2018). The edge of this late Pleistocene and early Holocene body of water is marked by several Late Epipalaeolithic and early Neolithic sites, several of which are substantial village sites (Richter et al. 2014). Excavations at the Natufian site of Shubayqa 1, extending about 5000 m² over a low terrace at the edge of the playa, reveal three main phases of occupation, with round stone-built and stone-paved structures, human burials, an abundance of ground stone mortars, chipped stone, botanical and faunal remains, and combustion features (Richter 2017). Reconstructing the environmental context of the site is greatly aided by the excellent preservation of archaeobotanical and faunal remains that include an abundance of gazelle and waterfowl, as well as the presence of tubers, cereals (*Hordeum spontaneum, Triticum* spp.), tamarix and a range of chenopods (Arranz-Otaegui et al. 2018; Jones et al. 2016; Richter 2017; Richter et al. 2016).

Substantial occupation of the playa edges in prehistory seems restricted to the Early Natufian and Early Neolithic, with little evidence of Late Natufian occupation. It is possible that settlement was at its greatest during wetter phases before and after the Younger Dryas, when water resources may have been notably lowered (Jones et al. 2016). It is interesting to note that Epipalaeolithic occupations of the Azraq Basin appear to shift northward around the Late Epipalaeolithic. Large Epipalaeolithic aggregation sites in the south and west Azraq Basin like Kharaneh IV and Jilat 6 were in-tensively occupied during the Early and Middle Epipalaeolithic periods and not occupied during the Late Epipalaeolithic. Late Epipalaeolithic sites begin to expand across the northern parts of the Azraq Basin, where Early and Middle Epipalaeolithic sites were rare. These Late Epipalaeolithic sites like Shubayqa 1 were smaller, had stone architecture, and large heavy-duty mortars for processing a wide variety of plants (Arranz-Otaegui et al. 2018; Richter 2017; Richter and Maher 2013).

Beyond the marsh: Wadi Jilat

A refined picture of the relationship between landscape change and pre-historic occupation of the Wadi Jilat is provided by the work of A. Garrard and colleagues who conducted a series of surveys and excavations at several Epipalaeolithic and Neolithic sites in the 1980s. An excellent and thorough summary of this work is presented in Garrard and Byrd (2013a), where the authors discuss the results of stratigraphic and lithic analysis from these sites. Occupied continuously throughout the late Pleistocene and into the Holocene, the Wadi Jilat was clearly a focus of intense occupation in the Epipalaeolithic. At Jilat 6, a site located on an alluvial or colluvial fan on the northern side of Wadi Jilat, occupation began during the Upper Palaeolithic. The earliest le-vels indicate marsh conditions that were likely fed by a spring. A shift toward seasonally arid conditions occurred and persisted through the Initial Epipalaeolithic. The presence of pedogenically altered loess with calcareous nodules during this period indicate seasonally arid phases with enough water and vegetation to allow for pedogenesis. The trend toward aridity continued into the Early Epipalaeolithic as aeolian sediment, unaltered loess deposits, and little evidence of organic matter were noted even during periods of intense occupation (Garrard et al. 1994; Garrard and Byrd 1992, 2013).

Aside from several smaller campsites and base camps, this area, located approximately 20 km due south from the Wadi Kharaneh, is the only other site of comparable size and density to Kharaneh IV – Jilat 6 (see Figure 5.1). Although only limited test excavations have been conducted at the site so far, a comparison of its multi-phase stratigraphy, lithic assemblages, and other features, including high densities of marine shell and possible evidence for structures, suggest a repeatedly occupied, prolonged-use aggregation site very similar to Kharaneh IV. And, like Kharaneh IV (discussed above), the dates for Middle Epipalaeolithic sites here are early in comparison to culturally comparable sites to the west and north (Richter et al. 2013).

Ongoing work in the north at Shubayqa 1 and to the west at the Early and Middle Epipalaeolithic site of Kharaneh IV hopes to shed light onto human occupation during these phases of the Epipalaeolithic outside the oasis. So far, intense occupation of both areas, at different times, suggests that parts of the basin were always favourable for settlement during the Epipalaeolithic. With the presence of the oasis serving at times to buffer against harsher, drier conditions elsewhere, the specific locations of refugia within the basin relate heavily to changing water availability and thus changed over time. For example, Kharaneh IV and Jilat 6 were preferred locales in the Early and early Middle Epipalaeolithic, while areas in the basalt desert to the north appear to have been more appealing in the Late Epipalaeolithic and early Neolithic.

Discussion: water as a dwindling resource linking past & contemporary land use

To return to the question posed at the beginning of the last section regarding the nature of water resources in the basin and their impact on human populations, a few scenarios can be suggested in light of the reconstructions presented above. These are best thought of as part of a palaeoenvironmental time series and, as such, are not mutually exclusive at any one point in time or across the basin (see Table 5.2). In one scenario, a central palaeolake might have been fed by surface runoff associated with a stronger Mediterranean cyclone and African or Indian monsoon, which preserved denser vegetation cover, competent to sustain productive soils and prevent highly erosive overland flow, resulting in higher infiltration rates and surface-groundwater connectivity, supporting an abundance of drought-tolerant flora and fauna year-round. Another scenario might present a network of wetlands of varying seasonal persistence – for instance, more robust wetlands in the north, fed by both surface and groundwater derived from a stronger Mediterranean cyclone, versus seasonal wetlands in the south, dependent on the water-table rise due to seasonal precipitation. A third scenario might describe a discontinuous patchwork of springs and associated marshes generated by local hydrogeological quirks and potentially threatened by headward extension of stream networks, tectonic shifts, and piracy of surface or groundwater divides, which might adversely affect recharge rates and water table elevation.

These three scenarios, and many gradations between, present significantly different opportunities for prehistoric human populations, who would have had varying landscape-management techniques and socioecological strategies available to their use to contend with these realities. We would expect quite different settlement and mobility strategies in each of these situations and would thus interpret archaeological traces of settlement differently as well. Even archaeological data from the Mediterranean littoral zone east to the Jordan Valley and escarpment should be considered in light of these palaeoenvironmental possibilities (along with those of other adjacent basins, such as al-Mujib, al-Hasa, and al-Jafr) since it is likely – depending on contemporary Lake Lisan-Dead Sea extent – that groups would have circulated throughout a considerable east-west range.

The possibility of a Pleistocene palaeolake at the centre of the Azraq Basin was first proposed by Bender (1974), and it has remained an active explanation for the high density of Palaeolithic sites noted here (Abed and Yaghan, 2000; Abed et al. 2008; Ames and Cordova 2015; Besançon and Sanlaville 1988; Cordova et al. 2013; Davies 2005; Garrard, Harvey, et al. 1985; Ginat et al. 2018; Jones and Richter 2011; Rech et al. 2017). More recent refinement of the palaeoenvironmental record of the basin has modified our picture from focus on a single palaeolake to a complex array of lakes, playas, and wetlands and to secondary questions regarding their spatial and temporal extent and persistence.

The deposition of marls at Kharaneh IV, containing an abundance of low-velocity, freshwater ostracods, matches reconstructed wetland contexts from the Azraq Oasis during the Epipalaeolithic (Jones and Richter 2011), confirming the presence of extensive and fluctuating marshland habitats at and around Kharaneh IV during the occupation of the site (Jones, Maher, Macdonald, et al. 2016). Surrounding this wetland (especially as it dried up) was a combination of open grasslands and parkland, making the immediate area rich in a wide variety of floral and faunal resources (Jones, Maher, Macdonald, et al. 2016; Ramsey et al. 2016; Ramsey and Rosen 2016). Intensive use of wetland and grassland resources is documented in the analysis of phytoliths from a variety of occupational contexts, including in association with several Early Epipalaeolithic structures (Ramsey et al. 2018). Being situated near both wetland and steppe/parkland resources allowed the inhabitants of Kharaneh IV to access reliable sedges and reed resources from the wetland while simultaneously exploiting more risky seasonal grasses and cereals from the steppe (Ramsey et al. 2016). Thus, it seems that the Early Epipalaeolithic occupants were living at the margins of these fluctuating wetlands, with ready and reliable access to fresh water and all of the flora and fauna it sustains; all in all, an attractive location for sustained settlement.

Movement, aggregation and dispersal on the Azraq palaeolandscape: past patterns of land use in relation to changing water availability

Although Epipalaeolithic groups depopulated Kharaneh IV sometime after 18,600 BP, the abandonment of the area was not permanent. Two Late Neolithic sites, KH2 and KH7, are located within several kilometres of Kharaneh IV in Wadi Kharaneh (Garrard and Byrd 2013), and Qasr Kharaneh is located less than a kilometre to the northeast (Harding 1959). Wadi Kharaneh has provided water to different groups for over 20,000 years. It currently contains a large water reservoir, constructed in 2011 and fed by rainwater and surface runoff, that is predominantly used by Bedouin herders and for water trucks to fill up to supply water to nearby farms. Intriguingly, there is an approximately 10,000 year absence in archaeological remains in Wadi Kharaneh between Middle Epipalaeolithic Kharaneh IV, and Late Neolithic sites KH2 and KH7 that includes periods of climatic "amelioration", such as the Bølling-Allerød interstadial and Early Holocene (e.g., see Maher, Banning, et al. 2011). This absence adds further complexity to the concept of climate amelioration and water availability, as the occupations in Wadi Kharaneh occur before and after these times. The archaeological record of Wadi Kharaneh demonstrates that the relationship between land use and water availability is not directly proportional; it appears that populations aggregated at Kharaneh IV during a time of relative water scarcity, not abundance.

However, when we step back and consider the larger Azraq Basin with the *longue durée* in mind, basin-wide shifts in settlement location, intensity and type become more apparent, and we can begin to make some preliminary suggestions about human decision-making and settlement choice within these changing environments (see also Richter and Maher 2013). There appears to be a concentration of sites within the Azraq wetlands during the Late Lower Palaeolithic and Middle Palaeolithic, and it seems that this area was central to migrating human populations, perhaps providing refuge to overwhelmingly arid conditions in surrounding regions. Notably, Upper Palaeolithic sites (beyond surface scatters) throughout the basin remain rare and seem restricted to permanent water sources, such as the marsh and major wadis like Wadi Jilat. The Uwaynid area, Wadi Jilat and Wadi Kharaneh were home to substantial and sustained Early and Middle Epipalaeolithic occupations, buffered from the LGM aridity by reliable water sources like rivers and wetlands. However, these southern and western portions of the basin are largely abandoned by 18,500 cal BP (until reoccupation in the Late Neolithic). Instead, the northern reaches of the basin, around Shubayqa, show an abundance of Late Epipaleolithic and Early Neolithic sites, suggesting large-scale movement of populations to these potentially better-watered locales. Late Neolithic occupations are sporadic around Kharaneh and Jilat, yet they are comparatively settled and dense in the eastern reaches of the basin at Maitland's Mesa and the Wissad Pools (Rowan et al. 2015).

Conclusions

The scenario of a single substantial palaeolake in the Azraq Basin provided an interesting explanation for the dense concentrations of the sites here, comparable to other areas in the region (e.g., Hasa). However, it was not able to account for the noted spatial and temporal shifts in site distributions. More recent refinement of the palaeoenvironmental record of the basin has modified our picture from focus on a single palaeolake to a complex array of lakes, playas, and wetlands and to secondary questions regarding their spatial and temporal extents, their persistence or disappearance in the Pleistocene and Holocene, and about varying human activities and strategies of sustainable land use within these dynamic environments.

It seems readily apparent that, like today, water availability was a crucial factor for past populations. Recent work at the Epipalaeolithic site of Kharaneh IV (and others nearby) indicates substantial settlement of this intensively used aggregation site at the LGM, yet by the start of the HS1, the occupants of Kharaneh IV reconsidered their use of this location for large-scale settlement and – as such aggregation sites soon disappear entirely from the region – perhaps even their very practice of aggregation as a socio-ecological strategy. In contrast to narratives that associate aggregation and "settling in" during the Epipalaeolithic of Southwest Asia with climatic amelioration during the Bølling-Allerød, local records indicate increased rates of evaporation, shrinking wetlands and disappearing resources (Jones, Maher, Richter, et al. 2016). Subsequent occupation of the Azraq Basin at the end of the Pleistocene and into the Holocene was heavily dependent on fluctuations in water availability in this newly arid landscape. From 20,000 years onwards, here water was an increasingly dwindling and precious re-source and greatly shaped how people engage with the landscape.

Once an economic hub, unique for its demographic, cultural, and natural diversity, the larger Azraq Basin has lost much of its water due to climatic and human development-related impacts (Jones, Maher, Richter, et al. 2016; Jones, Maher, Macdonald, et al. 2016; Jones Contreras 2016; Maher, Richter, et al. 2011). Since the 1990s, however, the people of Azraq have worked to reclaim and revitalize their wetlands, alongside a commensurate concern for associated cultural resources. In this climate, our work on human responses to changing landscapes at Kharaneh IV engages with equally complex and meaningful stewardship in Azraq today.

Acknowledgements

First and foremost, we thank our many different field crews from 2008 until 2019, which included international teams of students and specialists and Jordanian teams of students and professionals. We would like to gratefully acknowledge the support of the General Director of the Department of Antiquities and the invaluable assistance of our Department of Antiquities Representatives from 2008–2019. In particular, we would like to thank Ahmad Lash, who inspired us to think critically and make connections between water availability in the Azraq area today and in the past, and for facilitating bringing our work to the Azraq local community in so many meaningful ways. We have benefitted greatly from discussions with several colleagues working in the Azraq Basin, past and present, who have generously shared their research and time.

References

Abed, A. M. and R. Yaghan, 2000. On the paleoclimate of Jordan during the last glacial maximum. *Palaeogeography, Palaeoclimatology, Palaeoecology* 160: 23–33.

Abed, A. M., S. Yasin, R. Sadaqa, and Z. Al-Hawari, 2008. The paleoclimate of the eastern desert of Jordan during marine isotope stage 9. *Quaternary Research* 69 (3): 458–468.

Al-Kharabsheh, A., 2000. Ground-water modelling and long-term management of the Azraq basin as an example of arid area conditions (Jordan). *Journal of Arid Environments* 44: 143–153.

Allison, R. J., J. R. Grove, D. L. Higgitt, A. J. Kirk, N. J. Rosser, and J. Warburton, 2000. Geomorphology of the eastern Badia basalt plateau, Jordan. *Geographical Journal* 166 (4): 352–370.

Allison, R. J., D. L. Higgitt, A. J. Kirk, J. Warburton, A. Al-Homoud, B. Sunna, and K. White, 1998. Geology, geomorphology, hydrology, groundwater and physical resources. In *Arid Land Resources and Their Management: Jordan's Desert Margin*, edited by R. W. Dutton, J. I. Clarke, and A. Battikhi, pp, 21–46. Kegan Paul International, London.

Ames, C. J. H., and C. E. Cordova, 2015. Middle and Late Pleistocene Landscape Evolution at the Druze Marsh Site in Northeast Jordan: Implications for population continuity and hominin dispersal. *Geoarchaeology* 30: 307–329. https://doi.org/10.1002/gea.21516

Ames, C. J. H., A. Nowell, C. E. Cordova, J. T. Pokines, and M.l S. Bisson, 2014. Paleoenvironmental change and settlement dynamics in the Druze Marsh: Results of recent excavation at an open-air paleolithic site. *Quaternary International* 331: 60–73.

Ames, C. J. H., C. E. Cordova, K. Boyd, C. Schmidt, D. Degering, B. G. Jones, A. Dosseto, J. Pokines, A. Alsouliman, J. Beller, and A. Nowell. In press. Middle to Late Quaternary Palaeolandscapes of the central Azraq Basin, Jordan: Deciphering discontinuous records of human-environment dynamics at the arid margin of the Levant. *Quaternary International (Geoarchaeology from Mediterranean Desert Margins to Drylands).*

Amr, Z. S., and M. Abu Baker, 2004. Freshwater snails of Jordan. *Denisia* 14: 221–227.

Arranz-Otaegui, A., L. G. Carretero, M. N. Ramsey, D. Q. Fuller, and Tobias Richter, 2018. Archaeobotanical evidence reveals the origins of bread 14,400 years ago in northeastern Jordan. *Proceedings of the National Academy of Sciences* 115: 7925–7930.

Avni, Y. 2017, Tectonic and physiographic settings of the Levant. In *Quaternary of the Levant: Environments, Climate Change, and Humans*, edited by O. Bar-Yosef and Y. Enzel, pp. 3–16. Cambridge University Press, Cambridge.

Bajjali, W., and N. Abu-Jaber, 2001. Climatological signals of the paleogroundwater in Jordan. *Journal of Hydrology* 243 (1–2): 133–147.

Bard, Edouard, Frauke Rostek, Jean-Louis Turon, and Sandra Gendreau, 2000. Hydrological impact of Heinrich events in the subtropical northeast Atlantic. *Science* 289 (5483): 1321–1324.

Bar-Mathews, M., A. Ayalon, A. Kaufman, and G. J. Wasserburg, 1999. The Eastern Mediterranean paleoclimate as a reflection of regional events: Soreq Cave, Israel. *Earth and Planetary Science Letters* 166 (Journal Article): 85–95.

Bar-Matthews, M., and A. Ayalon, 2003. Climatic Conditions in the eastern Mediterranean during the Last Glacial (60–10 ky bp) and their relations to the Upper Palaeolithic in the Levant: Oxygen and Carbon Isotope Systematics of Cave Deposits. In *More Than Meets the Eye: Studies on Upper Palaeolithic Diversity in the Near East*, edited by A. Nigel Goring-Morris and Anna Belfer-Cohen, A, pp. 13–18. Oxbow Books, Oxford.

Bar-Matthews, M., and A. Ayalon, 2011. Mid-Holocene climate variations revealed by high-resolution speleothem records from Soreq Cave Israel and their correlation with cultural changes.*The Holocene* 21 (1): 163–171.

Bar-Matthews, M., A. Ayalon, and A. Kaufman, 1997. Late Quaternary paleoclimate in the eastern Mediterranean Region from stable isotope analysis of Speleothems at Soreq Cave, Israel. *Quaternary Research* 47: 155–168.

Bartov, Y., S. L. Goldstein, M. Stein, and M. Enzel, 2003. Catastrophic arid episodes in the eastern Mediterranean linked with the North Atlantic Heinrich events. *Geology* 31 (5): 439–442.

Belfer-Cohen, A., and O. Bar-Yosef, 2000. Early sedentism in the near east: A bumpy ride to village life. In *Life in Neolithic Farming Communities: Social Organization, Identity, and Differentiation*, edited by I. Kuijt, pp. 19–37. Kluwer Academic/Plenum Publishers, New York.

Bender, F., 1974. *Geology of Jordan. Vol. 7. Contributions to the Regional Geology of the Earth, Supplement*. Gebrüder Bornträger, Berlin.

Besançon, J., and P. Sanlaville, 1988. L'évolution géomorphologique du Bassin d'Azraq (Jordanie) depuis le Pléistocene Moyen. *Paléorient* 14 (2): 23–30.

Besançon, J., B. Geyer, and P. Sanlaville, 1989. Contribution to the study of the geomorphology of the Azraq Basin, Jordan. In *The Hammer on the Rock: Studies in the Early Palaeolithic of Azraq, Jordan*, edited by L. Copeland and F. Hours, pp. 7–63. British Archaeological Reports International Series 540. British Archaeological Reports, Oxford.

Betts, A. V. G., 1998. *The Harra and the Hamad: Excavations and Surveys in Eastern Jordan, Volume 1*. Sheffield Academic Press, Sheffield.

Byrd, B., and A. Garrard, 2017. The Upper and Epipalaeolithic of the Azraq Basin, Jordan. In *Quaternary of the Levant: Environments, Climate Change, and Humans*, edited by O. Bar-Yosef and Y. Enzel, pp. 669–678. Cambridge University Press, Cambridge.

Catlett, G. A., J. A. Rech, J. S. Pigati, M. Al Kuisi, S. Li, and J. S. Honke, 2017. Activation of a small ephemeral lake in southern Jordan during the last full glacial period and its paleoclimatic implications. *Quaternary Research* 88 (1): 98.

Clark, G. A. Coinman, and M. P. Neeley, 2001. The Paleolithic of Jordan in the Levantine Context. In *Studies in the History and Archaeology of Jordan VII*, edited by G. Bisheh, pp. 49–68. Department of Antiquities of Jordan, Amman.

Coinman, N. R., 1998. *The Archaeology of the Wadi al-Hasa, West-Central Jordan, Vol. 1: Surveys, Settlement Patterns and Paleoenvironments.* Arizona State University Press, Tempe.

Copeland, L., 1989. Analysis of the Paleolithic artifacts from the sounding of A. Garrard at C-Spring, Azraq; 1985 season. In *The Hammer on the Rock: Studies in the Early Palaeolithic of Azraq, Jordan*, edited by L. Copeland and F. Hours, pp. 325–390. British Archaeological Reports International Series 540. British Archaeological Reports, Oxford.

Copeland, L., 1991. The Late Acheulean Knapping-floor at C-Spring, Azraq Oasis, Jordan. *Levant* 23: 1–6. https://doi.org/10.1179/lev.1991.23.1.1

Copeland, L., and F. Hours, 1989. The Hammer on the Rock: Studies in the Early Palaeolithic of Azraq, Jordan. In *British Archaeological Reports International Series 540*, edited by L. Copeland and F. Hours, British Archaeological Reports, Oxford.

Cordova, C., 2007. *Millenial Landscape Change in Jordan: Geoarchaeology and Cultural Ecology.* University of Arizona Press, Tucson.

Cordova, C. E., G. O. Rollefson, R. Kaichgruber, P. Wilke, and L. Quintero, 2008. Natural and cultural stratigraphy of 'Ayn as-Sawda, al-Azraq Wetlands Reserve. 2007 excavation report and discussion of finds. *Annual of Department of Antiquities of Jordan* 52: 417–425.

Cordova, C. E., A. Nowell, M. Bisson, C. J. H. Ames, J. Pokines, M. Chang, and M. Al-Nahar, 2013. Interglacial and glacial desert refugia and the Middle Paleolithic of the Azraq Oasis, Jordan. *Quaternary International* 300: 94–110.

Danielson, J. J., and D. B. Gesch, 2011. Global Multi-resolution Terrain Elevation Data 2010 (GMTED2010). U.S. Department of the Interior, U.S. Geological Survey, Washington, D.C.

Davies, C. P., 2005. Quaternary paleoenvironments and potential for human exploitation of the Jordan Plateau desert interior. *Geoarchaeology* 20 (4): 379–400.

Dottridge, J., 1998. Water resources quality, sustainability and development. In *Arid Land Resources and Their Management: Jordan Desert Margin*, edited by R. W. Dutton, J. I. Clarke, and A. M. Battikhi, pp. 67–80. Kegan Paul International, London.

Drury, D., 1998. Baseline hydrochemical study of the Azraq Basin. In *Arid Land Resources and Their Management: Jordan's Desert Margin*, edited by R. W. Dutton, J. I. Clarke and A. Battikhi, pp. 81–85. Kegan Paul International, London.

El-Naqa, A., M. Al-Momani, S. Kilani, and N. Hammouri, 2007. Groundwater deterioration of shallow groundwater Aquifers due to overexploitation in Northeast Jordan. *CLEAN – Soil, Air, Water* 35 (2): 156–166.

Enzel, Y., and O. Bar-Yosef, 2017. *Quaternary of the Levant: Environments, Climate Change and Humans*. Cambridge University Press, Cambridge.

Enzel, Y., R. Amit, U. Dayan, O. Crouvi, R. Kahana, B. Ziv, and D. Sharon, 2008. The climatic and physiographic controls of the eastern Mediterranean over the late Pleistocene climates in the southern Levant and its neighboring deserts. *Global and Planetary Change* 60: 165–192.

Garrard, A., and N. P. Stanley Price, 1975–1977. A survey of prehistoric sites in the Azraq Basin, Eastern Jordan. *Paléorient* 3 (1): 109–126. https://doi.org/10.3406/paleo.1975.4192

Garrard, A., and B. Byrd, 1992. New dimensions to the epipalaeolithic of the Wadi el-Jilat in Central Jordan. *Paléorient* 18 (1): 47–62.

Garrard, A., and B. Byrd, 2013 *Beyond the Fertile Crescent: Late Palaeolithic and Neolithic Communities of the Jordanian Steppe. Volume 1: Project Background and the Late Palaeolithic - Geological Context and Technology. CBRL Levant Supplementary Series*. Oxbow Books, Oxford.

Garrard, A., D. Baird, and B. Byrd, 1994. The chronological basis and significance of the Late Palaeolithic and Neolithic sequence in the Azraq Basin, Jordan. In *Late Quaternary Chronology and Palaeoclimates of the Eastern Mediterranean*, edited by O. Bar-Yosef and R. Kra, pp. 177–199. Journal of Radiocarbon in association with the American School of Prehistoric Research, Preabody Museum, HarvardUniversity, Cambridge, Massachusetts.

Garrard, A., P. Harvey, F. Hivernel, and B. Byrd, 1985. The environmental history of the Azraq Basin. In *Studies in the History and Archaeology of Jordan II*, pp. 109–115. Department of Antiquities of Jordan, Amman.

Garrard, A., B. Byrd, P. Harvey, and F. Hivernel, 1985. Prehistoric environment and settlement in the Azraq Basin. A report on the 1982 survey season. *Levant* 17 (1): 1–28.

Garrard, A., D. Baird, S. Colledge, L. Martin, and K. Wright, 1994. Prehistoric environment and settlement in the Azraq Basin: An interim report on the 1987 and 1988 excavation seasons. *Levant* 26: 73–109.

Ghazleh, S., and S. Kempe, 2009. Geomorphology of Lake Lisan terraces along the eastern coast of the Dead Sea. *Geomorphology* 108: 246–263.

Ginat, H., S. Opitz, L. Ababneh, G. Faershtein, M. Lazar, N. Porat, and S. Mischke, 2018. Pliocene-Pleistocene waterbodies and associated deposits in southern Israel and southern Jordan. *Journal of Arid Environments* 148: 14–33.

Gleick, P. H., 2014. Water, Drought, Climate Change, and Conflict in Syria. *Weather, Climate and Society* 6: 331–340. http://dx.doi.org/10.1175/WCAS-D-13-00059.1

Goring-Morris, A. N., and P. Goldberg, 1990. Late Quaternary dune incursions in the southern Levant: Archaeology, chronology, and palaeoenvironments. *Quaternary International* 5: 115–137.

Goring-Morris, A. N., E. Hovers, and A. Belfer-Cohen, 2009. The dynamics of Pleistocene and Early Holocene settlement patterns and human adaptations in the Levant: An overview. In *Transitions in Prehistory: Essays in Honor of Ofer Bar-Yosef*, edited byJ. Shea and D. Lieberman, pp. 185–252. American School of Prehistoric Research Monographs. Oxbow Books, Oxford.

Grant, K. M., E. J. Rohling, M. Bar-Matthews, A. Ayalon, M. Medina-Elizalde, C. Bronk Ramsey, C. Satow, and A.P. Roberts, 2012. Rapid coupling between ice volume and polar temperature over the past 150,000 years. *Nature* 491 (7426): 744–747. https://doi.org/10.1038/nature11593

Hamdan, M., and G. Brook, 2015. Timing and characteristics of Late Pleistocene and Holocene wetter periods in the Eastern Desert and Sinai of Egypt, based on 14 C dating and stable isotope analysis of spring tufa deposits. *Quaternary Science Reviews* 130: 168–188.

Harding, G. L., 1959. *The Antiquities of Jordan*. Praeger, New York.

Henton, E., L. Martin, A. Garrard, A.-L. Jourdan, M. Thirlwall, and O. Boles, 2017. Gazelle seasonal mobility in the Jordanian steppe: The use of dental isotopes and microwear as environmental markers, applied to Epipalaeolithic Kharaneh IV. *Journal of Archaeological Science: Reports* 11: 147–158.

Higgitt, D., and R. Allison, 1998. Characteristics of the basalt boulder surfaces. In *Arid Land Resources and Their Management: Jordan's Desert Margin*, edited by D. Higgitt and R. J. Allison, pp. 171–182. Kegan Paul International, London.

Hill, J. Brent, 2001. The role of environmental degradation in long term settlement trends in Wadi al-Hasa, West-Central Jordan. In *Studies in the History and Archaeology of Jordan VII*, edited by Ghazi Bisheh, pp. 79–83. Department of Antiquities of Jordan, Amman.

Huckriede, R., and G. Wiesemann, 1968. Der jungpleistozäne Pluvial-See von El-Jafr und weitere Daten zum Quartär Jordaniens. *Geologica et Paleontologica* 2: 73–95.

Copeland, L., and F. Hours, editors, 1989. *The Hammer on the Rock: Studies in the Early Palaeolithic of Azraq, Jordan*. British Archaeological Reports International Series 540. British Archaeological Reports, Oxford.

Hunt, C., and A. N. Garrard, 1989. The 1985 excavation at C-Spring. In *The Hammer on the Rock: Studies in the Early Palaeolithic of Azraq, Jordan*, edited by L. Copeland and F. Hours, 319–324. British Archaeological Reports International Series. British Archaeological Reports, Oxford.

Ibrahim, K., I. Rabba, and K. Tarawneh, 2001. *Geological and mineral occurrences map of the northern Badia region, Jordan, scale 1: 250,000*. Natural Resource Authority, Amman.

Ilani, S., Harlavan, Y., K. Tarawneh, I. Rabba, R. Weinberger, K. Ibrahim, S. Peltz, and G. Steinitz, 2001. New K-Ar ages of basalts from the Harrat Ash Shaam volcanic field in Jordan: Implications for the span and duration of the upper-mantle upwelling beneath the western Arabian plate. *Geology* 29 (2): 171–174.

Iriarte, E., A. L. Balbo, M. A. Sánchez, J. E. González Urquijo, and J. J. Ibáñez, 2011. Late Pleistocene and Holocene sedimentary record of the Bouqaia Basin (central Levant, Syria): A geoarchaeological approach. *Comptes Rendus Palevol* 10 (Journal Article): 35–47.

Jones, M., and T. Richter, 2011. Palaeoclimatic and archaeological implications of Pleistocene and Holocene environments in Azraq, Jordan. *Quaternary Research* 76 (3): 363–372. https://doi.org/10.1016/j.yqres.2011.07.005

Jones, M., L. Maher, T. Richter, D. Macdonald, and L. Martin, 2016. Human-Environment Interactions through the Epipalaeolithic of Eastern Jordan. In *Correlation is not Enough: Building Better Arguments in the Archaeology of Human-Environment Interactions*, edited by D. Contreras, pp. 121–140. Routledge, New York.

Jones, M. D., L. A. Maher, D. A. Macdonald, C. Ryan, C. Rambeau, S. Black, and T. Richter, 2016. The environmental setting of Epipalaeolithic aggregation site Kharaneh IV. *Quaternary International* 396: 95–104.

Jones, M. D., N. Abu-Jaber, A. AlShdaifat, D. Baird, B. I. Cook, M. O. Cuthbert, J. R. Dean, M. Djamali, W. Eastwood, D. Fleitmann, A. Haywood, P. Kwiecien, J. Larsen, L. A. Maher, S. Metcalfe, A. Parker, C. Petrie, N. Primmer, T. Richter, N. Roberts, J. Roe, J. C. Tindall, E. Unal-Imer, and L. Weeks. 2019. 20,000 years of societal vulnerability and adaptation to climate change in southwest Asia. *Wiley Interdisciplinary Reviews: Water* 6 (2): e1330.

Kadosh, D., D. Sivan, H. Kutiel, and M. Weinstein-Evron. 2004. A Late Quaternary Paleoenvironmental sequence from Dor, Carmel Coastal Plain, Israel. *Palynology* 28 (Journal Article): 143–157.

Kelley, C. P., S. Mohtadi, M. A. Cane, R. Seager, and Y. Kushnir, 2015. Climate change in the Fertile Crescent and implications of the recent Syrian drought. *Proceedings of the national Academy of Sciences* 112 (11): 3241–3246. https://doi.org/10.1073/pnas.1421533112.

Kirkbride, J., 1989. Description of Stone Age sites at Azraq, Jordan. In *The Hammer on the Rock: Studies in the Early Palaeolithic of Azraq, Jordan*, edited by L. Copeland and F. Hours, 153–170. British Archaeological Reports International Series 540. British Archaeological Reports, Oxford.

Kushnir, Y., U. Dayan, B. Ziv, E. Morin, and Y. Enzel, 2017. Climate of the Levant: Phenomena and mechanisms. In *Quaternary of the Levant: Environments, Climate Change, and Humans*, edited by O. Bar-Yosef and Y. Enzel, pp. 31–44. Cambridge University Press, Cambridge.

Lehner, B., K. Verdin, and A. Jarvis, 2008. New global hydrography derived from spaceborne elevation data. *Eos, Transactions of the American Geological Union* 89(10): 93–94.

Lloyd, J. W., and M. H. Farag, 1978. Fossil ground-water gradients in arid regional sedimentary basins. *Groundwater* 16 (6): 388–392.

Macdonald, D. A., A. Allentuck, and L. A. Maher, 2018. Technological change and economy in the Epipalaeolithic: Assessing the shift from Early to Middle Epipalaeolithic at Kharaneh IV. *Journal of Field Archaeology* 43 (6): 437–456.

Maher, L. A., 2016. A Road well-travelled? Exploring terminal Pleistocene Hunter-gatherer activities, networks and mobility in Eastern Jordan. In *Fresh Fields and Pastures New: Papers Presented in Honor of Andrew M. T. Moore*, edited by M. Chazan and K. Lillios, pp. 49–75. Sidestone Press, Leiden.

Maher, L. A., 2017. Late Quaternary refugia: aggregations and palaeoenvironments in the Azraq Basin. In *Quaternary Environments, Climate Change and Humans in the Levant*, edited by O. Bar Yosef and Y. Enzel,pp. 679–689. Cambridge University Press, Cambridge.

Maher, L. A., 2019. Persistent place-making in prehistory: The creation, maintenance and transformation of an Epipalaeolithic landscape. *Journal of Archaeological Method and Theory* 25 (4): 998–1083. https://doi.org/10.1007/s10816-018-9403-1.

Maher, L. A., and M. Conkey, 2019. Homes for hunters? Exploring the concept of home at hunter-gatherer sites in Upper Palaeolithic Europe and Epipalaeolithic Southwest Asia. *Current Anthropology* 60 (1): 91–137.

Maher, L. A., E. B. Banning, and M. Chazan, 2011. Oasis or mirage? assessing the role of abrupt climate change in the prehistory of the Southern Levant. *Cambridge Archaeological Journal* 21 (1): 1–29.

Maher, L. A., T. Richter, and J. Stock, 2012. The Pre-Natufian Epipalaeolithic: Long-term behavioral trends in the Levant. *Evolutionary Anthropology* 21 (2): 69–81.

Maher, L. A., T. Richter, M. Jones, and J. T. Stock, 2011. The Epipalaeolithic Foragers in Azraq Project: Prehistoric landscape change in the Azraq Basin, Eastern Jordan. *CBRL Bulletin* 6: 21–27.

Maher, L. A., D. Macdonald, E. Pomeroy, and J. T. Stock. 2021. Life, death, and the destruction of architecture: Hunter-gatherer mortuary behaviors in prehistoric Jordan. *Journal of Anthropological Archaeology* 61: 101262. https://doi.org/10.101 6/j.jaa.2020.101262.

Maher, L. A., T. Richter, D. Macdonald, M. Jones, L. Martin, and J. T. Stock, 2012. Twenty thousand-year-old huts at a hunter-gatherer settlement in Eastern Jordan. *PLoS ONE* 7 (2): e31447. https://doi.org/10.1371/journal.pone.0031447.

Maher, L. A., D. A. Macdonald, A. Allentuck, L. Martin, A. Spyrou, and M. D. Jones, 2016. Occupying wide open spaces? Late Pleistocene hunter–gatherer activities in the Eastern Levant. *Quaternary International* 396: 79–94.

Martin, L., Y. Edwards, and A. Garrard, 2010. Hunting practices at an Eastern Jordanian Epipalaeolithic aggregation site: The case of Kharaneh IV. *Levant* 52 (2): 107–135.

McFadden, L. D., S. G. Wells, and M. J. Jercinovich, 1987. Influences of eolian and pedogenic processes on the origin and evolution of desert pavements. *Geology* 15 (6): 504–508.

EXACT, 1998. *Overview of Middle East water resources: Water resources of Palestinian, Jordanian and Israeli interest. Jordanian Ministry of Water and Irrigation, Palestinian Water Authority and Israeli Hydrological Service.* U.S. Geological Survey for the Executive Action Team, Middle East Water Data Banks Project, Amman.

Mischke, S., S. Opitz, J. Kalbe, H. Ginat, and B. Al-Saqarat, 2015. Palaeoenvironmental inferences from late Quaternary sediments of the Al Jafr Basin, Jordan. *Quaternary International* 382: 154–167.

Muheisen, M., 1983. La Préhistoire en Jordanie. Recherches sur l'Epipaléolithique. L'Example du Gisement de Kharaneh IV. Unpublished doctoral disserttion, l'Université de Bordeaux I, Bordeaux.

Muheisen, M., 1988a. The Epipalaeolithic phases of Kharaneh IV. In *The Prehistory of Jordan. The State of Research in 1986*, edited by A. Garrard and H.G.K. Gebel, 353–367. British Archaeological Reports 396. British Archaeological Reports, Oxford.

Muheisen, M., 1988b. Le Gisement de Kharaneh IV, Note Summaire Sur la Phase D. *Paléorient* 14: 265–269.

Muheisen, M., 1988c. Le Paléolithique et l'Epipaléolithique en Jordanie. Unpublished doctoral disserttion, l'Université de Bordeaux I, Bordeaux.

Nadel, D., 2003. The Ohalo II brush huts and the dwelling structures of the Natufian and PPNA sites in the Jordan Valley. *Archaeology, Ethnology and Anthropology of Eurasia* 13 (1): 34–48.

Nelson, J. B., 1973. *Azraq: Desert Oasis*. Penguin Press, London.

Nelson, J. B., 1985. Azraq - A case study. In *Studies in the History and Archaeology of Jordan II*, pp. 39–44. Department of Antiquities of Jordan, Amman.

Noble, P., 1998. Quantification of recharge to the Azraq Basin. In *Arid Land Resources and their Management: Jordan's Desert Margin*, edited by R. W. Dutton, J. I. Clarke and A. M. Battikhi, pp. 103–109. Kegan Paul International, London.

Nowell, A., C. Walker, C. E. Cordova, C. J. H. Ames, J. T. Pokines, D. Stueber, R. DeWitt, and A. S. A. Al-Souliman, 2016. Middle Pleistocene subsistence in the Azraq Oasis, Jordan: Protein residue and other proxies. *Journal of Archaeological Science* 73: 36–44.

Oguchi, T., K. Hori, and C. Oguchi, 2008. Paleohydrological implications of late Quaternary fluvial deposits in and around archaeological sites in Syria. *Geomorphology* 101: 33–43.

Orland, I. J., M. Bar-Matthews, A. Ayalon A. Matthews R. Kozdon T. Ushikubo and J. W. Valley, 2012. Seasonal resolution of Eastern Mediterranean climate change since 34 ka from a Soreq Cave speleothem. *Geochimica et Cosmochimica Acta* 89: 240–255.

Pokines, J. T., A. M. Lister, C. J. H. Ames, A. Nowell, and C. E. Cordova, 2019. Faunal remains from recent excavations at Shishan Marsh 1 (SM1), a Late Lower Paleolithic open-air site in the Azraq Basin, Jordan. *Quaternary Research* 91: 768–791. https://doi.org/10.1017/qua.2018.113.

Pollastro, R. M., A. S. Karshbaum, and R. J. Viger, 1999. Maps showing geology, oil and gas fields and geologic provinces of the Arabian Peninsula. *U.S. Geological Survey Open-File Report 97-470-B.* https://doi.org/10.3133/ofr97470B.

Rambeau, C., 2010. Palaeoenvironmental reconstruction in the Southern Levant: synthesis, challenges, recent developments and perspectives. *Philosophical Transactions of the Royal Society* 368: 5225–5248.

Ramsey, M. N., and A. M. Rosen, 2016. Wedded to wetlands: Exploring Late Pleistocene plant-use in the eastern Levant. *Quaternary International* 396: 5–19.

Ramsey, M., M. Jones, T. Richter, and A. Rosen, 2015. Modifying the marsh: Evaluating early Epipaleolithic hunter-gatherer impacts in the Azraq wetland, Jordan. *Holocene* 25 (10): 1553–1564. https://doi.org/10.1177/0959683615594240.

Ramsey, M. N., L. A. Maher, D. A. Macdonald, and A. Rosen. 2016. Risk, reliability and resilience: Phytolith Evidence for Alternative "Neolithization" Pathways at Kharaneh IV in the Azraq Basin, Jordan. *PLoS ONE* 11 (10): e0164081.

Ramsey, M. N., L. A. Maher, D. A. Macdonald, D. Nadel, and A. M. Rosen, 2018. Sheltered by reeds and settled on sedges: Construction and use of a twenty thousand-year-old hut according to phytolith analysis from Kharaneh IV, Jordan. *Journal of Anthropological Archaeology* 50: 85–97.

Rech, J., 2011. Millennial Landscape Change in Jordan: Geoarchaeology and Cultural Ecology. *Near Eastern Archaeology* 74 (4): 250–251. <Go to ISI>:// 000306412600008.

Rech, J. A., H. Ginat, G. A. Catlett, S. Mischke, E. W. Tully, and J. S. Pigati, 2017. Pliocene-Pleistocene water-bodies and associated deposits in southern Jordan and Israel. In *Quaternary Environments: Climate Change and Humans in the Levant*, edited byO. Bar-Yosef and Y. Enzel, 127–134. Cambridge University Press, Cambridge.

Richter, T., 2014. Margin or Centre? The Epipalaeolithic in the Azraq Oasis and the Qa'Shubayqa. In *Settlement, Survey and Stone: Essays on Near Eastern Prehistory in Honour of Gary Rollefson*, edited byB. Finlayson and C. Makarewicz, pp. 27–36. Ex Oriente, Berlin.

Richter, T. 2017. Natufian and Early Neolithic in the Black Desert, Eastern Jordan. In *Quaternary of the Levant: Environments, Climate Change, and Humans*, edited byO. Bar-Yosef and Y. Enzel, pp. 715–722. Cambridge University Press, Cambridge.

Richter, T., and L. Maher, 2013. The Natufian of the Azraq Basin: An Appraisal. In *Natufian Foragers in the Levant: Terminal Pleistocene Social Changes in Western*

Asia, edited byO. Bar-Yosef and F. Valla, pp. 429–448. International Monographs in Prehistory, Ann Arbor.

Richter, T., J. T. Stock, L. Maher, and C. Hebron, 2010. An Early Epipalaeolithic Sitting Burial from the Azraq Oasis, Jordan. *Antiquity* 84: 1–14.

Richter, T., M. Jones, L. Maher, and J. T. Stock, 2014. The Early and Middle Epipalaeolithic in the Azraq Oasis: Excavations at Ain Qasiyya and AWS-48. In *Jordan's Prehistory: Past and Future Research*, edited byM. Jamhawi, pp. 93–108. Department of Antiquities of Jordan, Amman.

Richter, T., A. Arranz-Otaegui, L. Yeomans, and E. Boaretto, 2017. High resolution AMS dates from Shubayqa 1, northeast Jordan reveal complex origins of Late Epipalaeolithic Natufian in the Levant. *Scientific reports* 7 (1): 1–10.

Richter, T., L. A. Maher, A. N. Garrard, K. Edinborough, M. D. Jones, and J. T. Stock, 2013. Epipalaeolithic settlement dynamics in southwest Asia: new radiocarbon evidence from the Azraq Basin. *Journal of Quaternary Science* 28 (5): 467–479.

Richter, T., S. Alcock, M. Jones, L. Maher, L. Martin, J. Stock, and B. Thorne, 2010. New light on Final Pleistocene settlement diversity in the Azraq Basin: some preliminary results from Ayn Qasiyah. *Paléorient* 35 (2): 49–68.

Richter, T., A. Arranz-Otaegui, E. Boaretto, E. Bocaege, E. Estrup, C. Martinez-Gallardo, G. A. Pantos, P. Pedersen, I. Sæhle, and L. Yeomans, 2016. Shubayqa 6: a new Late Natufian and Pre-Pottery Neolithic A settlement in north-east Jordan. *Antiquity* 90 (354).

Robinson, S. A., S. Black, B. W. Sellwood, and P. J. Valdes, 2006. A review of palaeoclimates and palaeoenvironments in the Levant and Eastern Mediterranean from 25,000 to 5000 years BP: setting the environmental background for the evolution of human civilisation. *Quaternary Science Reviews* 25: 1517–1541.

Rohling, E. J., K. M. Grant, A. P. Roberts, and J.-C. Larrasoaña, 2013. Paleoclimate Variability in the Mediterranean and Red Sea Regions during the Last 500,000 Years. *Current Anthropology* 54 (S8): S183–S201.

Rollefson, G. O., 1984. A Middle Acheulian surface site from Wadi Uweinid, Eastern Jordan. *Paléorient*10: 127–133. https://doi.org/10.3406/paleo.1984.4353.

Rollefson, G. O., 2000. Return to 'Ain el-Assad (Lion Spring), 1996: Azraq Acheulian Occupation in situ. In *The Archaeology of Jordan and Beyond: Essays in Memory of James A. Sauer*, edited by J. A. Sauer, L. E. Stager, J. A. Greene, and M. D. Coogan, pp. 418–428. Eisenbrauns, Winona Lake, Indiana.

Rollefson, G., D. Schnurrenberger, L. Quintero, R. P. Watson, and R. Low, 1997. Ain Soda and'Ayn Qasiya: New Late Pleistocene and Early Holocene Sites in the Azraq Shishan Area, Eastern Jordan. In *The Prehistory of Jordan II. Perspectives from 1997, Studies in Near Eastern Production, Subsistence, and Environment*, edited byH.G.K. Gebel, Z. Kafafi and G. O. Rollefson, pp. 45–58. Ex Oriente, Berlin.

Roskin, J., I. Katra, N. Agha, N. Goring-Morris, N. Porat, and O. Barzilai, 2014. Rapid anthropogenic response to short-term aeolian-fluvial palaeoenvironmental changes during the Late Pleistocene-Holocene transition in the northern Negev Desert, Israel. *Quaternary Science Reviews* 99: 176–192.

Rowan, Y., G. Rollefson, A. Wasse, W. Abu-Azizeh, A. C. Hill, and M. Kersel, 2015. The "land of conjecture": New late prehistoric discoveries at Maitland's Mesa and Wisad Pools, Jordan. *Journal of Field Archaeology* 40 (2): 176–189.

Salameh, E., K. Bandel, I. Alhejoj, and G. Abdallat, 2018. Evolution and Termination of Lakes in Jordan and Their Relevance to Human Migration from Africa to Asia and Europe. *Open Journal of Geology* 8 (12): 1113.

Shahbaz, M., and B. Sunna, 2000. *Integrated studies of the Azraq Basin in Jordan.* U.S. Department of Agriculture, Forest Service, Rocky Mountain Research Station, Fort Collins, Colorado.

Spyrou, A,. 2019. Meat Outside the Freezer: Drying, Smoking and Sealing Meat in Fat in an Epipalaeolithic Megasite in Eastern Jordan. *Journal of Anthropological Archaeology* 54: 84–101.

Steinel, A., K. Schelkes, A. Subah, and T. Himmelsbach, 2016. Spatial multi-criteria analysis for selecting potential sites for aquifer recharge via harvesting and infiltration of surface runoff in north Jordan. *Hydrogeology Journal* 24: 1753–1774. https://doi.org/10.1007/s10040-016-1427-6

Tansey, K. J., and A. C. Millington, 2001. Investigating the potential for soil moisture and surface roughness monitoring in drylands using ERS SAR data. *International Journal of Remote Sensing* 22 (11): 2129–2149.

Torfstein, A., S. L. Goldstein, M. Stein, and Y. Enzel, 2013. Impacts of abrupt climate changes in the Levant from Last Glacial Dead Sea levels. *Quaternary Science Reviews* 69: 1–7.

Wasse, A., Y. M. Rowan, and G. O. Rollefson, 2020. Flamingos in the Desert: how a chance encounter shed light on the 'Burin Neolithic' of eastern Jordan. In *Landscapes of survival: Pastoralist societies, rock art and literacy in Jordan's Black Desert*, edited by P. M. M. G. Akkermans, 79–102. Sidestone Press, Leiden.

Wells, S. G., J. C. Dohrenwend, L. D. McFadden, B. D. Turrin, and K. D. Mahrer, 1985. Late Cenozoic landscape evolution on lava flow surfaces of the Cima volcanic field, Mojave Desert, California. *Geological Society of America Bulletin* 96 (12): 1518–1529.

Yeomans, L. 2018. Influence of Global and Local Environmental Change on Migratory Birds: Evidence for Variable Wetland Habitats in the Late Pleistocene and Early Holocene of the Southern Levant. *Journal of Wetland Archaeology* 18 (1): 20–34. https://doi.org/10.1080/14732971.2018.1454702.

Zeuner, F. E., D. Kirkbride, and B. C. Park, 1957. Stone age exploration in Jordan, I. *Palestine Exploration Quarterly* 1957: 17–54.

6 Creating living predictive models of coastal palaeoenvironmental landscapes

Georgia, USA

Lindsey E. Cochran, Victor D. Thompson, and Bryan Tucker

Dynamic landscapes present distinct challenges to develop models of occupational histories and site selection that incorporate both the nature of human traditions as well as environmental change. Here, archaeological predictive modelling, or site suitability models, are defined as a set of techniques used to estimate "the location of archaeological sites or materials in a region, based either on a sample of that region or on fundamental notions concerning human behavior" (Kohler and Parker 1986, 400). Coastal zones and islands are subject to drastic changes since their initial occupations often with shifts in exposed lands as and sea levels have gradually, and sometimes abruptly shifted over time. The coast of Georgia, USA, bordering the Atlantic Ocean, is one such landscape that has experienced dramatic environmental shifts over time in terms of availed inhabitable land and shifts in estuarine resources (DePratter and Thompson, 2013; Turck and Alexander 2013). To account for such drastic environmental change, we test the applicability of three site suitability techniques to determine which method is most appropriate in a unique environment with an incredibly deep and dynamic history. We use an aggregate approach to this model, using sites from the total historical and near historical occupation of the Georgia coastal zone and sea islands (5000–100 YBP). These techniques are: (1) generalised linear modelling; (2) weighted overlay analysis; and (3) fuzzy logic. The aims of each technique are the same: to develop a regionally-specific multi-factor spatial decision support system to estimate the probability of identifying an archaeological site in a given area given the environmental characteristics and culture history of previously identified sites.

Coastal Georgia is part of the South Atlantic Bight, which extends from modern South Carolina to northern Florida. This region of the southeastern United States is a dynamic landscape, subject to natural forces of wind and tides, with sediments eroding and aggregating, forming and reshaping the coastline. The estuarine region of the Georgia coast contains incredibly productive and resource-rich estuaries, 14 major barrier islands and 1400

DOI: 10.4324/9781003139553-6

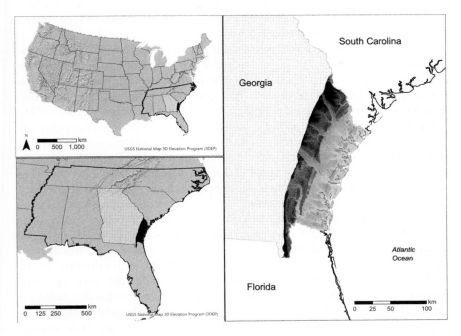

Figure 6.1 Area of interest, Georgia, USA. Sea Island ecoregion is highlighted on the right.

small marsh islands in the tidal area of the coast (Thompson and Turck 2010) (Figure 6.1).

These estuaries are surrounded by more than 160 km of coastline and about 5500 km of tidal shoreline. The region is characterised by a low overall elevation, especially around the smaller marsh islands, making it particularly vulnerable to increased storm frequency and amplitude associated with increases in sea level rise, both of which contribute to compressed sediments and sinking landforms (Anderson et al. 2017). The physical geology of the Georgia Sea Islands brings about interesting questions concerning how people used and moved across their landscape and how that landscape has changed since its initial settlement. Despite the constant habitation of the area over the last 5000 years, the consistent fluctuations of sea level in the Georgia coastal zone affect population densities over time. The most drastic change was at the end of the Pleistocene through the mid-Holocene. This sea level rise flooded the area, resulting in the formation of the barrier islands located on the Georgia coastline. By 6000 YBP, the sea level rise in the area began to slow. Based on archaeological and calibrated radiocarbon dates, a sea level regression likely occurred during the Late Archaic period (~3600 YBP) and the Early Woodland period (3100–2400 YBP). In addition to these larger-scale changes, seven small scale fluctuations occurred in the last 5000 years local to the Georgia coast and estuaries. Overall, however, sea level rise and fall affect the productivity of the marsh and estuarine ecosystems, which in turn restrict potential locations for resource acquisition.

The area between the Atlantic Ocean fronting barrier islands and the mainland are comprised of large expansive salt marshes and tidal estuaries that have been in constant flux. Smaller islands in these vast salt marsh estuaries serve as key structures in the ecosystem, making these prime site settlement locations for past peoples and cultures (Thompson et al. 2013).

The Georgia Coastal zone has a complex history of Native American settlement, with the earliest archaeologically visible occupation occurring almost 5000 years ago (Turck 2012; Turck and Thompson 2019). By 4500 years ago, the area saw the emergence of some of the first sedentary villages at shell ring sites along the coast, large circular deposits of shellfish representing sedentary occupations that relied on harvesting resources from the estuaries (Lulewicz et al. 2017; Sanger et al. 2020; Savarese et al. 2016; Thompson and Andrus 2011). With these villages also too emerged long-distance trade networks (Sanger et al. 2018) and the invention of pottery technologies, which were among the earliest in North America (Sassaman and Gilmore 2021). Archaeologists document a series of shifts in the settlement, with an abandonment of the previously established shell ring villages and a decreased reliance on shellfishing, beginning around 3800 years ago, largely thought to be responses to environmental changes, driven in part by sea level change (Ritchison et al. 2020; Turck and Thompson 2016, 2019; Thompson and Turck 2010). Around 3000 years ago, sea levels rebounded and once again, the archaeological record indicates a reliance on shellfish by villages along the coast (Thomas 2014). Finally, beginning around 1000 years ago, the coast saw immigration from the interior and the development and an increase in site density and size, along with an increase in maize agriculture by villages, yet fish and shellfish from the estuaries continued to play a large role in these economies (Reitz 2014; Thomas 2014; Thomas and Andrus 2008).

While a deep suite of resources drives past and present habitation location decisions (Kvamme 2020, 2006; Lulewicz et al. 2017; Verhagen and Gazenbeek 2006; Verhagen and Whitley 2020), there are certain key resources tied to the coastal landscape that heavily structure site location and habitation it seems, given the often long-term and repeated occupation of many sites over multiple time periods. As explored by Thompson et al. (2020), shellfish reefs and portions of the estuaries served as harvest areas over extended periods of time from 4500 through 500 years ago. Based on size classes from multiple sites up and down the coast, they argue that there were long term property rights systems that were established during early shell ring village times that continued until European colonisation (Thompson et al. 2020). This view of harvest territories or proprietorship (Letham et al. 2020) is also supported, in part, by the types of fishes that archaeologists also recover from middens at these sites (Reitz 2014). Thus, for the coastal zone, while there were major historical shifts, estuarine resources were also a key factor in determining site selection and village locations, albeit with some variation and different plants and animals (e.g., maize) were incorporated into these economies.

These estuarine resources were also critical for rice, sugar, and Sea Island cotton plantations on the Georgia coast. Because of the temporary illegality of slavery in 1735 by General James E. Oglethorpe, the Georgia plantation economy developed later than most. However, in 1751, the ban of slavery was reversed by royal decree, thus beginning the institution of slavery in the Province of Georgia and leading to massive efforts to grow sugar, indigo, and rice cultivation. These cultigens were grown in similar harvesting areas as their past counterparts.

Remnants of Indigenous habitation were repurposed often in the plantation era on the Georgia coast. Oyster shells sourced from past middens, rings, and mounds were used to construct buildings made of tabby, a type of cement comprised of equal parts of sand, lime, water, and oyster shell. Repurposed oysters were often repurposed in Sea Island cotton fields as a fertiliser or mixed with eggshells to be used as a whitewash on these historic tabby buildings. Landscape changes from the plantation era are clear today on aerial imagery, for instance, showing dikes for tidal rice agriculture, clearings for sugar and the later Sea Island cotton cultivation have permanently changed the coastal landscape, and ironically making these landscapes some of the first to be affected by modern sea level rise.

GIS and landscape archaeology

Geographic Information Systems (GIS) create graphical representations of the mathematics of landscapes; however, the software was not created to model the ways in which humans choose where to create materials that serve as evidence for archaeologists to interpret the ways in which past people "embodie[d] space and spatial relationships within a material world" (Gillings, Hacigüzeller, and Lock 2020:1). A natural landscape, or a space, is a physical, non-cultural setting in which placemaking, the human process of inscribing meaning to a given area, does not occur. Spaces or the culturally empty aspects of a landscape are the "physical dimensions or characteristics of architecture and landscape" (Heath 2010: 159). In contrast, a built or cultural landscape includes materials and ideational components of a person or group associated within a location, habitat, or place. Spatial thinking, an overall theme that subsumes the concepts of spatial analysis, "cannot be divorced from the interplay between quantitative and qualitative approaches, formal and informal aspects" (Gillings et al. 2020: 13). The accessibility of comprehensive and cohesive datasets like state site files are leading to research questions that are increasingly defined by movement patterns, characteristics of mobility, and settlement patterning.

Statistical methodologies do not adequately capture human intention, particularly not the reasoning of past peoples' movement, settlement strategies, or resource acquisition. Spatial archaeology works to create an interface between statistical methods to model past landscapes with anthropological, archaeological questions. Site suitability models are technical methods to identify landscapes that contain a suite of environmental and cultural characteristics that are most likely to contain evidence of past human activities. As these techniques improve, so too does the potential to use digital modelling techniques to access aspects of the past landscape that were otherwise unobservable.

A recent survey of space and spatiality in archaeology (Gillings et al. 2020) identified five relationships between the concept and action of past interactions between humans and space. They are, briefly: (1) time; (2) mobility; (3) daily practices that generate material culture; (4) the absence of people in a space; and (5) modern interpretation and representation. These themes of spatial relationships intersect with broad research questions about the long-term trajectories of human habitation along the Georgia coast. Here, we view historical ecology as a driving factor in the praxis of place-making. Each of the above spatial themes can be integrated into the major archaeological research questions about the long-term trajectories of human habitation along the Georgia coast, such as the emergence of pottery technology, the emergence of village societies, the transition from forager to farmer, long-term responses to sea level fluctuations, the impact of European colonialism, and the emergence of plantation economies and rapid adoption and eventual dissolution of slavery.

While suitability modelling, or statistical methods used to estimate the type and location of a potential cultural resource given known archaeological attribute data, cannot access these anthropological themes directly, the dialectical relationship of historical ecology is visible archaeologically. Each adoption of new technologies, be it pottery or Sea Island cotton, produced a new suite of materials visible in the archaeological record that are a fundamental dataset in spatial analyses. Likewise, social networks within and among emergent village societies led to new diagnostic trade goods, more robust and diagnostically variable material culture, and a larger footprint of a site on a landscape. As people shifted toward more sedentary lifeways but expanded in trade, eventually importing manufactured and exporting raw goods across the Atlantic World, social and material networks continued to expand, and cultural landscapes became more integrated. Each new use of space was predicated on a different use of environmental resources, as each development of place and reliance on new technologies or social developments continued to change that surrounding environment.

Archaeological insights generate spatial data that illustrate where people did or did not live, and these data are often time seriated. However, the majority of archaeological sciences conducted in the United States are on a by-project basis, which is excellent for recording sites and surveying land, but do not often focus on archaeological synthesis (Heilen 2020, 1). Site suitability modelling techniques must match the degree of refinement and visibility of past actions allowed by the archaeological data to avoid Type I and Type II statistical errors, untested or uninformed assumptions, and user bias. Ultimately, a dissolution of arbitrary boundaries between both past and present, integrating multiple nested scales of analysis, and building iterative models are the keys to assess more than simply habitat suitability for past people on the Georgia coast.

The ultimate problem is that statistical modelling techniques rarely address human autonomy or choice. Rather, many models are driven by calculating proximity analyses of cultural heritage sites to environmental resources like food sources and water. However, humans do not work within statistical boundaries, nor are distances to necessary resources the only reasons why people settled or moved through a certain area. In this chapter, we call for more conscious and careful implementation of statistical techniques while creating site suitability models. We integrate core concepts of historical ecology and landscape theory with site suitability modelling of at-risk cultural heritage resources to show how people have used certain places over time and which environmental resources were most critical. We illustrate that there is no uniform solution to modelling; the environmental context for human activity is a crucial and recursive relationship. Our goal is to determine which suitability modelling technique is most capable of capturing the nuance of the human experience for the Georgia Coast.

Using spatial archaeology to investigate historical ecology

Historical ecology is a theoretical derivative of primarily cultural ecology, evolutionary ecology, and behavioural ecology but differs from these parent predecessors in two fundamental ways (Thompson 2010: 2). First, historical ecology views humans as a keystone species, meaning that people are primary drivers of ecological change (Balée and Erikson 2006: 1, 3, 5). The changes that humans make to their surroundings can be intentional, such as clearing a forest or building a homestead; these are grouped under one of the seven foundational concepts in ecology, that of intentional ecosystem management. Conversely, these changes can be unintentional and subtle, such as reduction of prey due to overhunting, overfishing, or changing the chemical composition of soils as a result of long-term farming. At its core, historical ecology is based on materiality. Likewise, the second distinguishing aspect of historical ecology is fundamentally multi-scalar, taking the perspective that no action or reaction is spatially isolated (Szabó 2015), meaning that interpretations of located human interactions with the landscape must be situated within regional contextualisation.

The main outcome of an historical ecological perspective is to create a multi-disciplinary, regionalised synthesis of human-environmental interactions. Just as historical ecology fundamentally rejects isolationism, the paradigm also rejects concepts of human stagnation over time and place. The relationship between humans and their surroundings is a continuum.

Impacts of humans on their ecosystems is observable, especially within fragile but resource-rich ecological zones such as the southeastern seaboard and islands. In the southeastern United States, and Georgia specifically, recent studies demonstrate the importance of archaeology and paleobiology to understanding oyster reef loss and resilience and how Indigenous communities played a role in shaping these environments over time (Thompson et al. 2020). Modern Native American communities, along with other populations, such as the Gullah Geechee Nation, descendants of people who were enslaved on rice, indigo, and Sea Island cotton southeastern coastal plantations, still maintain a connection to vulnerable archaeological sites in the region. These resources, however, are eroding and dying at alarming rates. Over 92% of oyster reefs on the Georgia and South Carolina coasts have already been lost (Coen et al. 2011; Kirby 2004; Landis 2020;Ricka et al. 2016).

The spatiality of archaeology is not new nor revolutionary (Spaulding 1960), but it is both implicit and fundamental to any discussion of the archaeology of landscapes. In the more humanities-driven worlds, especially within history, a new "spatial turn" has been recognised over the last two decades (Blake 2004; Bodenhamer 2012; de Cunzo and Ernstein 2015; Hegemon 2003; Pouncett 2020; Warf and Arias 2008). In contrast to landscape histories, landscape archaeology is the study of the spatial relationship between past people and the space they occupied, or more specifically, past interactions between people and the bounded spaces in which cultural activities occurred (Branton 2009: 51–52). Concepts of space and place help archaeologists to consider a landscape as a whole by moving beyond the site to incorporate broader environmental and cultural areas of place-making (Anscheutz et al. 2001: 159). Landscapes do not belong to one person or population; rather, they are a diachronic compilation of human activity, in which individuals or groups react to those who preceded them, either implicitly or explicitly. Placemaking activities of humans alter the ecosystem in often permanent and observable ways.

Archaeological landscapes, considered as natural or cultural and the latter as either a space or a place, are terms that subsume the myriad and many ways in which people experience their world. Landscapes, which are always nested, materialises the relationship between people and the area they occupy, all the while essentially imprinting aspects of themselves, of variable physicality and durability (e.g., forest clearing or debitage), onto that landscape (Delle 1999, Anscheutz 2001). A landscape is an area bounded naturally (e.g., island coastlines) or culturally (e.g., boundaries of a village) in which human activity occurred. At a fundamental level, each landscape has different "meaning[s] to a discrete group of people at a defined time and place" (Ashmore 2004; Branton 2009: 53; Knapp and Ashmore 1999). These landscapes, considered by Kosiba and Bauer (2013, 63) as a "mélange of places, practices, and concepts through which people experience and perceive their environment." Most broadly, places are created through human practices or are the outcome of the social processes of space; both are influenced by the daily activities of people, their activities, and cultural context (Ashmore 2004).

The process of place-making results in a material record and a portion of those places are recorded as cultural heritage sites. The historical ecological dialectic drives the site suitability modelling involving the recursive and reoccurring relationship between human behaviour, local and regional ecological niches, and the landscape that contains these aspects. The ultimate goal of suitability modelling is to access exactly how past humans changed their landscape and how they, in turn, were shaped by the environment. Developments in predictive modelling techniques and creative applications of biological and ecological suitability statistics bring archaeologists closer to accessing a more holistic interpretation of past movements, social interactions, and relationships with the environment, especially in a landscape as dynamic and challenging as the Georgia coast.

An environment known for its remote islands provides challenges for modelling. While the area is flat, an input parameter that commonly ranks as one of the most important settlement and site suitability factor-there are many other landscape features that significantly affect how people use the landscape around them, creating places from spaces.

Digital synthesis of archaeological spaces

Predictive models are estimates of past human behaviour that must always be interpreted as flawed. Archaeological science demands that datasets are, by their nature, incomplete, and establishing how representative a sample of sites to a certain area is challenging, but new statistical approaches are helping to address this problem. But despite these flaws with site suitability modelling, there is no better way to analyse a large area of multitemporal space to estimate where people used to live and why.

Hodder and Orton (1976), in Spatial Analysis in Archaeology, commented on the necessity of accumulating a "slow collection of large bodies of reliable data." Spatial archaeology, subsumed within spatial thinking, makes explicit two key factors of the use of digital data today: The recycling of spatial data and the basic necessity of accessible, reliable data organised within databases (Gupta 2020). Most critically, archaeologists can use the large datasets collected through the state site files and integrate those points into software created for biology and ecology. The detailed Georgia state site files, nested within the Georgia Natural and Historical Resources Geographic Information System (GNAHRGIS), tracks the sites location, cultural association, temporal association, and chronologies of cultural resources within the state.

Types of models

With roots in the New Archaeology of the 1970s, the premise of quantitative site locational methodology is that human behavioural patterns are observable and predictable (Vita-Finzi and Higgs 1970). Therefore, the locations of daily activities can be statistically estimated from these patterns. These tools are generally considered to be heuristic devices and take either a data-driven or theory-driven approach to estimating past human behaviour (Verhagen 2006; Verhagen and Gazenbeek 2006; Verhagen and Whitley 2020). Data-driven modelling approaches are commonly used in CRM as a planning aid to create a cost/benefit assessment of disturbing cultural heritage sites during construction projects (Heilen 2020; Verhagen and Whitley 2020). These models are deeply rooted in environmental parameters that are interpreted to influence the choice of settling in a particular location. Parameters always include slope, distance to water, soil type, and geology (Branton 2009; Duncan and Beckman 2000; Elith et al. 2011; Harris 2013). The confluence of these variables are extrapolated to other areas with the same environmental scenario that has not been surveyed, and thus labelling those as high priority areas to survey or an area likely to contain cultural heritage sites. Data-driven models are often processed with logistic regressions, a popular statistical technique to fit a prediction curve to a particular set of observations.

Theory-driven models, in contrast, tend to weight certain variables as more important than others. These models are more question-based and less exploratory, as aspects of the model are assumed. A theory-driven model makes certain assumptions about settlement patterning strategies.

Both theory-driven and data-driven models are subject to flaws and pitfalls, as summarised by Leusen et al. (2005) and further developed by Verhagen and Whitley (2020, 233). These issues can be summarised as problems with: (1) the incomplete nature of archaeological datasets; (2) historically variable data collection and archaeological survey methods; (3) environmentally driven analyses; (4) tension between limited temporal control in archaeological datasets and reliance of modern environmental datasets; (5) modern bias while interpreting levels of significance of palaeoenvironmental resources to past people; and (6) problematic statistical validation methods and a near-complete lack of ground truthing models. Predictive modelling has been often criticised for prioritising these environmentally driven datasets because the datasets are modern and reductionist. These inputs, most often derivations of Digital Elevation Models, are used most frequently because they are easy to access, easy to implement, and require little to no interpretation after post-processing.

Though requiring more interpretation and cognitive outputs from the mapmaker, culturally driven variables are accessible and often derived from Euclidian distance or raster proximity calculations. Suitability models nearly always include nearness indices to roads, horizontal and vertical distance to water sources, groundwater, certain soils and geological formations (e.g., lithic quarries), and distance to other archaeological sites. Just as with environmental variables, these cultural variables must be implemented carefully and with due consideration. The region and scale of analysis are critical when determining which variables are needed and important. Furthermore, much of the published literature about suitability modelling is based in ecology, where the science of site prediction is very well developed but lacks time depth and the influence of choice and the logic (or absence of) involved in humans' decision-making processes that result in settlement or materiality. Therefore, when implementing ecological modelling techniques, the spatial archaeologist must carefully interpret the appropriateness of those techniques to archaeological questions.

Tensions between archaeological data and statistical methods

Common tensions between archaeological datasets and statistical approaches lie in the structure of data collection, and the type of research questions asked. First, archaeological sites are polygons but must be represented as a point for statistical analyses, so sites with large and small polygon footprints contribute the same amount to the model (Mink et al. 2006). Second, archaeological surveys are conducted mostly as cultural and environmental surveys prior to construction efforts. Overall collections of survey points in a state system are not collected on a grid; they are collected around areas that need frequent surveys, such as military bases, water reservoirs, highways, and large cities, thus skewing the data sample (Aiello-Lammens et al. 2015; Engemann et al. 2015; Fithian et al. 2015; Zizka et al. 2020). Third, many states collect presence-only data, or the location of known sites rather than areas that have been surveyed but were found not to contain artefacts or features, so spatial analyses are reinventing the wheel of site/non-site designations to a certain degree. Fourth, the significance of environmental variables is highly dependent on the total environment. Slope, for example, carries little weight in the site settlement decision making process on the Georgia coast, but in the North Georgia mountains, the slope is extremely significant. In contrast, distance to a deltaic or non-deltaic hydrologic system is quite significant on the coast but wholly irrelevant elsewhere. Environmental variability makes the size of analysis in a suitability model critical. Finally, the most frequent questions in archaeology are centred around chronologies and movement (for an excellent discussion of stationary model populations as an apriori assumption, see Crema 2020). Creating theory-based models with time sensitivity will be an important area for future development in the world of spatial archaeology.

Georgia Sea Island Suitability Modeling Project

The purpose of the Georgia Sea Island Suitability Modeling Project is to determine general locations where cultural heritage sites are more likely to be identified, given the environmental and cultural attributes attached to each known archaeological site that is documented in GNAHRGIS. As a liberal experimental model, the Georgia Sea Island Suitability Modeling Project incorporates sites from all time periods into the same model to determine which resources were, in general, most important to past people and groups. These data are incorporated into the model as x, y point data on a presence, absence level. We apply three common techniques, summarised in Table 6.1. To determine which is appropriate to synthesise past uses of spaces (as presence/absence) in an ever-changing palaeoenvironment. The goals of the model are to determine: (1) which environmental suite of characteristics most significantly contributes to a high likelihood of containing an archaeological site; (2) which type of modelling technique best captures the interface between nature and culture on the Georgia Sea Islands; and (3) are the models capable of integrating geological landforms where past people were likely to have lived? The following section introduces, discusses the pros and cons of implementing each model type, and presents geoprocessing techniques and resulting rasters of the AOI.

Each of the models was run with a suite of input parameters, summarised in Table 6.2. Input variables derived from a digital elevation model (DEM) were not particularly useful due to the flatness of the coastline. Instead, environmental variables like distance calculations to the shoreline, estuary, mainland, or major hydrological source were prioritised to estimate critical areas of past resource acquisition. Landcover, geologic, and soil categories included important habitat implications of characteristics people have found desirable, such as dense habitation barrier islands but a near-total absence of habitation on the marsh or lagoonal facies as visually estimated from known points in GNAHRGIS.

Logistic regression: generalised linear model (GLM)

Logistic regression is used to model the probability of the presence or absence of an event. This statistical model uses a logistic function to model a binary dependent variable, in this case, the existence of an archaeological site at a given spatial coordinate. The independent or predictor variables are the environmental and cultural variables like slope or distance to a water source. Independent variables can be categorical or continuous, making the model flexible because it allows for data with arbitrary distributions. The output of an ordinary logistic regression is a value estimation of an unknown quantity, known as a response variable. A generalised linear model (GLM) is a non-normal form of the parametric (bell-shaped) general linear model. The GLM is a data-driven suitability model.

Table 6.1 Summary of generalised linear models, weighted overlay analysis, and fuzzy logic.

Model type	Premise	Data manipulation	Pros	Cons
Generalised Linear Model	Generalisation of ordinary linear regression that allows for response variables with a non-normal distribution	• Binary • Presence-absence Records of Dependent Variable	• Good for experimental models • Data do not need to be parametric • Broad application	• All input parameters contribute equally • Oriented towards mathematics and statistics • Relies on R and GIS for processing
Weighted Overlay Analysis	Statistically model importance of eligibility criteria of input parameters	• Range of scores • Rescaled and re-weighted rasters	• Straightforward implementation • Requires little data manipulation • User chooses significance contribution of each input parameter	• Requires significant background research to determine how to weigh each input parameter • Low statistical accountability • Interpretation statistically programmed leading to unconscious bias
Fuzzy Logic	Statistically models imprecision and uncertainty	• Binary • Rescaled and re-weighted rasters	• Excellent for adapting to data incompatibility • Rejects assumptions of past landscapes • Statistically models uncertainty • Can be weighted	Some categorical data are not scalable for the question at hand, i.e., aspect

Table 6.2 Data input into each of the three models, including per cent contribution of each input variable to each model.

Data	Processing type	Data Type	GLM	WOA	Fuzzy
Hydrology	Euclidean Distance	Continuous	10	5	10
Deltaic	Euclidean Distance	Continuous	10	10	10
Non Deltaic	Euclidean Distance	Continuous	10	10	10
Mainland	Euclidean Distance	Continuous	10	10	10
Ecoregions	Euclidian Distance	Continuous	10	10	10
Landcover	Visual estimation of site densities	Categorical	10	5	10
Soils	Visual estimation of site densities	Categorical	10	10	10
Slope	0°–2°=Best; >10°=Worst	Categorical	10	5	10
Geology	Visual estimation of site densities	Categorical	10	10	10
GNAHRGIS	Presence	Binary	10	25	10

* Euclidian distance is used here rather than the raster proximity tool because: (1) it is less computationally intensive and takes seconds rather than minutes to process; and (2) the output is more general, which is appropriate given uncertainties in the specific makeup of the past landscape and human reactions to the past landscape.

** Visual estimations of site densities per category were used in conjunction with USGS category definitions. Important characteristics included drainage qualities, overall porosity, proclivity for sustainable agricultural growth, and USGS scores of quality for settlement or farming.

Traditionally, site suitability models have been run using regression analyses in the form of a generalised linear model. The GLM uses both categorical and continuous variables to model the likelihood of the presence or absence of an archaeological site at a particular location using a maximum likelihood estimation. Environmental and archaeological data are cleaned and standardised in an ArcGIS environment, though an R workspace is also commonly used. The area of interest is divided into cells (a 30 m cell was used here) that are coded as a presence point or a pseudo-absence point. Ultimately, all independent variables are joined into a single.csv with presence and pseudo-absence points coded as "0" or "1", representing "no site" and "site". In suitability models, the fundamental purpose of the model is to compare the variability of independent variables at background pseudo-absence points versus those at site presence points, measured through coefficients, z-values, and p-value. One way that statistical significance is measured is through regression coefficients; the more negative the coefficient, the less likely the presence of a site. A z-value is the regression coefficient divided by its standard error. To obtain coefficients, probabilities, and z-values, the.csv is moved to R and processed using the "glm" function. Results from R are re-imported to GIS and processed further in the raster calculator. The GLM of the Georgia coast captured geological and environmental differences, with the data input, the model poorly predicted the location of archaeological sites (Figure 6.2). Despite this being an environmentally-driven

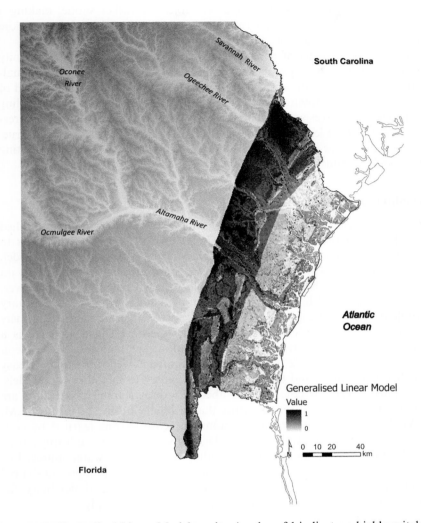

Figure 6.2 Generalised Linear Model results. A value of 1 indicates a highly suitable area for archaeological sites, whereas a value of 0 indicates a less suitable area for sites. The greyscale DEM to the left is the outline of Georgia, with major river basins that drain into the project area labelled. The drastic break between values in the centre of the image reflect remnants of palaeo-shorelines.

site suitability model, the main benefit of the GLM was creating observation points of what datasets were well suited to capturing elements of the coastline. This particular iteration of the GLM was ill-suited for a test area like the Georgia coast, where people favoured areas subject to frequent environmental change rather than a stable palaeoenvironment that is easily visible using modern imagery datasets. The input variables are rather subtle, making a blanket statistical approach like the data-driven GLM useful to better explore data quality and parameters.

Overall, the benefits of using a GLM are in its simplicity and ubiquity among modelling strategies. Additionally, the data do not require normalisation techniques as in other regression-based analyses. While the world of spatial analyses is moving away from GLM to more theory-based modelling strategies, like Bayesian modelling, Forest-based classification and regression, and machine learning techniques, GLM approaches are tried and true. Because GLM are common across many disciplines, verifying and testing the model is intuitive and accessible.

Weighted overlay analysis

Weighted overlay analyses (WOA) are one of the most commonly applied tools for multicriteria problems like site suitability. WOA allows users to determine how much each input criteria layer weighs into the overall model. User-defined inputs are especially useful for a problem like site suitability on the Georgia coast where slope, for example, is nearly meaningless. However, it serves as a useful buffer between habitable landforms and steep estuary shorelines when viewing DEMs. Soil types, on the other hand, are extremely useful in determining site suitability on the Georgia coast, as there is a significant difference between the non-agricultural marsh sands of the Leon series versus fine Norfolk sands, which are compact, well-draining, and easy to cultivate (Latimer and Bucher 1911). In contrast to the other two models, a weighted overlay is considered a liberal model, meaning that results are presented as a range of scores rather than a binary output as in the GLM.

WOA allows the user to apply common sense lessons learned by conducting Phase I surveys, such as adages about reading landforms, belt-buckle slope measurements, and eyeballing distance to a water source. For better or worse, GIS has no "beltbuckle as a measure of slope suitability" function. However, implementing a WOA gives statistical flexibility for these real-world observations. The general formula is simply:

Overall Score = [Criteria Score 1 $*$ Weight 1] + [Criteria Score 2 $*$ Weight 2]

A result of the general statistical flexibility inherent in a WOA is that the user input is critical to the overall success of the model. The overall problem must be very clearly defined because the ranking system between input criteria is entirely based on the user. Each category is individually rescaled then reclassified according to this scale of settlement "goodness". In sum, each layer is individually selected, rasterised, and normalised as part of the basic data preparation for the WOA, then rescaled and reclassified according to the WOA parameters for the individual suitability analysis.

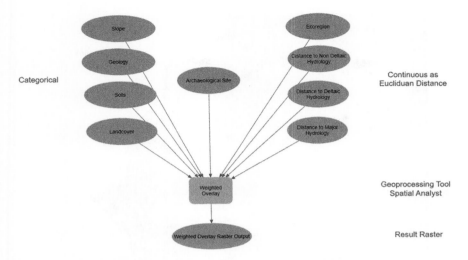

Figure 6.3 Annotated Modelbuilder process for the weighted overlay analysis.

The ranking must be tailored to the question at hand. Similarly, the quality of spatial data that represent the criteria is important. Because each input will be ranked against the other input criterion layers, the cell size must be identical, data within each layer comparable, and each layer rankable. Each layer, be it continuous or categorical, must be broken into submodels, in this case on a scale of 10, with 1 being least good for human habitation and 10 being the best possible option for human habitation. Though the data required significant pre-processing in the form of standardisation and reclassification, the actual implementation of the WOA was straightforward (Figure 6.3).

This theory driven model required significant user input for each category of every parameter input. The hands-on nature of the weighted statistic provided levels of control over the model that was well-suited for application to archaeological problems. However, such high user input demands significant background research to accurately determine – beyond visual estimates – how to divide and reallocate each data point. As with Munsell values, much of these determinations are in the eye of the beholder, and without good record keeping can lead to biases buried deep in the model. The WOA of the Georgia coast did an excellent job of capturing the locations of archaeological sites. As shown in Figure 6.4, programming a training sample of 75% of known archaeological sites that contribute 25% to the model was well suited to capture the prime locations of past activities.

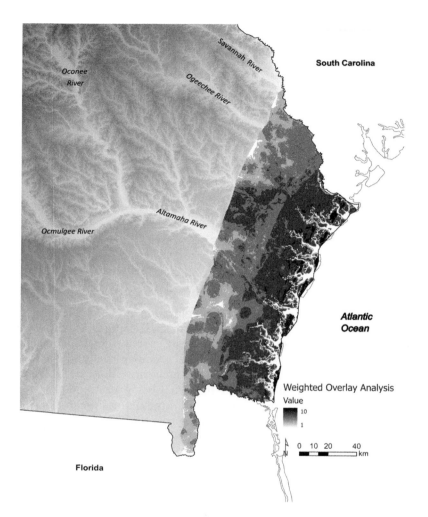

Figure 6.4 Weighted Overlay Analysis Result Raster. A value of 10 indicates a highly suitable area for archaeological sites, whereas a 1 indicates a less suitable area for sites. The location and environment around existing sites contributed 25% to this model.

Modelling site suitability with fuzzy logic

The Georgia coast is physiographically unique and has changed drastically over the last 10,000 years. The nature of a waterway is to change. Estuaries carve into palaeosols of the barrier islands, and sands from northern beaches reallocate to the southern ones until they have no more to erode and are gone forever. While we know the general geological profile of the coastline and general physiographic and ecological changes over time, scientists are still learning the specific mechanism of these changes and various environmental profiles throughout the geological history of the coast. However, site suitability modelling techniques rely on input environmental data, so the "unknowns" of the past compromise the accuracy and integrity of the model. Calling for cognisant and careful quality control by improving data cleaning and tidying and especially version control within database usage, Gupta (2020) argues that the nature of the ways in which data are collected in the field, shared, and reproduced are changing drastically due to our "digital data-rich environment" (Gupta 2020: 18).

Spatially fuzzy sets statistically model these inherent unknowns (Fusco and de Runz 2020). Rather than attempting to upwardly standardise data, or the process of unifying heterogeneous data, which inherently leads to reductionism and data loss, fuzzy logic works "within imperfection" (Fusco and de Runz 2020, 169). Imperfection can be attributed to one or more of four general divisions: incompleteness, uncertainty, vagueness, and ambiguity (Fisher 2005; Fisher et al. 2006). The archaeological record is by its nature incomplete, as archaeologists can only access past activities with information via available materials. Only a certain amount of information is known, and only a certain about can ultimately be known. Though that information is drawn from a wide range of materials, our interpretation is recursively driven by previously synthesised information. Furthermore, the source of information within those syntheses may be compromised due to measuring errors, transformed cartographic coordinate systems, varying standardisation protocol, or even regional differences in typologies, leading to uncertainty in datasets (Mink et al. 2009).

Modern environmental wetland demarcations, among other classified variables, can be imprecise. Much of archaeological data and data interpretation is vague, be it through imprecise language, boundaries, interpretations of data that have not been ground truthed, typological classifications, or regionalised assumptions about chronologies. Ambiguity is a type of uncertainty that combines imprecision and uncertainty, for "when, exactly, is a house a house? The questions always revolve around the threshold value of some measurable parameter or the opinion of some individual" (Fisher 2005: 7). The boundaries of archaeological inquiry are not limited to material typologies, and especially not in suitability modelling. Palaeoecological boundaries are just as significant as the boundaries set in material studies but are perhaps vaguer because the scale of data is at odds. Palaeoecological reconstructions are generally based around a region or micro-region, but the scale of analysis of an archaeological site is driven by the materials within the site itself.

Fundamental to the concept of fuzzy logic, rather than force incompatible accumulations of heterogeneous data, which leads to imperfections in newly generated datasets, analyses, results, and interpretations, we turn to fuzzy data. The process of fuzzification essentially allows us to work from within imperfection rather than around it, taking into consideration the effects of environmental change over time on our interpretation of modern data sets. Other than linear binary dataset conversions, the specific methods and data type constraints are very similar to the WOA, except rather than data being tied to the spatiality of a specific boundary, the boundaries themselves are essentially pixelated.

Overall, the concept of the fuzzy spatial data set appears to hold the most potential to best create suitability models with general, rather than specific, palaeoenvironmental datasets. However, the reality of the model presented in Figure 6.5 favours certain overlapping environmental characteristics

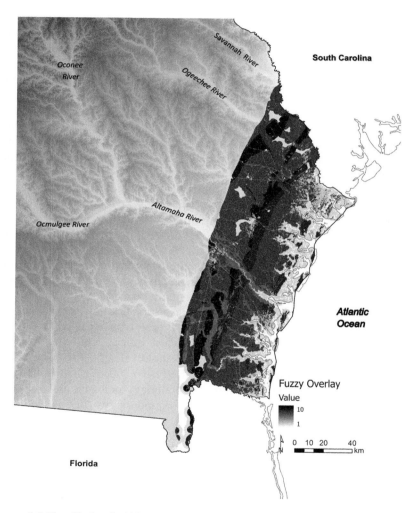

Figure 6.5 Fuzzified suitability model of the Georgia coastline. A value of 10 indicates a highly suitable area for archaeological sites, whereas a 1 indicates a less suitable area for sites.

rather than culturally driven environmental characteristics. While this model adequately captured site suitability on the islands and marshes, areas that have not been thoroughly surveyed archaeologically and without critical environmental resources like major rivers (i.e., bottom left) are likely biased against site identification. This particular model serves as a reminder of the fallible nature of suitability modelling. The fuzzified model below is not weighted, giving credence equally to all ten input variables. Though the theory behind the suitability model is most appropriate for this question at hand, the results of the fuzzified model are lacking. However, future research should consider a fuzzified weighted overlay model with added time seriation. Ultimately, a cost path analysis between high probability areas along the coast separated by time period and with added materials analysis (i.e., stamp patterns via network analysis) would be an excellent way to apply the methods presented here with movement-based questions about the history of the Georgia coast.

Future directions

Geospatial analyses in archaeology are driven by data collected by someone else, entered into a system created by someone else, and using a primary methodology most often developed for an entirely different field of study. Because of the "hands off" nature of geospatial analyses, it is important for the user to recognise that "spatial analysis exists at the interface between the human and the computer, and both play important roles" (de Smith et al. 2018). Overall, the models are contingent upon the data input. Major issues are seen via the tensions resulting from asking archaeological research questions and answering using modern datasets. Given the results presented between the GLM, WOA, and Fuzzy logic suitability models, the WOA preformed the best for the Georgia Coast. Part of this analysis is also the inherent measurement of resource continuity over time. On the Georgia Coast, in particular, past people targeted similar resources, living spaces, and travelled through similar corridors. Future models in areas with more diverse and temporally divergent resource acquisition should re-run models by time rather than this cluster approach.

Methods like WOA and Fuzzy logic analyses promise ways to ease tensions between statistical laws and archaeological research questions because they allow users to integrate the breadth of archaeological and anthropological knowledge into the statistical methodology. As suitability models become less environmentally and more culturally driven, archaeologists are better able to access, via GIS models, distinctions between areas of the Georgia coast likely to have been both spaces and places to past people throughout the total history of the coast.

Ultimately, the study of space is the study of mobility. How did people use, adapt to, and change what was around them? Why did people settle in one area over another? What routes did people take to trade, to move, to inhabit or leave? However, to understand movement, migration, and trade at a site-specific scale rather than a regional scale using suitability and predictive modelling, geological, ecological, and environmental changes that took place over the last 10,000 years must be more holistically developed, as "being human embodies space and spatial relationships within a material world" (Gillings et al. 2020: 1).

The methods presented here highlight the need for continuous investigation into innovative ways to estimate ways to assess past human movement, reasoning, and intention behind the point. As it stands, much of modern suitability modelling is reductionist and environmentally deterministic. However, this is simply a reflection of the tensions between statistical laws and the archaeological data at hand. We propose that future research and development of suitability models begin with a reassessment of the archaeological data, focusing on collecting, collating, and synthesising radiometric, palaeoethnobotanical, chemical sourcing studies for clay-based materials, to name a few. If these datasets can be accurately integrated into spatial datasets, then archaeologists can begin to refine our suitability modelling techniques to more accurately estimate past human behaviour, trade networks, group, and personal identities.

The technology to understand how past people moved through and changed their landscape exists due to Georgia's robust and well-developed state site file system. Creating accessible and reasonably accurate palaeoenvironmental time-sliced reconstructions in a DEM format will allow detailed estimates of broad-scale landscape use over time. The Georgia coast is a perfect illustration of the ever-changing landscape. People have always adapted to the unchangeable spaces around them, like the shape of an island or the structure of the tide, while concurrently changing what can be changed, such as methods of resource acquisition like culling certain competitive resources while encouraging the growth of another.

Acknowledgements

We would like to express our thanks to the editor for arranging an outlet for publication in light of conference cancellations due to COVID-19. We thank our colleagues who offered vital information and support, as well as the hospitality of our proxy work-from-home colleagues, particularly Richard B. Descoteaux in Swanzey, New Hampshire.

References

Aiello-Lammens, Matthew E., Robert A. Boria, Aleksandar Radosavljevic, Bruno Vilela, and Robert P. Anderson, 2015. SpThin: An r package for spatial thinning of species occurrence records for use in ecological niche models. *Ecography* 38 (5): 541–545. https://doi.org/10.1111/ecog.01132.

Anderson, David G., Thaddeus G. Bissett, Stephen J. Yerka, Joshua J. Wells, Eric C. Kansa, Sarah W. Kansa, Kelsey Noack Myers, R. Carl DeMuth, and Devin A. White, 2017. Sea-level rise and archaeological site destruction: an example from the Southeastern United States Using DINAA (Digital Index of North American Archaeology). *PLoS ONE* 12(11). https://doi.org/10.1371/journal.pone.0188142.

Anscheutz, K. F., R. H. Wilshusen, and C.L. Scheick, 2001. An Archaeology of Landscapes: Perspectives and Directions. *Journal of Archaeological Research* 9: 157–201.

Ashmore, Wendy, 2004. Social archaeologies of landscape. In *A Companion to Social Archaeology*, 255–271. Blackwell Publishing, Malden, Massachusetts.

Balee, William, and Clark L. Erikson, 2006. Time, Complexity, and Historical Ecology. In *Time and Complexity in Historical Ecology: Studies in the Neotropical Lowlands*, edited by Balee William and Clark L. Erikson, pp. 1–17. Columbia University Press, New York.

Blake, Emma, 2004. Space, spatiality and archaeology. In *A Companion to Social Archaeology*, 230–254. Blackwell Publishing, Malden, Massachusetts.

Bodenhamer, David J., 2012. Beyond GIS: Geospatial technologies and the future of history. In *History and GIS: Epistemologies, Considerations and Reflections*, edited by Alexander von Lünen and Charles Travis, pp.1–13. Springer, Netherlands.

Branton, Nicole, 2009. Landscape approaches in historical archaeology: The archaeology of places. In *International Handbook of Historic Archaeology*, edited by Theresa Majewski and David Gamister, pp. 51–65. Springer, New York.

Coen, Loren D, Nancy Hadley, Virginia Shervette, and Bill Anderson, 2011. Managing Oysters in South Carolina: A Five year program to enhance/restore shellfish stocks and reef habitats through shell planting and technology improvements. Charleston, South Carolina. www.dnr.sc.gov.

Crema, Enrico R., 2020. Non-stationary and local spatial analysis. In *Archaeological Spatial Analysis: A Methodological Guide*, edited by Mark Gillings, Piraye Hacigüzeller, and Gary Lock, pp. 155–168. Routledge, London.

Cunzo, Lu Ann de, and Julie H. Ernstein, 2015. Landscapes, ideology and experience in historical archaeology. In *The Cambridge Companion to Historical Archaeology*, edited by Dan Hicks and Mary C. Beaudry, pp. 255–270. Cambridge University Press, Cambridge. https://doi.org/10.1017/CCO9781139167321.014.

Delle, James A., 1999. The landscapes of class negotiation on coffee plantations in the blue mountains of Jamaica: 1790–1850. *Historical Archaeology* 33: 136–158.

DePratter, Chester, and Victor D. Thompson, 2013. Past shorelines of the Georgia Coast. In *Life Among the Tides: Recent Archaeology on the Georgia Bight.*, edited by Victor D. Thompson and David Hurst Thomas, pp. 145–167. Anthropological Papers of the American Museum of Natural History, New York.

Duncan, Richard B, and Kristen A Beckman, 2000. The application of GIS predictive site location models within Pennsylvania and West Virginia. In *Practical Applications of GIS for Archaeologists: A Predictive Modelling Toolkit*, pp. 37–61. Taylor and Francis, New York. http://000-0748408304.

Elith, J., S. J. Phillips, T. Hatsie, M. Dudik, Y. A. Chee, and C. J. Yates, 2011. A statistical explanation of MaxEnt for ecologists. *Diversity and Distributions* 17 (1): 43–57.

Engemann, Kristine, Brian J. Enquist, Brody Sandel, Brad Boyle, Peter M. Jørgensen, Naia Morueta-Holme, Robert K. Peet, Cyrille Violle, and Jens

Christian Svenning, 2015. Limited sampling hampers "big data" estimation of species richness in a tropical biodiversity hotspot. *Ecology and Evolution* 5 (3): 807–820. https://doi.org/10.1002/ece3.1405.

Fisher, P. F. 2005. Models of uncertainty in spatial data. In *Geographical Information Systems: Principles, Techniques, Management and Applications*, edited by P. A. Longley, Michael F. Goodchild, D. J. Maguire, and D. W. Rhind, pp. 191–205. Wiley, Hoboken, New Jersey.

Fisher, P. F., A. Comber, and R. Wadsworth, 2006. Approaches to Uncertainty in Spatial Data. In *Fundamentals of Spatial Data Quality*, edited by A Morris and S Kokhan, pp. 167–186. ISTE, London.

Fithian, William, Jane Elith, Trevor Hastie, and David A. Keith, 2015. Bias correction in species distribution models: pooling survey and collection data for multiple species. *Methods in Ecology and Evolution* 6 (4): 424–438. https://doi.org/10.1111/2041-21 0X.12242.

Fusco, Johanna, and Cyril de Runz, 2020. Spatial fuzzy sets. In *Archaeological Spatial Analysis: A Methodological Guide*, edited by Mark Gillings, Piraye Hacıgüzeller, and Gary Lock, pp. 169–191. Routledge, London.

Gillings, Mark, Piraye Hacıgüzeller, and Gary Lock, 2020. Archaeology and spatial analysis. In *Archaeological Spatial Analysis: A Methodological Guide*, edited by Mark Gillings, Piraye Hacıgüzeller, and Gary Lock, pp. 1–16. Routledge, New York.

Gupta, Neha. 2020. Preparing archaeological data for spatial analysis. In *Archaeological Spatial Analysis: A Methodological Guide*, edited by Mark Gillings, Piraye Hacıgüzeller, and Gary Lock, pp. 17–40. Routledge, New York.

Harris, Matthew, 2013. Pennsylvania Department of Transportation Archaeological Predictive Model Set Task 1: Literature Review Contract #335I01 Archaeological Predictive Model Set. URS Corporation, Burlington New Jersey. https://www.penndot.gov/ProjectAndPrograms/Cultural%20Resources/Documents/state-wide-archaeological-predictive-model-volume-1.pdf

Heath, Barbara A., 2010. Space and Place within Plantation Quarters in Virginia, 1700–1825. In *Cabin, Quarter, Plantation: Architecture and Landscapes of North American Slavery*, edited by Clifton Ellis and Rebecca Ginsburg, pp. 157–176. Yale University Press, New Haven, Connecticut.

Hegemon, Michelle, 2003. Setting theoretical egos aside: Issues and theory in North American Archaeology. *American Antiquity* 68: 213–244.

Heilen, Michael, 2020. The role of modeling and synthesis in creative mitigation. *Advances in Archaeological Practice* 8 (3): 263–274. https://doi.org/101017/aap.2020.23.

Hodder, Ian, and C. Orton, 1976. *Spatial Analysis in Archaeology*. Cambridge University Press, Cambridge.

Kirby, Michael Xavier, 2004. Fishing down the Coast: Historical expansion and collapse of oyster fisheries along Continental Margins. *Proceedings of the National Academy of Sciences of the United States of America* 101 (35): 13096–13099. https://doi.org/10.1073/pnas.0405150101.

Knapp, Bernard, and Wendy Ashmore, 1999. Archaeological Landscape: Constructed, conceptualized ideational. In *Archaeologies of Landscape: Contemporary Perspectives*, edited by Wendy Ashmore and Bernard Knapp, pp. 1–32. Blackwell Publishers, Oxford.

Kohler, Timothy A., and S. C. Parker, 1986. Predictive models for archaeological resource location. In *Advances in Archaeological Method and Theory, Volume 9*, edited by Michael B Schiffer, pp. 397–452. Academic Press, New York.

Kosiba, Steve, and Andrew M. Bauer, 2013. Mapping the political landscape: Toward a GIS analysis of environmental and social difference. *Journal of Archaeological Method and Theory* 20 (1): 61–101. https://doi.org/10.1007/s10816-011-9126-z.

Kvamme, Kenneth L. 2006. There and back again: Revisiting archaeological locational modeling. In *GIS and Archaeological Site Location Modeling*, edited by Mark L Mehrer and Konnie L Wescott, pp. 2–35. Taylor and Francis, New York.

Kvamme, Kenneth L. 2020. Analysing regional environmental relationships. In *Archaeological Spatial Analysis: A Methodological Guide*, edited by Mark Gillings, Piraye Hacıgüzeller, and Gary Lock, pp. 212–230. Routledge, New York.

Landis, Josh, 2020. Climate Change, Development Batter Mississippi Delta Oysters. PBS NewsHour Weekend. 8 February 2020. https://www.pbs.org/newshour/show/climate-change-development-batter-mississippi-delta-oysters

Latimer, W. J., and Floyd S. Bucher, 1911. Soil Survey of Chatham County, United States Department of Agriculture National Resources Conservation Service: Georgia, pp. 1–556. https://www.nrcs.usda.gov/Internet/FSE_MANUSCRIPTS/georgia/chatham GA1911/chathamGA1911.pdf

Letham, B., A. Martindale, and K. M. Ames, 2020. Endowment, investment, and the transforming coast: Long-term human-environment interactions and territorial proprietorship in the Prince Rupert Harbour, Canada. *Journal of Anthropological Archaeology* 59: 1–22 (101179). https://doi.org/10.1016/j.jaa.2020.101179

Leusen, M. Van, and H. Kamermans, 2005. Predictive *Modeling for Archaeological Heritage Management: A Research Agenda*. Rijksdienst voor het Oudheidkundig Bodemonderzoek, Amersfoort.

Lulewicz, Isabelle H., Victor D. Thompson, Justin Cramb, and Bryan Tucker, 2017. Oyster paleoecology and native american subsistence practices on Ossabaw Island, Georgia, USA. *Journal of Archaeological Science: Reports* 15 (October): 282–289. https://doi.org/10.1016/j.jasrep.2017.07.028.

Mink, Phillip B., B. Jo Stokes, and David Pollack. 2006. Points vs. polygons: A test case using a statewide geographic information system. In *GIS and Archaeological Site Location Modeling*, edited by Mark L Mehrer and Konnie L Wescott, pp. 200–219. Taylor and Francis, New York.

Mink, P. B., M. J. Ripy, Keiron Bailey, H. B. Box, and Ted Grossardt, 2009. Predictive Archaeological Modeling Using GIS-Based Fuzzy Set Estimation. *ESRI International Users Conference*, San Diego, California, http://proceedings.esri.com/library/userconf/proc09/uc/papers/pap_1495.pdf

Pouncett, John, 2020. Spatial approaches to assignment. In *Archaeological Spatial Analysis: A Methodological Guide*, edited by Mark Gillings, Piraye Haciguzeller, and Gary Lock, pp. 192–211. Routledge, New York.

Reitz, Elizabeth J., 2014. Continuity and Resilience in the Central Georgia Bight (USA) Fishery between 2760 BC and AD 1580. *Journal of Archaeological Science* 41: 716–731.

Ricka, Torben C., Leslie A. Reeder-Myers, Courtney A. Hofmana, Denise Breitburg, Rowan Lockwood, Gregory Henkes, Lisa Kellogg, et al., 2016. Millennial-scale sustainability of the Chesapeake Bay Native American Oyster Fishery. *Proceedings of the National Academy of Sciences of the United States of America* 113 (23): 6568–6573. https://doi.org/10.1073/pnas.1600019113.

Ritchison, Brandon T., Victor D. Thompson, Isabelle Lulewicz, Bryan Tucker, and John A. Turck, 2020. Climate Change, resilience, and the native american fisher-hunter-gatherers of the Late Holocene on the Georgia Coast, USA. *Quaternary International*, 584:82–92. August. https://doi.org/10.1016/j.quaint.202 0.08.030

Sanger, Matthew C., I. R. Quitmyer, C. E. Colaninno, N. Cannarozzi, and D. L. Ruhl, 2020. Multiple-proxy seasonality indicators: An integrative approach to assess shell midden formations from Late Archaic Shell Rings in the Coastal Southeast North America. *The Journal of Island and Coastal Archaeology* 15 (3): 333–363.

Sanger, Matthew C., Mark A. Hill, Gregory D. Lattanzi, Brian D. Padgett, C. S. Larsen, Brendan J. Culleton, and Douglas J. Kennett, 2018. Early metal use and crematory practices in the American Southeast. *Proceedings of the National Academy of Sciences* 115 (33): E7672–E7679.

Sassaman, Kenneth E., and Z. I. Gilmore, 2021. When edges become centred: the ceramic social geography of Early Pottery Communities of the American Southeast. *Journal of Anthropological Archaeology* 61:1–16 (101253).

Savarese, Michael, Karen J. Walker, Shanna Stingu, William H. Marquardt, and Victor Thompson, 2016. The Effects of shellfish harvesting by aboriginal inhabitants of Southwest Florida (USA) on productivity of the Eastern Oyster: Implications for Estuarine Management and Restoration. *Anthropocene* 16: 28–41. https://doi.org/10.1016/j.ancene.2016.10.002.

de Smith, M. Michael F. Goodchild, and P. Longley, 2018. *Geospatial Analysis: A Comprehensive Guide*. 6th edition. spatialanalysisonline.com.

Spaulding, A., 1960. The dimensions of archaeology. In *Essays in the Science of Culture in Honor of Leslie A. White*, edited by G. E. Doles and R. L. Cameiro, pp. 437–456. Crowell, New York.

Szabó, Péter, 2015. Historical Ecology: Past, present and future. *Biological Reviews* 90 (4): 997–1014. https://doi.org/10.1111/brv.12141.

Thomas, David Hurst, 2014. The Shellfishers of St. Catherine's Island: Hardscrabble foragers of farming beachcombers? *Journal of Island and Coastal Archaeology* 9: 169–182.

Thomas, David Hurst, and C. F. T. Andrus, 2008. *Native American Landscapes of St. Catherines Island, Georgia*. Anthropological Papers of the American Museum of Natural History, Number 88. American Museum of Natural History, New York.

Thompson, Victor D., and John A. Turck, 2010. Island archaeology and the Native American Economies (2500 BC–AD 1700) of the Georgia Coast." *Journal of Field Archaeology* 35 (3): 283–297.

Thompson, Victor D., and C. F. T. Andrus, 2011. Evaluating mobility, monumentality, and feasting at the Sapelo Island Shell Ring Complex. *American Antiquity* 76 (2): 315–343. https://doi.org/10.7183/0002-7316.76.2.315.

Thompson, Victor D., John Turck, and Chester DePratter, 2013. Cumulative actions and the historical ecology of islands along the Georgia Coast. In *The Archaeology and Historical Ecology of Small Scale Economies*, edited by Victor D. Thompson and James C Waggoner Jr., pp. 79–85. University Press of Florida, Gainesville.

Thompson, Victor D., Torben Rick, Carey J. Garland, David Hurst Thomas, Karen Y. Smith, Sarah Bergh, Matt Sanger, et al. 2020. Ecosystem stability and native American Oyster Harvesting along the Atlantic Coast of the United States. *Science Advances* 6 (28):1–8.

Turck, John A., 2012. Where were all of the coastally adapted people during the Middle Archaic Period in Georgia, USA? The *Journal of Island and Coastal Archaeology* 7 (3): 404–424.

Turck, John A., and Clark R. Alexander, 2013. Coastal landscapes and their relationship to human settlement on the Georgia Coast. In *Life Among the Tides: Recent Archaeology on the Georgia Bight.*, edited by Victor D. Thompson and David Hurst Thomas, pp. 168–189. Anthropological Papers of the American Museum of Natural History New York.

Turck, John A., and Victor D. Thompson, 2019. Human-environmental dynamics of the Georgia Coast. In *The Archaeology of Human-Environmental Dynamics on the North American Atlantic Coast*, edited by Leslie A. Reeder-Myers, John A. Turck, and Torben C. Rick, pp. 164–198. University Press of Florida, Gainesville.

Verhagen, Philip, 2006. Quantifying the qualified: the use of multicriteria methods and Bayesian statistics for the development of archaeological predictive models. In *GIS and Archaeological Site Location Modeling*, edited by Mark L Mehrer and Konnie L Wescott, pp. 176–199. Taylor and Francis, New York.

Verhagen, Philip, and Michiel Gazenbeek. 2006. The use of predictive modeling for guiding the archaeological survey of Roman Pottery Kilns in the Argonne Region (Northeastern France). In *GIS and Archaeological Site Location Modeling*, edited by Mark L Mehrer and Konnie L Wescott, pp. 411–423. Taylor and Francis, New York.

Verhagen, Philip, and Thomas G Whitley. 2020. Predictive spatial modelling. In *Archaeological Spatial Analysis: A Methodological Guide*, edited byMark Gillings, Piraye Hacıgüzeller, and Gary Lock, pp. 231–246. Routledge, New York.

Vita-Finzi, C., and E. S. Higgs, 1970. Prehistoric economy in the Mount Caramel Area of Palestine: Site catchment analysis. *Proceedings of the Prehistoric Society* 36: 1–37.

Warf, Barney, and Santa Arias, 2008. *The Spatial Turn: Interdisciplinary Perspectives.* Routledge, New York.

Zizka, Alexander, Alexandre Antonelli, and Daniele Silvestro, 2020. Sampbias, a method for quantifying geographic sampling biases in species distribution data. *Ecography* 44(1):25–32, October, ecog.05102. https://doi.org/10.1111/ecog.05102

7 The Maya domestic landscape and household resilience at Actuncan, Belize: a reconstruction and modern implications

Kara A. Fulton and David W. Mixter

The first thing you notice are the cows. Glassy-eyed, they stand and lay, chewing on cow grass chosen for its nutritional value or relaxing in the shade of the occasional palm tree standing solitary over broad grassy fields. Here we come up the hill, heading to work. Archaeologists in heavy work boots tromp along the paths through the grass etched by the cows' daily migrations (Figure 7.1). Our route is determined by theirs. Cow heads perk up, now standing at attention and staring at these outsiders from beyond the fence. Intruders in their home. The cows gather, some shuffling, others moving more quickly as they form their herd. Now they are off. Sprinting together, the herd wheels around trampling the earth, crushing grass, and compacting clay. Hopefully, they stay off the ancient mounds – the now well-flattened remains of long-abandoned Maya houses – this time.

Today in western Belize, the cows form the landscape (Figure 7.2). For locals in the rural landowning class, herds kept for dairy and beef underpin some household livelihoods (Robinson 1985). Because of the local and international demand for beef, those with substantial land around their homesteads are able to leverage the land into an economically sustainable living. To do so, landowners cut down the forest and replace traditional *milpa* agriculture with fields of imported grasses. The cows then compact the land, pressing down the clay alluvium and flattening the microtopography formed by ancient Maya mounds as they kick and dislodge stones that once formed platform edges.

As humans frequently do, these modern ranchers are leveraging the resources around their homes for economic stability. They reshape the land around their homes, building a landscape that suits their needs and aspirations. The continued success of the choices they make anchors the financial stability of the household and determines its resilience. Yet, by virtue of having land to shape into a sustaining landscape, these farmers have greater potential to secure their livelihoods.

Many of our young friends who live landless in Belize's towns talk about the need to acquire land where they can build a house and milpa, securing stability for their future. Without land, they are dependent on the poor job market and expensive food prices at the market. With land, they have corn

DOI: 10.4324/9781003139553-7

Figure 7.1 The modern site setting of Actuncan. Photograph by Lisa J. LeCount.

Figure 7.2 Modern inhabitants of Actuncan with Maya mounds and archaeologists in the background. Photograph by Lisa J. LeCount.

for tortillas, vegetables to diversify their meals, a yard where they can raise chickens, and a roof of their own over their heads. In other words, they have a home that sustains and resources that fill their fundamental needs to fall back on when the fickle local economy fails. The local economy is driven by the booms and busts of foreign tourists, but household resilience is anchored to the land.

These contemporary land use choices and the resulting transformations to the landscape are only the most recent examples in a long history of human settlement in the region. In the area where our case study is located, there exists a modern cattle ranch. The land beneath this ranch was first settled by sedentary populations over 3000 years ago and was the site of a continuously occupied Maya village, known as Actuncan, from around 1000 BC to AD 900. Actuncan's residents transformed the landscape, carving public plazas from the clay and building pyramids with imported stone. Like the modern ranchers, the Maya of Actuncan chose plots of land, built houses, and then built their livelihoods off the land. Some built their houses in town, along the edges of the broad plazas, while others settled outside in communities where they lived on larger plots of land.

Yet, over the long occupation of Actuncan, the people living in these houses were not equally resilient. At the time of the town's founding as a formally planned community between 300 and 200 BC, households laid claim to plots of land and established their homes. While the placement of a house may be intended to maximise control over optimal croplands, rich pastures, and abundant forage lands, social considerations and constrictions – land tenure rules, existing property ownership, the desire to live near or away from others, the desire to live near or away from centres of power – often drive people to choose where they live. Similarly, the personalisation of the landscape imbues homes with individual character but is also reflective of status, economic roles, social aspirations, and political position. With these considerations in mind, some placed their houses along the edges of the town's public plazas, anchoring themselves to the political authority and cosmopolitan vibrance located in the centre. These houses grew in size over time in an apparent attempt to match the aesthetic of the town's increasingly tall and elaborate political and religious buildings.

Other people chose to settle around the outskirts of the centre, building compounds surrounded by rich, alluvial soils with space to spread out. They built airy homes with multiple buildings surrounding central patios. They had plenty of spaces for gardens and maintained farmland immediately adjacent to their homes. They even had space designated for special purposes, including stone tools and pigment production areas. While the people on the outskirts of town sacrificed proximity to the centre, they were able to use their land to diversify their economic strategies.

Eventually, tough times came to Actuncan. Around AD 300, just past the onset of the Early Classic period (AD 250–600), construction slowed in Actuncan's centre, resources became scarce, routine maintenance suffered, public buildings and households were gradually abandoned, and some buildings seem to have been torn down and burned. Yet, despite clear evidence of impoverishment, some houses remained occupied during Actuncan's centuries-long downturn. Those households that persevered were rewarded when Actuncan again flourished, eventually regaining its status as a local political centre in the 9th century AD. These resilient households anchored that resurgent community.

Why were some of Actuncan's households resilient while others were quickly abandoned? As the Maya community occupied Actuncan for over two millennia, they created a layered palaeolandscape inscribed by the physical markers and by-products of their lives and actions. The land was cleared, hilltops were levelled, and stone buildings were constructed. Unintentionally, erosion likely affected hillslopes. At the same time, people's actions imbued these landscapes with social significance. We argue that these long-term processes of landscape formation and signification at the community and household scales contributed to the differential resilience of different households.

Drawing on a case study from the pre-Columbian Maya settlement of Actuncan, Belize, this study reflects on the palaeolandscape of homes and how the choices made in the establishment and development of domestic landscapes impact the resilience of households. In the following pages, we will first outline our contextualisation of space and place on the landscape, particularly considering the areas where people live and conduct their everyday activities. Then, we discuss the settlement on which we focus our case study, the site of Actuncan, Belize. We situate Actuncan within the natural, political, and social landscape of the region. We then compare the long-term resilience of people living in two types of homes – patio-focused groups along the edge of the settlement and individual houses located in the settlement's monumental core. Based on these comparisons, we argue that a household's level of resilience to shifting political structures is tied in part to its capacity to be self-sustaining and its investment in social prestige. Understanding the connections between self-sufficiency, conspicuous consumption, and resilience has implications for how we approach current and future issues of changing landscapes.

Natural, political, and social landscapes

What gives a place meaning?

The physical environments in which people reside and perform everyday tasks often affect the meanings and values associated with the activities themselves (Hendon 2010; Lawrence and Low 1990). Humans define the spatial boundaries of activity spaces in a variety of ways, including the construction of architecture and identifying landmarks on the natural landscape (Low 2000). Spaces are not passive backdrops for human actions.

Rather, they are active participants that constrain, regulate, and inspire the creation of social meaning by providing the context for human activities (e.g. Hendon 2010; Lawrence and Low 1990; Robin 2002). The relationship between society, culture, and space is interactive; people create space and are, in turn, influenced by it (Lawrence and Low 1990).

Although similar ideas exist, we draw on de Certeau's (1984) definitions of the concepts of "space" and "place" in relation to everyday life. He argued that places represent a realm of possibilities with which people can interact; places – as with social structure – can both constrain and enable the actions of the individual. Space, on the other hand, is "the multifaceted experiences of being and doing in a place which may or may not conform to previously constructed and/or conceived meanings and which continue to constitute and reconstitute the meanings of places" (Robin 2002, 249). However, it is important to bear in mind that space and place are interrelated and cannot be separated from one another outside of our own abstract conceptions. Space gives place meaning through human experience and interaction. For example, a house is simply an architectural creation (a place), but it is transformed into a space by those who inhabit it and their activities. Further, shared activities, such as cooking, farming, or tool making, produce social memories and attach meanings to places, which foster a common awareness and sense of self (Hendon 2010; McAnany 2013). These shared activities are often connected to shared group identities, which can connect a place or kind of place to ideas that integrate people at various social levels (Burke 1989; Gillespie 2000; Gillis 1994; Goff 1992; Yaeger 2000a; 2000b; 2003a; 2003b). As such, daily activities and their engagement with the material world entangle social meaning, values, and relationships (Wells and McAnany 2008). Humans transform places – a realm of possibilities with which people can interact – into spaces imbued with meaning through human experience and engagement. The sociocultural production of space affects the way that people operate on an everyday basis and, in turn, is influenced by human activities.

We would like to take a moment here to briefly acknowledge the diversity of ways scholars have approached the ideas of "space" and "place" as we have presented. Many well-known scholars have explored human relationships with social space from different angles, including Bourdieu (1977), Durkheim (1995 [1915]), Foucault (2012 [1977]), Giddens (1984), Lefebvre (1991 [1974]), Mauss (1979 [1906]), Rabinow (1989), and Soja (1989) to name only a few. Many recent researchers have based their work on these theorists. Yet, as with many disciplines, archaeologists do not abide by unified definitions for the concepts we discuss. While some have defined these notions in a similar manner to what we have proposed—with "place" representing the physical, measurable surroundings and "space" referring to cultural meaning (e.g. Morris 2004; Steadman 2016) – others have used the terms "space" and "place" in reverse (e.g. Aucoin 2017; Janz 2005). Further, some scholars have preferred entirely different terminology to denote physical places/spaces that hold meaning, such as "social landscapes" (e.g. Ucko and Layton 1999), or simply "landscapes" (e.g. Anschuetz, Wilshusen, and Scheick 2001). Although we do not have the room here to delve into all of these approaches, we wanted to acknowledge them as they are related to the themes we discuss in this chapter.

How are cities planned?

One way that humans create meaningful space is through the intentional planning of cities. Researchers have documented tremendous social and spatial variability in the organisation of modern and past cities (Isendahl and Smith, 2013; Marcus and Sabloff 2008; Sinclair et al. 2010). This variability stems from a multitude of local factors, including (but not limited to) the underlying natural landscape, the distribution of local resources, preferred modes of transportation and social interaction, and cultural norms of land tenure or possession. Cities do not normally emerge as unintended consequences from a series of actions but, rather, are "actively and intentionally created" (Cowgill 2004, 258). Exploration into the social construction and organisation of settlements can help us understand how cities form and how residents integrate and/or differentiate themselves from one another (Cowgill 2004).

One way to understand the social underpinnings of different cities is to investigate the logic behind settlement planning and design. The reasoning behind planning gives the built environment meaning at many levels and has a dialectical relationship with activities conducted in and around places (Rapoport 1988, 1990). The degree and kinds of planning are defined by local cultural constraints and forms of political authority. The evaluation of features such as common building orientation, orthogonal street layout, and the standardisation and spatial patterning of architectural forms can be used to evaluate the role of political entities and local cultural norms in planning (Smith 2007). By exploring the interplay of urban environments and human behaviours, researchers can attempt to understand in detail who occupies cities, how and why people organise their environment in different ways, and how people's lives are structured by the urban spaces they have inherited (Smith 2007).

In the case of the ancient Maya, monumental centres follow common orientations and are composed of public buildings that fit a standard inventory. Occasionally, hieroglyphic texts indicate that specific public buildings were commissioned by and dedicated to specific rulers, pointing clearly to the central role political authorities played in the design and layout of public space (Stuart 1998). However, the residential sectors of Maya communities tend to be sprawling and, at first glance, seem less ordered, perhaps pointing to less centralised control over the neighbourhood and household layouts. Indeed, evidence indicates that Maya land tenure systems were determined when a settlement was first established and that land tenure rights were assiduously maintained through time (LeCount et al. 2019; McAnany 2013; Thompson and Prufer 2021). Later arrivals were often forced to occupy less desirable plots of land, and, consequently, they could be limited in their access to resources or disadvantaged by the lack of opportunity to improve their land over previous centuries (Ashmore et al. 2004; Robin et al. 2012; Thompson and Prufer 2021). The durability of Maya land tenure likely made it difficult for later leaders to enact formal urban planning because, in the context of these strong norms, it would have been challenging to muster the political capital required to displace established households.

What makes a home?

At a more localised level, humans create meaningful space through the formation of a house and the maintenance of a home. Here, we would like to differentiate between a house, household, and home as distinct, though interrelated, concepts bounded by social organisations and space. We consider a house to be the physical structure(s) in which people live. Household refers to the people living within the house structures. Home is the meaning given to the house by the household. Hodder (1990, 1992) argues that when people began to build permanent structures in which to live, they began to differentiate between human space and the "other", the natural world. By bringing certain activities into the living space, such as food preparation activities and burial of the dead, people imbue structures with meaning. In other words, the physical house structure becomes transformed into a home by the everyday activities the people conducted within the house. As such, a home represents the cultural beliefs of the household while the house provides an environment for social life (Watkins 2004; Wilson 1988). We recognise that there may be additional complexity added to this model—that members of a household may live in more than one house (e.g., household groups or transhumance) or people living in one house may not consider themselves members of the same household (Carter 1984; Wilk and Rathje 1982). Nonetheless, the houses that we excavate can often be thought of as a combination of functional places as well as representations of generations of social identities inscribed on the landscape.

The idea that the everyday practices of the household create and maintain the meaning of a home is influenced by the work of Giddens (1979, 1984) and Bourdieu (1977, 1990), both of whom emphasised the actions of individuals and how they simultaneously manifest and constitute social and political structures. Bourdieu (1977, 1990) argued that individuals are variably aware of the rules within their society, and the level of knowledge of these rules informs their everyday actions. Going further into everyday practices, Giddens (1979) and Acabado (2018) suggested that human agency reproduces these structures imperfectly through time and states that unintended consequences of a person's actions modify future intended actions.

Building on these ideas, Yaeger (2000a, 2000b, 2003a, 2003b) argued that human behaviours and actions contribute to the creation and expression of shared social identity. Archaeologically speaking, spatial patterns of material remains can be indicative of specific household activities. Furthermore, households may have cooperated with each other and, through shared activities, had a sense of shared identity (Kent 1987; Rapoport 1990).

For the Maya, the socially constituted home should not be thought of as limited to just the house. Maya daily activities move fluidly between the interior and exterior living spaces that compose the Maya house lot and include courtyard patios, home gardens, rubbish disposal zones, auxiliary activity areas, and often in-field agricultural zones (Killion 1990, 1992; Robin 2002). Indeed, reserving space for outdoor activities, gardens, and agricultural zones is often provided as the best explanation for the dispersed organisation of Maya cities (Isendahl 2012). With the benefit of strong cultural protections of land tenure, Maya families often lived on the same plots of land for long periods of time. They would have developed collective identities and deep attachments to these places around their collective memory of everyday activities and explicit periodic events marking ancestral ties (e.g., Hendon 2010; McAnany 2013). But not all households attached their idea of home to the same kinds of everyday activities, special events, or practices of affiliation. In her work at Chan, Robin (2013) shows how Maya farming communities contained diverse sets of actors whose resilience can be related to everyday life as manifested in household economies, social networks, and political entanglements. At Chan, the long-tenured farming households and households with economic specialisations survived the longest when difficult times arrived while relatively newly arrived farmers and those who strongly affiliated with short-lived political actors quickly abandoned their homes (Robin 2013: 168–173). In general, we argue that among the Maya, daily practices that anchored deep emotional and economic attachments to residential space created household resilience, while domestic landscapes that prioritised political signalling tended to be vulnerable during social and environmental disruptions. In the section that follows, we focus on how one common kind of Maya household – smallholding farmers – built resilience as they built their homes.

Maya smallholders: what makes a home resilient?

Despite their urban setting, many Maya were smallholders – households reliant on farming land passed down through familial inheritance as their primary source of food production (Fisher 2014; Pyburn 1998, 2008; sensu Netting 1993). As defined by Netting (1993: 2), smallholders leverage their long-term control of land through ownership or land tenure rights into economic autonomy and self-determination at the household level. Smallholding households make individual decisions about how to utilise the land around their homes. They modify their land strategically through economic planning and everyday life, changing the land to meet their needs. Smallholder autonomy and self-determination is a source of resilience: It allows them to quickly pivot in their agricultural and other household economic production practices to social, ecological, and market changes (Netting et al. 1989; Stone et al. 1990). The diversity of production practices also allows smallholders to support themselves and their household when necessary, especially when support or resources from elsewhere are not available. Additionally, because smallholders are dependent on their land, they are invested in the quality of the land. They don't pollute or overuse the soils because they and their descendants are dependent on long-term productivity. Smallholders' sustainable approach to working their land and moulding the house and land into a productive home also builds less tangible emotional attachments to the land as the home becomes instilled with meaning and the memory of a specific household's history. This attachment is then magnified by special events that reify ancestral attachments, such as the Maya practice of burying ancestors beneath houses, which periodically reinforces a family's claim to a plot of land and solidifies their intention to stay connected to that land for generations (McAnany 2013).

However, it is important not to equate smallholder autonomy with independence or egalitarianism. They are not undifferentiated and unorganised, as Marx (1999: Chapter VII) insinuated by calling them "homologous magnitudes" and "a sack of potatoes." Smallholders are not peasants defined primarily by their lack of access to wealth and power (Netting 1993: 20–21). As much as smallholders utilise their autonomy to control their household economies, they still need to negotiate access to and maintenance of common goods. Things like water rights and regulations over access to property require the collaborative establishment and enforcement of rules and norms. Additionally, smallholders rely on each other for larger, collaborative infrastructure projects like building and maintaining *aguadas* (systems for freshwater collection and storage) or irrigation systems.

Furthermore, because smallholders make many decisions independently, households typically have varying levels of economic success that translate into hierarchies based on wealth and perceived achievement. However, unlike within rigid hierarchies anchored in class, smallholders can be upwardly mobile if new economic strategies lead to greater household prosperity. Further, because smallholders often need to cooperate, equality is not amplified in day-to-day interactions as it is among "farmer-owners and merchants, government officials and professionals, and landless laborers" (Netting 1993: 12). Smallholders move up and down the economic scale, but they do not stratify into landowners and subservient labourers over time (McGuire and Netting 1982). The result of smallholder autonomy, self-determination, and limited collaboration is an "economically efficient, environmentally sustainable, and socially integrative" (Netting 1993: 27) household unit that is able to adapt to the changing needs of population change, transformations in the natural landscape, and changes in larger economic structures. Beyond resilience, smallholding is a form of ecological resistance (sensu Langlie 2018a; see also Acabado 2018) as autonomy and self-sufficiency allow smallholders to resist incorporation into dependent relationships with states. By limiting their integration into political organisations, smallholders are able to buffer themselves against political instability.

So, what makes a home? For smallholders, the home is imbued with meaning by daily routines and everyday practice. These activities could include playing games and cooking meals or placing furniture and assembling decor. However, the home is also anchored to the household's economic base and the attachment of identity to activities in and around the house. In the United States of America today, Latin American immigrants start their own small farms not because they are a path to wealth and even financial stability – stability comes when the land is owned without debt – but because farming is integral to their sense of what a home is, a sense derived from their lives and heritage from before they migrated (Minkoff-Zern 2019). For smallholders, the agricultural economic base becomes part of the household's identity, heritage, and sense of home.

Case Study: Actuncan, Belize

Group 1, Actuncan, ca. AD 150[1]

It is mid-morning on a sunny day near the end of the dry season. Standing at the entrance to your house, one of several pole and thatch structures surrounding your household's central patio. Children play on the patio: one, about two years old, carries around a small wooden dog, another, around twelve, sits in the corner on a cloth practising making tools. Planting season comes soon, and you will need plenty of spare hoes for mounding the earth before planting the maize, beans, and squash.

Across the patio, your spouse is repairing the roof of the chert knapping hut. Harvested last night, the bay leaf is in the perfect season for this repair. These efforts are important to your continued affluence. Your family has benefited from making surplus tools and trading them at the market. While your fields ensure economic stability, the trade brings you some wealth: small bits of jade, an occasional painted pot, labour to help haul stone when it is time to renovate the house. These repairs are an investment in your future, but they also bring you the pride that comes with having a well-organised, well-maintained house.

To the right is your parents' house. They moved over to the smaller building when you started having children. Across the patio is a small shrine to your ancestors. Under the patio floor, your grandparents and their parents lay. They are buried in stone line crypts accompanied by their fancy painted pots. Living with your parents and over your older ancestors reminds you daily of the family tie to this place. They invested in its growth and in the fertility of the surrounding land for generations, ensuring it would be there for you. You belong here.

You walk west, out of the patio and beyond the buildings of your home. Downhill you see the browning stalks of last year's now-harvested maize crop. The cobs are now in storage, resting above ground in a covered corn crib. These will provide you with tortillas until next year's crop and beyond if needed. You have enough dried maize to not go hungry even if the rains don't come next year. Further downhill, you see your family's raised garden. These were made decades ago by your grandparents. The stone foundations have endured and provide you with small plots to grow vegetables year-round. These gardens were purposefully placed along the outlet of the community's aguada reservoir, where they can be irrigated by water stored from the wet season. Further downhill is your neighbour's garden, positioned to recycle water after it passes from your garden. Though it is adjacent to your house, the land below the aguada is shared with your neighbours so that each household may take advantage of the year-round gardening. You appreciate the convenience of being so close. The fields and gardens provide your household with plenty of food. It is also comforting to know that you can always trade the valuables purchased with stone tools in the market if either food source fails.

You look south into Actuncan, your home village. Beyond the edge of your house plot stretches the market. Quiet at the moment, it will bustle tomorrow on market day. Beyond that, at the southern edge of the market, the massive home of a noble family looms over the marketplace. Rather than pole and thatch, their house is built of stone, complete with a vaulted stone roof. The house is stuccoed and painted bright red, proclaiming their wealth. A decorative frieze depicting the Maize God looms over their market stall, which is built directly into the base of their home. Your house is drab by comparison with clay daub walls, its thatch roof, thin layer of white stucco, and lack of sculptural adornments. You can't help but be jealous of their fortune. Their house is located right on the market, and they clearly have wealth. Yet, you know they don't have farmland or space for substantial gardens. They are dependent on Actuncan's ruler for much of their wealth and are tied to his success. Yes, they feed on your corn and your vegetables, provided as tribute. But what if the ruler fails? What if their alliances prove fickle?

Actuncan is a pre-Columbian Maya settlement located in western Belize (Figure 7.3). Containing about 14 ha of civic and domestic buildings, the settlement is similar in size to neighbouring centres and exemplifies a low-density urban centre as defined by Smith (2010). Following early excavations by Gann (1925: 48–93) in 1924 and research into the site's monumental architecture by McGovern (2004) from 1992 to 1994 for his PhD dissertation, the Actuncan Archaeological Project (AAP), led by Dr. Lisa LeCount, has undertaken sustained research at Actuncan since 2001. The following reconstructions of Actuncan's urban development, political trajectory, and household occupation histories are a product of McGovern's original mapping effort, his establishment of a baseline chronology for the site, and intensive excavations of households and monumental public buildings

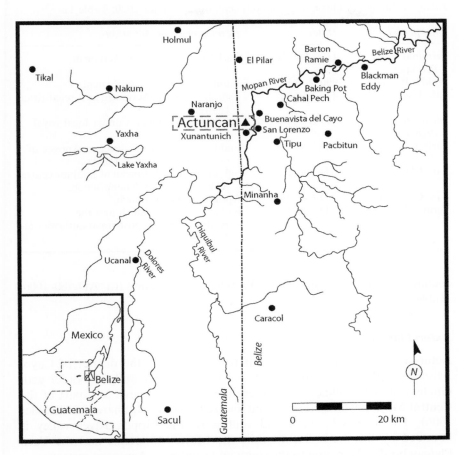

Figure 7.3 Map of the Belize River valley, showing the location of Actuncan (map from LeCount 2004, dotted lines added by the authors).

undertaken by the AAP. The details of these excavations can be found in AAP field reports, which are cited liberally in the sections that follow.

Actuncan was originally settled around 1000 BC, around the time local people began to make ceramics and build permanent architecture (LeCount et al. 2017). Prior to this, population levels in the area were likely low. Though pollen cores point to land clearance prior to 1000 BC, the early mobile pre-ceramic groups in Belize at that time left few material remains for archaeologists to find beyond a few rock shelter sites and isolated chert points (Brown et al. 2011; Jones 1994; Lohse 2010; Prufer et al. 2019). The human occupation of the region at this time remains poorly understood. Actuncan expanded from a small, sparsely populated village into a planned urban centre during the Late Preclassic period (ca. 400–150 BC) and was

Table 7.1 Summary chronology of Actuncan (after Mixter 2020: Table 1).

Time period	Dates	Developments at Actuncan
Terminal Early Preclassic	1100–900 BC	Earliest occupation of Actuncan.
Middle Preclassic	900–400 BC	Limited evidence of settlement.
Late Preclassic	400–150 BC	Establishment of Actuncan as a regal-ritual centre.
Terminal Preclassic	150 BC–AD 250	Apogee of Actuncan's under local royal leadership.
Early Classic	AD 250–600	Failure of Actuncan's leaders. Apogee of nearby Buenavista del Cayo.
Late Classic I	AD 600–670	Subjugation of Actuncan as a minor centre. Ruled by nearby Xunantunich.
Late Classic II	AD 670–780	Apogee of Xunantunich.
Terminal Classic	AD 780–1000	Failure of Xunantunich and the establishment of post-royal authority at Actuncan.

continuously occupied through the Terminal Classic period (ca. AD 780–1000) (Table 7.1).

Natural landscape

Actuncan is located in the Mopan River valley, in the district of Cayo, Belize, near the modern town of San Jose de Succotz and the border with Guatemala. The archaeological literature defines this area as a part of the Central Maya lowlands (Hammond and Ashmore 1981; Sharer and Traxler 2006). To the north, west, and southwest, the Mopan River valley is bounded by the Petén Basin, while the Maya Mountains and the Vaca Plateau border the region to the south and southeast. Actuncan rests upon a T–3 alluvial terrace west of the Mopan River, a tributary of the Belize River (Figure 7.4). The Mopan River, along with many other tributaries and minor streams of the Belize River, flows all year and would have been beneficial to prehispanic residents of the region as an abundant and consistent supply of freshwater (Fedick 1988; Smith 1998). Additionally, this waterway allows direct access to the Caribbean coast providing an ideal transportation route for goods and/or people to the ocean (Cap 2015). Likely due to swamps and low plains upstream that moderate flow downstream, the water levels of the Mopan River tend to rise slowly, even with heavy rains. The water levels rarely rise beyond the floodplain and return to their normal positions gradually (Yaeger 2000a).

The rugged upland terrain of the Mopan River valley includes enclosed depressions and residual hills with grey and brown soils of the Toledo Beds that are predominantly calcareous mollisols and vertisols (Beach et al. 2003; Day 1993; King et al. 1992). Underlying the area is a karst belt that runs

Figure 7.4 The Mopan River adjacent to Actuncan. Photograph by John H. Blitz.

through central Belize. The Cretaceous to Early Tertiary period limestone substrate is composed primarily of calcium, barium, and strontium (Day 1993). The soils are mostly acidic with a range of 5.5–6.5 on the pH scale and are often less than 0.5 m thick. However, probes at Actuncan have reached depths of more than 3 m, thus limestone bedrock was not readily available as a source for construction material. The soils underlying Actuncan consist of the Yalbac sub suite of the Yaxa soil suite (Birchall and Jenkins 1979). These clay-rich soils are low in potassium and phosphates, essential nutrients for plant growth, but are satisfactory for growing grain crops, including maize. The nearby Melinda suite soils are more nutrient-rich, being located on the Mopan River floodplain, and are ideal locations for agriculture (Birchall and Jenkins 1979; Cap 2015). Within the Yalbac sub suite, the Piedregal series, on which Actuncan is situated, is one of the most productive soil series in terms of potential productivity using prehispanic agricultural techniques (Fedick 1988). In other words, Actuncan's households were well-situated to produce abundant harvests if the rain cooperated.

Local stone resources at Actuncan include limestone and chert. Limestone cobbles used to create small objects (e.g., manos and bark beaters), and for construction fill would likely have come from the Mopan River (Cap 2015). The nearest limestone outcrop suitable for cut limestone blocks used in architecture is south of Actuncan, near Xunantunich. Chert is abundant in this portion of the Mopan River Valley, though the alluvial terraces of the Mopan River generally lack outcrops. However, two outcrops (one designated as a quarry) have been identified in the vicinity of Actuncan (Cap 2015; Horowitz 2018). For the people of Actuncan, raw chert would have been an abundant and difficult to control resource. Architectural quality limestone, on the other hand, was only available at a distance. In times of scarcity, Actuncan's residents do not appear to have ready access to architectural material and instead relied on reused blocks and tamped earth floors rather than floors made of lime plaster.

Actuncan resides within a subtropical zone that is marked by a dry season (from January to May) and a wet season (from June to December). Annual rainfall is between 2000 and 2400 mm, with the majority of precipitation falling in June at the beginning of the wet season. Temperatures do not vary much throughout the year; average temperatures are 25° C (77° F), though they are slightly cooler in fall and winter (Day 1993; Furley and Newey 1979). During summer and fall, tropical storms are common with occasional hurricanes (Yaeger 2000a). The fickle rains were a source of uncertainty for pre-Columbian Maya, as they are today. The Maya were dependent on the rains coming at the right time in the right amount for a successful crop. Maya markets, like the one hypothesised to have been located in Actuncan's Plaza H (Keller 2012), buffered against local crop failures. Additionally, to account for the variable rains, Actuncan's residents built a large *aguada* that was used to ensure sufficient drinking water and could be used to irrigate the raised gardens built by the Actuncan Maya along its outlet (Heindel 2019).

Political landscape

Actuncan's occupation as an urban centre spans approximately 1400 years from the Late Preclassic period (ca. 400–150 BC) to the Terminal Classic period (ca. AD 780–1000). Throughout this time, Actuncan as a whole had residential continuity while regional political authority shifted between three settlements: Actuncan, Buenavista del Cayo, and Xunantunich. Actuncan was established in the Late Preclassic period as a planned political and ceremonial centre containing monumental civic constructions including an E-Group – an early Maya architectural form that is often interpreted as commemorative solar shrines – and a triadic temple group – a common form of Maya monumental architecture with pyramids on three sites of a rectangular plaza (McGovern 2004; Mixter 2020). Additionally, Stela 1, a carved limestone stela found in the triadic temple group, is evocative of the well-known San Bartolo murals (Taube et al. 2010) and points to the early adoption of divine rulership at the site (Fahsen and Grube 2005; McGovern 1992). Several residential structures and household groups were also founded during this time. Although excavations support the occupation of the region prior to the Late Preclassic period, early constructions were buried when the entire ridgetop was levelled in preparation for the construction of the planned urban centre. During the Terminal Preclassic period (150 BC–AD 250), Actuncan continued to expand and rose to political prominence in the Mopan River valley under divine rulership. It is at this time that household groups became more firmly established in the urban centre, a trend that mirrors growing regional populations and, perhaps, reflects the growing attractiveness of life within the local capital (LeCount 2004). During the Early Classic period (AD 250–600), local authority shifted from Actuncan to Buenavista del Cayo (Ball and Taschek 2004; LeCount and Yaeger 2010a, 2010b; Leventhal and Ashmore 2004; McGovern 2004). At this time, Actuncan was likely just a village within a broader polity rather than a locus of political authority, albeit one with large crumbling public architecture. This period also marks an interval where many urban households at Actuncan were abandoned and ritually terminated as evidenced by burning and the scattering of burned artefacts within and around these homes (Mixter et al. 2014; Simova et al. 2015). At the same time, a two-hundred-year hiatus in large-scale civic construction began. During the Late Classic I period (AD 600–670), Buenavista del Cayo's prominence was eclipsed by the rapid rise of Xunantunich (possibly under the sponsorship of Naranjo), and Actuncan shifted to become subordinate to Xunantunich (Awe et al. 2020; LeCount and Yaeger 2010a; Mixter et al. 2013). The connection between Xunantunich and Actuncan is evident in a causeway built to connect the two settlements (Ashmore 1998). During the Late Classic II period (AD 670–780), the Mopan River valley continued to be dominated politically by Xunantunich's kings. The importance of Actuncan as a subject site and its corresponding lack of autonomy are marked by renewed occupation of Preclassic residential structures and an absence of intensive investment in public spaces (McGovern 2004; Mixter 2020; Mixter et al. 2014; Mixter et al. 2013). During the Terminal Classic period (AD 780–1000), the authority of Xunantunich's king decline while Actuncan likely gained independence and may have politically reorganised under the rule of a governing council (LeCount and Yaeger 2010a; Mixter 2016, 2017). Actuncan was continuously occupied through the Terminal Classic period and remained a vibrant population centre during the decline of Xunantunich. Population decline does not occur at Actuncan until the end of the Terminal Classic period.

Social landscape

The research presented here centres on the archaeological investigations of nine residential units at Actuncan that likely represent two distinct social strata: smallholding commoners and elite households (Figure 7.5). Epigraphic evidence indicates that the Maya had distinct social classes with elites identified by courtly titles in locations with a robust preserved written record

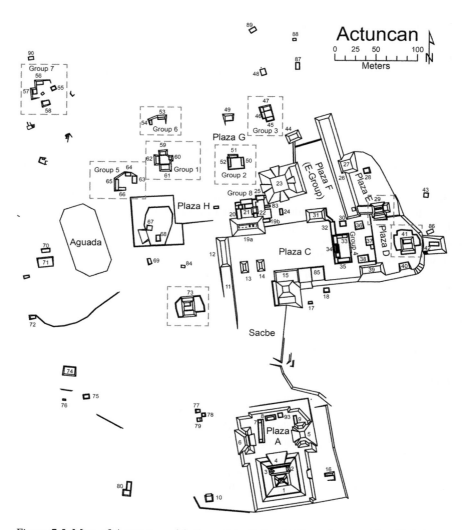

Figure 7.5 Map of Actuncan with the residential units from the text identified (map from Mixter and LeCount 2020, dotted lines added by the authors).

(Houston and Stuart 2001; Jackson 2013). Though we do not have textual evidence from Actuncan, the distinctiveness of the two forms of domestic architecture points to a clear difference in social status. At Actuncan, and in much of the Classic Maya world, status appears to be separate from wealth. Some of the commoner households discussed here appear to have been quite wealthy at some points in time. Elite households, on the other hand, appear to

Figure 7.6 Kite photograph oriented north of Group 1 (centre) and Group 5 (right-hand edge) during excavation. Note the organisation of buildings within Group 1. Four low platforms surround a central patio, as is typical of patio-focused residential groups at Actuncan. Plaza H, Actuncan's hypothesised market, extends south from Group 1 towards Structure 73 (obscured by the forest) at the far end of the plaza. The site's aguada is the low area visible in the upper right part of the image. Photograph by Chester P. Walker.

be more tightly connected both physically and economically to Actuncan's political leaders.

Six of these residential units (Groups 1, 2, 3, 5, 6, and 7) are patio-focused groups (Figure 7.6) located in the northern part of the settlement, referred to as the Northern Settlement, that formed the anchor for five smallholder households (Groups 1 and 6 are likely a single household). Groups 1, 5, and 7 date to the initial founding of Actuncan as a political centre during the Late Preclassic period. Their founders claimed prime agricultural land near the Actuncan's aguada and maintained control of that land throughout Actuncan's occupation (LeCount et al. 2019). Fulton (2019) has previously argued that these households likely cooperated to some extent, with Group 1 serving as a focal residence (sensu Gonlin 1993) for the neighbourhood where ancestors from all groups were buried in later time periods. Similarly, these households seem to have shared access to an array of common goods, including improved agricultural land around the aguada and raised garden plots (Heindel 2019; LeCount et al. 2019). Later, Groups 2 and 3 were established in a more tightly packed zone along a ridge to the east of Group 1. While it is unclear if these households participated in sharing resources with the earlier patio groups, their positioning clearly does not impede on the landed resources secured by the longer-tenured households. However, by virtue of being outside the site's planned monumental core, these households had access to land, though it may have been lower quality and smaller in quantity than the land farmed by earlier settlers. Test pitting to the north of Group 3 identified an array of special purpose platforms associated with specialised production activities, including pigment and stone tool production (LeCount et al. 2019: Supplementary Text 1). As smallholders, these households leveraged the productive capacity of their individual and collective control of land. Their location on house lots, their continued access to common goods, and the land tenure norms that protected the land around their houses provided opportunities for these households to amass wealth during Actuncan's prosperity and provided a modicum of stability in difficult times.

The remaining three residences are large single house mounds, likely inhabited by elite members of the community (Figure 7.7). Each of these houses is located along the boundaries of one of Actuncan's main plazas. Two of these residences, Structures 29 and 41, are located in the eastern part of the settlement core-forming edges of Plaza D, while the third elite residence, Structure 73, is located on the southern edge of Plaza H west of Actuncan's formal causeway (or *sacbe*) that connects the northern and southern portions of the settlement. These structures are each broad platforms greater than 2 m tall in height with lower, attached terraces. Around the edges of these houses, the terraces could have been used as segregated work areas, similar to those reported from elite Maya residences elsewhere (Folan et al. 2001; Inomata and Triadan 2000). These platforms would have supported superstructures that provided private inside space. Excavations on Structure 41 found evidence of a demolished masonry superstructure (Mixter et al. 2014), and it is likely that Structure 73 and 29 also had masonry superstructures during

Figure 7.7 Structure 41 and the adjoining Structure 40 prior to excavation. This photograph shows the scale of Actuncan's large urban houses. Photograph by Lisa J. LeCount.

Actuncan's peak. All of these buildings featured red-painted stucco apron moulding on the facade, and stucco collected from Structure 41 included fragments from broken painted stucco reliefs. The decoration and public location of these houses reflect a peak of wealth and influence. These houses are the inverse of the inward-focused patio groups occupied by Actuncan's commoners. They were ostentatiously decorated stages, where the occupants would engage in their daily life on terraces and platforms elevated above Actuncan's public plazas in full view of passersby. Furthermore, it is not clear that these households had access to house lots and land the way the commoner households did. Structure 29 is integrated into the layout of Actuncan's public architecture. Structure 41 is connected to auxiliary buildings but is at the edge of a relatively steep slope. Structure 73's control of land is not clear based on current data. Regardless, these households seem to have tied themselves through their positioning and implied social roles to Actuncan's political and economic fortunes. They were not smallholders with the agricultural resources to be self-sufficient during tough times. This lack of resilience is evident when their history is reviewed. What follows is a social history of Actuncan's households, which highlights the comparative resilience of local households across Actuncan's history.

Group 1 was first constructed as a large patio-focused group during the Late Preclassic period when Actuncan was established (Rothenberg 2012). Group 1 is the oldest and largest patio-focused group within the Actuncan core, a legacy that likely contributed to shifts in its function evident in later time periods. Groups 5 and 7 were founded shortly after, during the Terminal Preclassic period (Hahn 2012; Simova 2012), a time when Group 1 was also continuously occupied and repeatedly renovated (Mixter et al. 2014). Group 1 appears to have served as an important location for burials beginning in the Terminal Preclassic period, marked by the presence of three single interments in stone crypts placed within the patio space dating to this time (LeCount and Blitz 2005). The presence of these early burials supports interpretations that Group 1 was a founding household in the Northern Settlement.

Although excavations have not yet uncovered the earliest construction phases of the three elite residences, documented construction phases date to at least the Terminal Preclassic period. We suspect their founding corresponds to the establishment of the civic plan near the end of the Late Preclassic period. Excavations indicate that the three elite residences were occupied throughout the Terminal Preclassic period, with evidence of small construction projects that would be expected in domestic space (Mixter et al. 2014; Nordine 2014; Simova 2015).

All three elite residences were abandoned during the Early Classic period. Termination rituals consisting of densely layered broken ceramics were encountered on the Early Classic surfaces of Structures 41 and 73 (Mixter et al. 2014; Simova et al. 2015). The trajectories of these three elite residences appear to be closely tied to the fate of Actuncan's divine rulers. The elite residences were relatively stable throughout the Preclassic period. However, when Actuncan's political prominence declined with the rise of Buenavista del Cayo, the Actuncan elite families abandoned their homes. In contrast, the commoner groups do not appear to be as affected by the political shift as Groups 1, 5, and 7 continued to be occupied during the Early Classic period. Ongoing modifications of both Groups 1 and 5 occurred during the Early Classic period, including the placement of plaster patio floors and the enlargement of architecture. There is also evidence of occupation at Groups 2 and 3 during this time, though excavations of these household groups have been limited.

Groups 1, 2, 3, 5, and 7 all show evidence of continued occupation into the Late Classic I period. At Group 1, there is a hiatus in construction, though the residence was still occupied. The lack of renovation likely points to diminished access to labour and resources. In contrast, Group 5 contains evidence of continuous architectural modification and stability. The disparity in construction at this time may reflect differences in the social strategies adopted by these household groups. We hypothesise that Group 5's construction reflects its relative prosperity in comparison to Group 1's limited access to resources (Mixter et al. 2014). Perhaps residents of these households associated themselves with different aspiring leaders during the local contests for power, leading to Group 5's prosperity when their allies were victorious – and Group 1's poverty when their allies lost. Structure 73 was reoccupied at this time, though it was likely used for ritual purposes (Simova et al. 2015).

During the Late Classic II period, Groups 1, 2, 3, 5, and 7 were continuously occupied, and Group 6 was founded. Additionally, architectural modification at Group 1 began again with the expansion of building platforms and alterations throughout the group. Throughout the Late Classic period, the patio adjacent to the eastern structure of Group 1 served as a burial ground for many individuals over several generations, including possible revered ancestors. The manner in which individuals were buried and extensively disturbed (Freiwald et al. 2014) suggests that the space itself held more importance as a ritual location rather than recognition of the individual graves. New generations or, possibly, new inhabitants drew on the long-term occupation and history of the household to anchor their ritual practices in space. Activity analysis suggests that Group 6 served as an extension of Group 1, rather than its own household, and may have been used on special occasions in connection with ritual practices (Fulton 2015). These ritual investments speak to a renewed focus on Group 1, the first household group constructed at Actuncan, as an ancestral location. During this time, many commingled burials were also placed within a newly constructed platform of Structure 73. With the absence of domestic debris, it is likely that Structure 73 was not used as a residence but, rather, as a special purpose ritual structure (Simova et al. 2015). Structures 29 and 41 both appear to be reoccupied and renovated during the Late Classic II period, though the exact nature of reoccupation remains unclear. There are minor and major construction events that occurred at Structure 29. Structure 41's modifications are more complex and include the placing of a seated burial at its summit, creation of a two-level summit platform, and construction of three low benches (Freiwald et al. 2014; Mixter et al. 2014).

Prior to the Terminal Classic period, Groups 2 and 3 were abandoned, but the other commoner households continued to be occupied throughout this time period. Burials continued to be placed in the patio of Group 1 in addition to significant enlargement of the southernmost structure, Structure 61, which faces the public plaza to the south. Increasing the size of Structure 61 may have been an attempt to amplify the grandeur of the group when viewed from this plaza. The residents were likely trying to emphasise their perceived sociopolitical importance to regain or continue their prestige within the community. At Group 5, architectural modifications were more widespread and included modifying the group's alignment from 10° W of N to nearly exactly North. This realignment is unusual, as most residential structures at Actuncan maintained their alignment through time. The realignment may indicate instability within the lineage and customs of Group 5's residents, perhaps speaking to the reoccupation of this location by new residents or the affiliation of the residents with newly introduced identities. Although both Groups 1 and 5 contain evidence of modification, the residents of Group 5 constructed plaster floors, whereas the residents of Group 1 constructed earthen floors. This difference may indicate that the residents of Group 5 had more consistent access to resources than the residents of Group 1. However, activity analysis from this time suggests that Group 1 served as a locus for distinct activities, including burials and, possibly, affiliative ritual practices connected to ancestral landscape use. Based on this evidence Fulton (2019) has argued that Group 1 was a "focal residence," similar to what has been documented in the Copan Valley. Turning to the former elite houses, at Structure 73, there is clear evidence of ceremonial activities at the summit during the Terminal Classic period, suggesting that it continued to serve as a special purpose ritual structure. Structures 29 and 41 still appear to be occupied, but the nature of their occupation remains unclear. Both platforms had been stripped of their cut stone facades by this point, so occupants using the summit of these platforms were literally residing on ruins. Ultimately, in terms of domestic function, the patio groups along the edge of the settlement outlived the individual houses located in the settlement's monumental core.

Group 1, Actuncan, ca. AD 650 (see endnote 1)

It is late afternoon near the beginning of the harvest season. You are thankful to be surrounded by maize stalks taller than your head as you walk uphill from your small raised garden towards your home. Times have been difficult recently. Your family has always farmed, but your ancestors used to trade tools and crafts at the market and had ready access to building materials. In those days, Actuncan was a capital, home to the royal court and a burgeoning class of political elites. Then, tribute and labour flowed into Actuncan, facilitating the construction of grand and beautiful buildings in the town centre. The main regional market where your ancestors traded their wares was located here, just to the south of your house.

You reach your household and enter the patio. Recently, the plaster surface has worn thin. The overlords at Xunantunich have limited your access to the local limestone quarries, so you won't be able to repair the floor. You will soon be living on dirt. Yet, you still have your home and your land. You and your neighbours work together to regulate water use from the local aguada and to manage other community resources. You have reliable food and shelter for your family, even if many comforts are limited - you feel fortunate to have that. These gardens and fields have been worked by generations of your ancestors, now buried under your feet, and will be worked for generations to come.

You walk from your patio to the south around the back of your house and look out towards downtown Actuncan. Stones are strewn about the old plazas, tumbled from the city's once-grand now-deteriorating public architecture. Glancing south across the old marketplace, your eyes linger on one platform. Once the home of one of Actuncan's grand families, this building has fallen to ruin. Its once beautiful painted, sculpted facades have fallen, leaving behind a bare pile of stones. All of the old elite homes are abandoned now. Their families were too concerned with politics and staying close to the royal court. Without their own fields, they couldn't stay at Actuncan without royal patronage. The only reminder of their position in the community are the crumbling remains of their grand homes.

Yes, you are happy for your fields. Someday, Actuncan may be grand again, but until then, your family will be fine.

Discussion

Considering the ancient landscape of Actuncan, the commoner and elite households demonstrate different trajectories. The prosperity of elite re-sidences was tightly tied to their affiliation with Actuncan's leaders. As such, occupation of these residences directly reflects shifts in regional political authority. Most notably, they were established during the Late Preclassic period when Actuncan was a regional capital and abandoned during the Early Classic period when regional power shifted to another location. In contrast, several commoner households, which first settled on the land during the founding of the Late Preclassic city, maintained residential continuity into the Terminal Classic period despite regional political shifts. However, not all commoner households experienced equal prosperity or resilience. By the end of Actuncan's occupation, the pattern is clear: The oldest commoner households (Groups 1, 5, and 7) proved most resilient, surviving well after the residents of most households at Actuncan and in the

Mopan River valley departed. These households continually invested in building their houses, transformed their land to improve its usefulness and created a meaningful domestic space connected to their family's history and prosperity. In contrast, Groups 2 and 3, the relatively young household groups, did not last as long. They likely occupied lower quality land to start with and had substantially less time to invest in that place before the difficult time arrived in the Mopan River valley at the end of the Classic period. The pattern of abandonment at Actuncan in which the oldest, most entrenched households tended to be the last ones abandoned is typical for the Mopan River valley region (Ashmore, Yaeger, and Robin 2004) and is evocative of the logical outcome of the Principle of First Occupancy observed elsewhere in the Maya world: Those who settle in an area first are able to occupy the prime plots of land (McAnany 2013). Those who arrive later settle on less ideal plots of land are the first to feel the effects of resource stress.

The oldest smallholder households were buoyed by their control over the domestic landscape granted to them by local land tenure rules. They had access to prime agricultural land for gardens and in fields and plenty of space to place auxiliary activity areas where they could undertake additional economic production activities. We have evidence, for example, that Group 1 had an active chert tool production area. These households improved their land, improving drainage and building raised gardens to increase their productivity. Furthermore, the land tenure rules encouraged collaboration in the use of common goods, such as water from the aguada. Cooperation between commoner households is evident in burial practices and the establishment of Group 1 as a focal location for ancestors. For the households in Actuncan's Northern Settlement, their landscape provided them with basic nutritional needs. Additionally, their landscape created a social ethos of collaboration fostered by the proximity of three households living near each other, forced to share resources for over 1000 years.

In contrast, Actuncan's urban houses were less resilient. Despite being founded contemporaneously with Actuncan's resilient smallholders, the elite urban houses were abandoned and fell into disrepair along with the rest of Actuncan's monumental core when the centre's political fortunes turned. This abandonment indicates that these households' resilience was tied to those political entities. These households do not seem to have been smallholders. More likely, they were involved in the administration of the polity or production for the luxury economy as later Maya courtiers often were (i.e., Inomata 2001). Their homes reflect different priorities and different daily lives. During their peak, these households would have been reliant on other means – tribute or markets – to meet their basic nutritional needs. When Actuncan failed as a political seat, these households were abandoned.

The connection of these households to Actuncan's political and economic fortunes is evident in their position within Actuncan's urban landscape. These ostentatiously decorated houses were located directly on Actuncan's major plazas and intentionally placed to be part of the action. However, they did not have access to their own plots of land and were reliant on the state and/or local economy for their resilience. As a result, these households suffered most when Actuncan's political power and, subsequently, its economy collapsed. Like Actuncan's civic and religious buildings, these households crumbled. Their once beautiful sculptural facades fell apart and were left strewn about. Evidence indicates that Structure 41 was burned as Actuncan fell into disrepair, perhaps indicating that these elites may have been targets of the political struggle that brought Actuncan's rulers down. Their connection to Actuncan's political leaders and the position of these houses within Actuncan's urban core limited the resilience of these households and made them vulnerable to the vagaries of the region's dynamic political environment.

Current and future issues of changing landscapes

Writing in January of AD 2021, the productivity of the domestic landscape has never seemed more relevant. Reflecting from our perspectives in the states of Texas and New York in the United States, the year AD 2020 can only be seen as a challenging time. Bound to our homes by the global COVID-19 pandemic, we have watched as long-simmering tensions around racial inequalities spawned widespread street protests in the United States and rising authoritarian sentiments challenged political institutions across the Global North. Without other outlets, we have invested in our homes. Many have built home gardens as both a pastime – the kind of everyday practice that makes a space a home – and a locus of small-scale food production following the shock of empty grocery stores at the beginning of the pandemic. These contemporary household gardens resemble the so-called "victory gardens" that emerged in the early 1900s in response to food shortages during World War I and followed a common pattern in which people turn to the land for self-sufficiency during periods of crisis (Langlie 2018b; Langlie and Arkush 2016).

Yet only households with access to land may take advantage of this kind of supplemental gardening. In the past, smallholders at Actuncan did not suddenly change their farming and gardening practices during difficult times. Their resilience resulted from long-term investment in the land around their homes. In contrast, urban elite households were abandoned when the political tides turned, and their occupants could no longer depend on patronage or favourable market conditions to trade for foodstuffs. In this case, the powerful were victims of their political connections.

However, it is much more common for the landless poor to suffer the most when times are difficult. Unlike smallholders, the landless poor are dependent on the labour market for their income. With little or no time and space to tend their own crops and little else of value to trade for crops, the landless poor have no buffer against a failing economy.

For our friends and colleagues in Belize, 2020 was disastrous. The global pandemic shut down the tourism industry leaving many in western Belize without employment, while two major hurricanes, driven in part by global climate change, caused major flooding that filled houses with mud and ruined crops for those who had small *milpa* fields. Families with land who have been able to invest in its productivity, like the cattle ranchers discussed at the beginning of this chapter, have some buffer. These families still have beef to sell on the international market. However, without incomes or land, most Belizeans are hungry and suffering.

For the past few centuries, land ownership has been elusive for most Belizeans. During the 18th and 19th centuries, land policies enacted under Spanish and British colonial rule effectively alienated most Belizeans from both land and agriculture. During the 19th century, the British Honduras (as Belize was then called) territory was viewed primarily through an extractive lens as a source of hardwoods and forest products. African slaves were brought to British Honduras to work in mahogany production and other forestry industries, while indigenous Maya populations were subjected to constantly changing land tenure rules throughout the 19th century that left Maya land officially in control of the British Crown. Private land was consolidated in large-scale landholdings, half of which was controlled by a single company, the British Estate Company, by 1880 (Iyo et al. 2003, 146). Even after the end of slavery in 1938, ordinary Belizeans had little access to land and were tied as labourers to large foreign companies through debt peonage schemes such as the Truck and Advance systems (Iyo et al. 2003, 147). Furthermore, treaties between the British and Spain banned agriculture in Belize for parts of the 19th century, a policy that tied Belize to the British Empire by ensuring Belizeans dependence on imported foodstuffs, which were distributed through company commissaries at a profit (Petch 1986, 1005). The inability to turn to agriculture tied labourers to forestry work even if they were able to get access to land. These restrictions loosened in the second half of the 19th century and the first half of the 20th century, but forestry monopolies were just replaced by large scale industrial sugar cane, banana, coconut, and eventually orange plantations (Petch 1986). The work changed, but the conditions of working Belizeans largely did not.

Despite calls by politicians to "Give Every Belizean a Piece of Land" (United Democratic Party (UDP) slogan cited in Iyo et al. 2003: 159), efforts to create equitable access to land have continued to be hindered by the lack of a cohesive policy framework (Iyo et al. 2003: 162) and the historical legacy of colonial property rights concentrated in a few hands (see Petch 1986). Our young friends in San Jose Succotz, Belize talk about how they are working so they can get their land to build their house. But a postage stamp with a house on it doesn't provide the same security that the farms and gardens of Actuncan's smallholders did. In the meantime, they continue to live with their parents or rent while they work and try to save.

In the face of famine, the people of Belize must rely on different back-stops: Public institutions and collaborative support. The Belize government is offering expanded unemployment benefits to help those whose jobs have been affected by COVID-19 (Central Information and Technology Office, Belize 2020). Additionally, extended communities within and beyond Belize have come together to collect funds for flood relief and ongoing food support. While these are only short term-solutions, community collaboration is a major contributor to long-term resilience in the region, in the present as much as it was in the past. For pre-Columbian Maya smallholding farmers, collaboration on infrastructure projects paired with long-term investment in household land improved resilience. While we hesitate to make specific recommendations for modern-day communities, we do suggest that the smallholders of Actuncan provide a proven model of successful households occupied for over a millennium. Ultimately, we see evidence in the past and in the present that both natural and social landscapes contribute to the resilience of communities. Understanding how people maintain their homes over generations (or recognising strategies that were not successful) can inform how modern and future households can flourish when faced with major challenges and changing landscapes.

Acknowledgements

First of all, this article is inspired by our friends and collaborators in the Cayo District, Belize. Actuncan sits on the property of the Galvez and Juan families, and we are fortunate that they allow us to disturb their cows as we hike up to the site each day. Azucena Galvez, in particular, hosts us at Clarissa Falls and provides our small archaeological project with more than food and shelter. She makes our camp a home. We were assisted in the field and laboratory by many dedicated individuals from San Jose de Succotz. Their hard work, collaboration, and enthusiasm have contributed to many years of successful field seasons. We are indebted to them for their work as well as their friendship.

Our interpretations of a large amount of data collected by the Actuncan Archaeological Project since 2001, and we are indebted to the many people who made this work possible. First, thank you to Dr Lisa LeCount, who invited us both to work at Actuncan in 2010 and who has continued to support our research at the site ever since. Thanks to the Belize Institute of Archaeology, particularly directors Drs. John Morris and Jaime Awe, who have provided us with permission to undertake work in Belize. Our interpretations are based on excavations directed and reported on by a large number of colleagues. Thanks, in particular, go to Caroline Antonelli, Dr John Blitz, Lauren Bussiere, Dr Carolyn Freiwald, Dr Theresa Heindel, Kelsey Nordine, and Bobbie Simova, who directed excavations within the household contexts discussed in this paper. Thanks also to Dr Christian Wells for guidance with analysis. This research was funded by the National

Science Foundation (BCS-0923747 to LeCount and BCS-1418480 to Fulton), the National Geographic Society Committee for Exploration and Research (9279-13 to LeCount), the University of Alabama, and the Washington University Graduate School of Arts and Sciences. Thanks to Dr Mike Carson for inviting us to participate in this volume and to Dr Mike Carson, Dr BrieAnna Langlie, and anonymous reviewers for feedback on the article concept and the manuscript.

Note

1 Although there is deep theoretical literature on the social impact of landscapes, the abstract language of social theory often belies how social connections are built due to the rich experience humans feel interacting with these landscapes. Just re-read the previous sentence if you doubt the loss of emotion and perception that can take place when academics transform human experience into words. The two fictionalised narratives presented in this text are intended to place you into the landscape to gain back some of that loss. Hopefully, these help you conceptualise the deep connection to the place that Maya dwellings likely symbolised. These reconstructions are based largely on excavations at and around Group 1, on Structure 73, and within Actuncan's Plaza H described by Antonelli and Rothenberg (2011), Chambers-Koenig (2013), Craiker (2013), Freiwald et al. (2014), LeCount and Blitz (2001), Heindel (2015, 2019), Keller (2012), LeCount et al. (2005), Millar (2016), Rothenberg (2012), Simova (2012, 2015), and Simova et al. (2014).

References

Acabado, Stephen, 2018. Zones of refuge: Resisting conquest in the Northern Philippine Highlands through environmental practice. *Journal of Anthropological Archaeology* 52: 180–195. https://doi.org/10.1016/j.jaa.2018.05.005

Anschuetz, Kurt F., Richard H. Wilshusen, and Cherie L. Scheick, 2001. An archaeology of landscapes: Perspectives and directions. *Journal of Archaeological Research* 9 (2): 157–211.

Antonelli, Caroline, and Kara A. Rothenberg, 2011. Excavations at Group 1. In *Actuncan Early Classic Maya Project: Report of the Third Season*, edited by Lisa J. LeCount and Angela H. Keller, pp. 15–32. Report on file at the Belize Institute of Archaeology. Belmopan.

Ashmore, Wendy, 1998. Monumentos Politicos: Sitio, Asentamiento, y Paisaje Alrededor de Xunantunich, Belice. In *Anatomia de Una Civilizacion: Aproximaciones Interdisciplinarias a La Cultura Maya*, edited by Andrés Ciudad Ruiz, Yolanda Fernández Marquínez, José M. García Campillo, A. Josefa Iglesias Ponce de León, Alfonso L. García-Gallo, and Luis T. Sanz Casto, pp. 161–183. Publicaciones de La SEEM 4. Sociedad Española de Estudios Mayas, Madrid.

Ashmore, Wendy, Jason Yaeger, and Cynthia Robin, 2004. Commoner sense: Late and terminal classic social strategies in the Xunantunich Area. In *The Terminal Classic in the Maya Lowlands: Collapse, Transition, and Transformation*, edited by Arthur A. Demarest, Prudence M. Rice, and Don S. Rice, pp. 302–323. University Press of Colorado, Boulder.

Aucoin, Pauline McKenzie, 2017. Toward an anthropological understanding of space and place. In *Place, Space and Hermeneutics*, edited by Bruce B. Janz, pp. 395–412. Contributions to Hermeneutics 5. Springer International Publishing, New York. http://dx.doi.org/10.1007/978-3-319-52214-2_28

Awe, Jaime J., Christophe Helmke, Diane Slocum, and Douglas Tilden, 2020. Ally, client, or outpost? evaluating the relationship between Xunantunich and Naranjo in the Late Classic Period. *Ancient Mesoamerica* 31 (3): 494–506. https://doi.org/1 0.1017/S095653612000036X

Ball, Joseph W., and Jennifer T. Taschek, 2004. Buenavista Del Cayo: A short outline of occupational and cultural history at an Upper Belize Valley Regal–Ritual Center. In *The Ancient Maya of the Belize Valley: Half a Century of Archaeological Research*, edited by James Garber, pp. 149–167. University Press of Florida, Gainesville.

Beach, Timothy, Sheryl Luzzadder-Beach, Nicholas Dunning, and Vernon Scarborough, 2003. Depression soils in the lowland tropics of Northwestern Belize: Anthropogenic and natural origins. In *The Lowland Maya Area: Three Millennia at the Human-Wildland Interface*, edited by Arturo Gómez-Pompa, Michael F. Allen, Scott L. Fedick, and Juan J. Jiménez-Osornio, pp. 139–174. Hawthorn Press, Binghamton.

Birchall, C. J., and R. N. Jenkins, 1979. The Soils of the Belize Valley, Belize. Overseas Development Administration, Land Resources Development Centre, Surbiton, UK.

Bourdieu, Pierre, 1977. *Outline of a Theory of Practice*. Cambridge University Press, Cambridge.

Bourdieu, Pierre, 1990. *The Logic of Practice*. Stanford University Press, Stanford.

Brown, M. Kathryn, Jennifer Cochran, Leah McCurdy, and David W. Mixter, 2011. Preceramic to postclassic: A brief synthesis of the occupation history of group E, Xunantunich. *Research Reports in Belizean Archaeology* 8: 209–219.

Burke, Peter, 1989. History as Social Memory. In *Memory: History, Culture and the Mind*, edited by Thomas J. Butler, pp. 97–113. Basil Blackwell, Oxford.

Cap, Bernadette, 2015. Classic Maya Economies: Identification of a Marketplace at Buenavista Del Cayo, Belize. Unpublished doctoral dissertation, University of Wisconsin, Madison.

Carter, Anthony T., 1984. Household Histories. In *Households: Comparative and Historical Studies of the Domestic Group*, edited by Robert McC. Netting, Richard R. Wilk, and Eric J. Arnould, pp. 44–83. University of California Press, Berkeley.

Central Information and Technology Office, Belize, 2020. The COVID-19 Unemployment Relief Program, Belmopan, Belize. https://web.archive.org/web/20210201221222/https://www.covid19.bz/the-covid-19-unemployment-relief-program

Certeau, Michel de, 1984. *The Practice of Everyday Life*. University of California Press, Berkeley.

Chambers-Koenig, Emma, 2013. Excavations in the West Plaza. In *Actuncan Early Classic Maya Project: Report of the Fifth Season*, edited by Lisa J. LeCount, pp. 109–117. Report on file at the Belize Institute of Archaeology, Belmopan.

Cowgill, George L., 2004. Origins and development of urbanism: Archaeological perspectives. *Annual Review of Anthropology* 33: 525–549.

Craiker, Krystal, 2013. Things That fall through the screen: Microartifact analysis of the West Plaza at Actuncan. In *Actuncan Early Classic Maya Project: Report of the Fifth Season*, edited by Lisa J. LeCount, pp. 145–154. Report on file at the Belize Institute of Archaeology, Belmopan.

Day, M., 1993. Resource use in the tropical Karstlands of Central Belize. *Environmental Geology* 21 (3): 122–128. https://doi.org/10.1007/BF00775295.

Durkheim, Emile, 1995. *The Elementary Forms of the Religious Life.* Translated by Karen E. Fields. Free Press, New York.

Fahsen, Federico, and Nikolai Grube, 2005. The origins of Maya writing. In *Lords of Creation: The Origins of Sacred Maya Kingship*, edited by Virginia M. Fields and Dorie Reents-Budet, pp. 74–79. Scala, London.

Fedick, Scott L., 1988. Prehistoric Maya Settlement and Land Use Patterns in the Upper Belize River Area, Belize, Central America. Unpublished doctoral dissertation, Arizona State University, Tempe.

Fisher, Chelsea, 2014. The Role of Infield Agriculture in Maya Cities. *Journal of Anthropological Archaeology* 36 (December): 196–210. https://doi.org/10.1016/j.jaa.2014.10.001

Folan, William J., Joel D. Gunn, and María del Rosario Domínguez Carrasco, 2001. Triadic Temples, Central Plazas, and Dynastic Palaces: A diachronic analysis of the Royal Court Complex, Calakmul, Campeche, Mexico. In *Royal Courts of the Ancinet Maya, Volume 2: Data and Case Studies*, edited by Takeshi Inomata and Stephen D. Houston, pp. 223–265. Westview, Boulder, Colorado.

Foucault, Michel, 2012. *Discipline and Punish: The Birth of the Prison.* Translated by Alan Sheridan. Knopf Doubleday Publishing Group, New York.

Freiwald, Carolyn, David W. Mixter, and Nicholas Billstrand, 2014. Burial practices at Actuncan, Belize: A seated burial and ongoing analysis from the 2001–2013 field seasons. *Research Reports in Belizean Archaeology* 11: 95–109.

Fulton, Kara A., 2015. Community Identity and Social Practice during the Terminal Classic Period at Actuncan, Belize. Unpublished doctoral dissertation, University of South Florida, Tampa.

Fulton, Kara A., 2019. Community identity and shared practice at Actuncan, Belize. *Latin American Antiquity* 30 (2): 266–286. https://doi.org/10.1017/laq.2019.18

Furley, P. A., and W. W. Newey, 1979. Variations in plant communities with topography over tropical limestone soils. *Journal of Biogeography* 6 (1): 1–15. https://doi.org/10.2307/3038149

Gann, Thomas W. F., 1925. *Mystery Cities: Exploration and Adventure in Lubaantun.* Charles Scribner's Sons, New York.

Giddens, Anthony, 1979. *Central Problems in Social Theory: Action, Structure, and Contradiction in Social Analysis.* University of California Press, Berkeley.

Giddens, Anthony, 1984. *The Constitution of Society: Outline of the Theory of Structuration.* University of California Press, Berkeley.

Gillespie, Susan D., 2000. Maya "Nested Houses": The ritual construction of place. In *Beyond Kinship: Social and Material Reproduction in House Societies*, edited by Rosemary A. Joyce and Susan D. Gillespie, pp. 135–160. University of Pennsylvania Press, Philadelphia.

Gillis, John, R., editor, 1994. *Commemorations: The Politics of National Identity.* Princeton University Press, Princeton.

Gonlin, Nancy, 1993. Rural Household Archaeology at Copán, Honduras: A Thesis in Anthropology. Unpublished doctoral dissertation, Pennsylvania State University, State College.

Hahn, Lauren D., 2012. The 2011 Excavations at Group 5: Structures 64 and 65. In *Actuncan Early Classic Maya Project: Report of the Fourth Season*, edited by

Lisa J. LeCount and John H. Blitz, pp. 119–138. Report on file at the Belize Institute of Archaeology, Belmopan.

Hammond, Norman, and Wendy Ashmore, 1981. Lowland Maya Settlement: Geographical and chronological frameworks. In *Lowland Maya Settlement Patterns*, edited by Wendy Ashmore, pp. 19–36. University of New Mexico Press, Albuquerque.

Heindel, Theresa, 2015. The 2014 Lithic analysis: Chert debitage from structures 41, 73, 21 A and B, and 59. In *Actuncan Early Classic Maya Project: Report of the Seventh Season*, edited by Lisa J. LeCount, pp. 115–150. Report on file at the Belize Institute of Archaeology, Belmopan.

Heindel, Theresa, 2019. Plotting, Planting and Prospering: Ancient Maya Agricultural Production and Water Management at Actuncan, Belize. Unpublished doctoral dissertation, University of California, Riverside. https://search.proquest.com/docview/2382636307/abstract/91C3F4471BF94DC5PQ/1

Hendon, Julia A., 2010. *Houses in a Landscape: Memory and Everyday Life in Mesoamerica*. Illustrated Edition. Duke University Press Books, Durham, North Carolina.

Hodder, Ian, 1990. *The Domestication of Europe*. Basil Blackwood, Oxford.

Hodder, Ian, 1992. *Theory and Practice in Archaeology*. Routledge, London.

Horowitz, Rachel A., 2018. Uneven lithic landscapes: Raw material procurement and economic organization among the Late/Terminal Classic Maya in Western Belize. *Journal of Archaeological Science: Reports* 19 (June): 949–957. https://doi.org/10.1016/j.jasrep.2017.01.038

Houston, Stephen D., and David Stuart, 2001. Peopling the Maya Court. In *Royal Courts of the Classic Maya, Volume 1: Theory, Comparison, and Synthesis*, edited by Takeshi Inomata and Stephen D. Houston, pp. 54–83. Westview Press, Boulder, Colorado.

Inomata, Takeshi, 2001. The power and ideology of artistic creation: Elite craft specialists in Classic Maya Society. *Current Anthropology* 42 (3): 321–349.

Inomata, Takeshi, and Daniela Triadan, 2000. Craft production by Classic Maya Elites in Domestic Settings: Data from rapidly abandoned structures at Aguateca, Guatemala. *Mayab* 13: 57–66.

Isendahl, Christian, 2012. Agro-urban landscapes: The Example of Maya lowland cities. *Antiquity* 86 (334): 1112–1125. https://doi.org/10.1017/S0003598X00048286

Isendahl, Christian, and Michael E. Smith, 2013. Sustainable agrarian urbanism: The low-density cities of the Mayas and Aztecs. *Cities* 31: 132–143. https://doi.org/10.1016/j.cities.2012.07.012

Iyo, Joe, Patricia Mendoza, Jose Cardona, Armin Cansino, and Ray Davis, 2003. Belize: Land policy, administration and management in Belize. In *Land in the Caribbean: Issues of Policy, Administration and Management in the English-Speaking Caribbean*, edited by Allan N Williams, pp. 141–174. Terra Institute, Mount Horeb, Wisconsin.

Jackson, Sarah E., 2013. *Politics of the Maya Court: Hierarchy and Change in the Late Classic Period*. University of Oklahoma Press, Norman.

Janz, Bruce, 2005. Walls and borders: The range of place. *City & Community* 4 (1): 87–94. https://doi.org/10.1111/j.1535-6841.2005.00104.x

Jones, John G., 1994. Pollen evidence for early settlement and agriculture in Northern Belize. *Palynology* 18: 205–211.

Keller, Angela H., 2012. Everyday practices in Maya Civic Centers: Preliminary results of testing in the West Plaza at Actuncan, Belize. *Research Reports in Belizean Archaeology* 9: 39–50.

Kent, Susan, editor, 1987. *Method and Theory for Activity Area Research: An Ethnoarchaeological Approach.* Cambridge University Press, Cambridge.

Killion, Thomas W., 1990. Cultivation intensity and residential site structure: An ethnographic examination of peasant agriculture in the Sierra de Los Tuxtlas, Veracruz, Mexico. *Latin American Antiquity* 1: 191–215.

Killion, Thomas W., 1992. Residential ethnoarchaeology and ancient site structure: Contemporary farming and prehistoric settlement agriculture at Matacapan, Veracruz, Mexico. In *Gardens of Prehistory: The Archaeology of Settlement Agriculture in Greater Mesoamerica*, edited by Thomas W. Killion, pp. 119–149. University of Alabama Press, Tusaloosa.

King, R. B., I. C. Baille, and T. M. B. Abell, 1992. *Land Resource Assessment of Northern Belize. Volume 1 of 2.* Natural Resources Institute Bulletin 43. Chatham, UK.

Langlie, BrieAnna S., 2018a. Building ecological resistance: Late Intermediate Period Farming in the South-Central Highland Andes (CE 1100–1450). *Journal of Anthropological Archaeology* 52 (December): 167–179. https://doi.org/10.1016/j.jaa.2018.06.005

Langlie, BrieAnna S., 2018b. Gardening for Victory: War Gardens in the Ancient Andes. Presentation at the 83rd Meeting of the Society for American Archaeology, April 2018. Washington, D.C.

Langlie, BrieAnna S., and Elizabeth N. Arkush, 2016. Managing Mayhem: Conflict, environment, and subsistence in the Andean Late Intermediate Period, Puno, Peru. In *The Archaeology of Food and Warfare: Food Insecurity in Prehistory*, edited by Amber M. VanDerwarker and Gregory D. Wilson, pp. 259–289. Springer International Publishing, Cham, Swittzerland. https://doi.org/10.1007/978-3-319-18506-4_12

Lawrence, Denise L., and Setha M. Low, 1990. The Built environment and spatial form. *Annual Review of Anthropology* 19: 453–505.

LeCount, Lisa J., 2004. Looking for a needle in a haystack: The Early Classic Period at Actuncan. *Research Reports in Belizean Archaeology* 1: 27–36.

LeCount, Lisa J., and John H. Blitz, 2001. *Actuncan Early Classic Project: Report of the First Season.* Report on file at the Belize Institute of Archaeology, Belmopan.

LeCount, Lisa J., and John H. Blitz, 2005. The Actuncan Early Classic Maya Project: Progress report on the second season. *Research Reports in Belizean Archaeology* 2: 67–77.

LeCount, Lisa J., and Jason Yaeger, 2010a. A brief description of Xunantunich. In *Classic Maya Provincial Politics: Xunantunich and Its Hinterlands*, edited by Lisa J. LeCount and Jason Yaeger, pp. 67–78. University of Arizona Press, Tucson.

LeCount, Lisa J., and Jason Yaeger, 2010b. Placing Xunantunich and its hinterland settlements in perspective. In *Classic Maya Provincial Politics: Xunantunich and Its Hinterlands*, edited by Lisa J. LeCount and Jason Yaeger, pp. 337–369. University of Arizona Press, Tucson.

LeCount, Lisa J., John H. Blitz, and Rebecca Scopa Kelso, 2005. *Actuncan Early Classic Project: Report of the Second Season.* Report on file at the Belize Institute of Archaeology, Belmopan.

LeCount, Lisa J., David W. Mixter, and Borislava S. Simova, 2017. Preliminary thoughts on ceramic and radiocarbon data from Actuncan's E-Group Excavations. In *Actuncan Early Classic Maya Project: Report of the Ninth Season*, edited by Lisa J. LeCount, pp. 27–48. Report on file at the Belize Institute of Archaeology, Belmopan.

LeCount, Lisa J., Chester P. Walker, John H. Blitz, and Ted C. Nelson, 2019. Land tenure systems at the Ancient Maya Site of Actuncan, Belize. *Latin American Antiquity* 30 (2): 245–265. https://doi.org/10.1017/laq.2019.16

Lefebvre, Henri, 1991. *The Production of Space.* Translated by Donald Nicholson-Smith. Wiley, New York.

le Goff, Jacques, 1992. *History and Memory. Translated by Elizabeth Claman* and *Steven Rendall.* Columbia University Press, New York.

Leventhal, Richard, and Wendy Ashmore, 2004. Xunantunich in a Belize Valley Context. In *The Ancient Maya of the Belize Valley: Half a Century of Archaeological Research,* edited by James Garber, pp. 168–179. University Press of Florida, Gainesville.

Lohse, Jon C., 2010. Archaic origins of the Lowland Maya. *Latin American Antiquity* 21 (3): 312–352.

Low, Setha M., 2000. *On the Plaza: The Politics of public space and culture.* 1st Edition. University of Texas Press, Austin.

Marcus, Joyce, and Jeremy A. Sabloff, editors, 2008. *The ancient city: New perspectives on urbanism in the old and new world.* School for Advanced Research Press, Santa Fe, New Mexico.

Marx, Karl, 1999 [1985]. *The Eighteenth Brumaire of Louis Bonaparte.* Translated by Saul K. Padover and Progress Publishers. Marxists Internet Archive, Gunzenhausen, Germany. https://www.marxists.org/archive/marx/works/1852/18th-brumaire/

Mauss, Marcel, 1979. *Seasonal Variations of the Eskimo: A Study in Social Morphology.* Translated by J. J. Fox. Routledge, London.

McAnany, Patricia A., 2013. *Living with the Ancestors: Kinship and Kingship in Ancient Maya Society.* Revised Edition. Cambridge University Press, Cambridge.

McGovern, James O., 1992. Study of Actuncan (Cahal Xux). In *Xunantunich Archaeological Project: 1992 Field Season,* edited by Richard Leventhal, pp. 74–83. Report on file at the Belize Institute of Archaeology, Belmopan.

McGovern, James O., 1993. Survey and excavation at Actuncan. In *Xunantunich Archaeological Project: 1993 Field Season,* edited by Richard M. Leventhal, pp. 100–127. Report on file at the Belize Institute of Archaeology, Belmopan.

McGovern, James O., 2004. Monumental Ceremonial Architecture and Political Autonomy at the Ancient Maya City of Actuncan, Belize. Unpublished doctoral dissertation, University of California, Los Angeles.

McGuire, Randall, and Robert McC. Netting, 1982. Leveling peasants? The maintenance of equality in a Swiss Alpine Community. *American Ethnologist* 9 (2): 269–290. https://doi.org/10.1525/ae.1982.9.2.02a00040

Millar, Jane E., 2016. Ground truthing magnetic anomalies at Actuncan: The Second Season. In *Actuncan Early Classic Maya Project: Report of the Eighth Season,* edited by Lisa J. LeCount and David W. Mixter, pp. 52–68. Report on file at the Belize Institute of Archaeology, Belmopan.

Minkoff-Zern, Laura-Anne, 2019. *The New American Farmer: Immigration, Race, and the Struggle for Sustainability.* Massachusetts Institute of Technology Press, Cambridge, Massachusetts.

Mixter, David W., 2016. Surviving Collapse: Collective Memory and Political Reorganization at Actuncan. Unpublished doctoral dissertation, Washington University, Saint Louis, Missouri.

Mixter, David W., 2017. Political change expressed in public architecture: The terminal Classic Maya Civic Complex at Actuncan, Belize. *Research Reports in Belizean Archaeology* 14: 65–75.

Mixter, David W., 2020. Community resilience and urban planning during the Ninth-Century Maya Collapse: A Case Study from Actuncan, Belize. *Cambridge Archaeological Journal*, 219–237. https://doi.org/10.1017/S095977431900057X

Mixter, David W., Thomas R. Jamison, and Lisa J. LeCount, 2013. Actuncan's Noble Court: New insights into political strategies of an enduring center in the Upper Belize River Valley. *Research Reports in Belizean Archaeology* 10: 91–103.

Mixter, David W., Kara A. Fulton, Lauren Hahn Bussiere, and Lisa J. LeCount, 2014. Living through collapse: An analysis of Maya Residential modifications during the Terminal Classic Period at Actuncan, Cayo, Belize. *Research Reports in Belizean Archaeology* 11: 55–66.

Mixter, David W., and Lisa J. LeCount, 2020. *Actuncan Archaeological Project: Report of the Third Season*, Report on file at the Belize Institute of Archaeology, Belmopan.

Morris, David, 2004. *The Sense of Space*. State University of New York Press, Albany.

Netting, Robert McC., 1993. *Smallholders, Householders: Farm Families and the Ecology of Intensive, Sustainable Agriculture*. Stanford University Press, Stanford.

Netting, Robert McC., M. Priscilla Stone, and Glenn D. Stone, 1989. Kofyar cash-cropping: Choice and change in indigenous agricultural development. *Human Ecology* 17 (3): 299–319.

Nordine, Kelsey, 2014. The 2013 excavations at Structure 29. In *Actuncan Early Classic Maya Project: Report of the Sixth Season*, edited by Lisa J. LeCount, pp. 155–169. Report on file at the Belize Institute of Archaeology, Belmopan.

Perez, Don C., 2011. Mapping Actuncan. In *Actuncan Archaeological Project: Report of the Third Season*, edited by Lisa J. LeCount and Angela H. Keller, pp. 10–14. Report on file at the Belize Institute of Archaeology, Belmopan.

Petch, Trevor, 1986. Dependency, land and oranges in Belize. *Third World Quarterly* 8 (3): 1002–1019.

Prufer, Keith M., Asia V. Alsgaard, Mark Robinson, Clayton R. Meredith, Brendan J. Culleton, Timothy Dennehy, Shelby Magee, et al., 2019. Linking Late Paleoindian stone tool technologies and populations in North, Central and South America. *PLOS ONE* 14 (7): e0219812. https://doi.org/10.1371/journal.pone.0219812

Pyburn, K. Anne, 1998. Smallholders in the Maya Lowlands: Homage to a garden variety ethnographer. *Human Ecology* 26 (2): 267–286. https://doi.org/10.1023/A:1018770907863

Pyburn, K. Anne, 2008. Pomp and circumstance before Belize: Ancient Maya commerce and the New River Conurbation" In *The Ancient City: New Perspectives on Urbanism in the Old and New World*, edited by Joyce Marcus and Jeremy A. Sabloff, pp. 247–272. School for Advanced Research, Santa Few, New Mexico.

Rabinow, Paul, 1989. *French Modern: Norms and Forms of the Social Environment*. University of Chicago Press, Chicago.

Rapoport, Amos, 1988. Levels of meaning in the built environment. In *Cross-Cultural Perspectives in Nonverbal Communication*, edited by Fernando Poyatos, pp. 317–336. Hogrefe, Toronto.

Rapoport, Amos, 1990. Systems of activities and systems of settings. In *Domestic Architecture and the Use of Space*, edited by Susan Kent, pp. 9–20. Cambridge University Press, Cambridge.

Robin, Cynthia, 2002. Outside of houses: The practices of everyday life at Chan Nòohol, Belize. *Journal of Social Archaeology* 2 (2): 245–268.

Robin, Cynthia, 2013. *Everyday Life Matters*. University Press of Florida, Gainesville.

Robin, Cynthia, Andrew R. Wyatt, Laura Kosakowsky, Santiago Juarez, Ethan Kalosky, and Elise Enterkin, 2012. A Changing cultural landscape: Settlement survey and GIS at Chan. In *Chan: An Ancient Maya Farming Community*, edited by Cynthia Robin, pp. 19–41. University Press of Florida Gainesville.

Robinson, G. M., 1985. Agricultural change in the Belize River Valley. *Caribbean Geography* 2 (1): 33–44.

Rothenberg, Kara A., 2012. The 2011 excavations at Group 1. In *Actuncan Early Classic Maya Project: Report of the Fourth Season*, edited by Lisa J. LeCount and John H. Blitz, pp. 33–48. Report on file at the Belize Institute of Archaeology, Belmopan.

Salberg, Daniel J., 2012. Mapping Actuncan during the 2011 Season. In *Actuncan Early Classic Maya Project: Report of the Fourth Season*, edited by Lisa J. LeCount and John H. Blitz, pp. 25–32. Report on file at the Belize Institute of Archaeology, Belmopan.

Sharer, Robert J., and Loa P. Traxler, 2006. *The Ancient Maya*. Stanford University Press, Stanford.

Simova, Borislava S., 2012. Test excavations in Actuncan households. In *Actuncan Early Classic Maya Project: Report of the Fourth Season*, edited by Lisa J. LeCount and John H. Blitz, pp. 139–162. Report on file at the Belize Institute of Archaeology, Belmopan.

Simova, Borislava S., 2015. Ceramic analysis and the related construction sequence of structure 73. In *Actuncan Early Classic Maya Project: Report of the Seventh Season*, edited by Lisa J. LeCount, pp. 105–114. Report on file at the Belize Institute of Archaeology, Belmopan.

Simova, Borislava S., Carolyn Freiwald, and Nicholas Billstrand, 2014. Continuing excavations at Structure 73. In *Actuncan Early Classic Maya Project: Report of the Sixth Season*, edited by Lisa J. LeCount, pp. 49–74. Report on file at the Belize Institute of Archaeology, Belmopan.

Simova, Borislava S., David W. Mixter, and Lisa J. LeCount, 2015. The social lives of structures: Ritual resignification of the cultural landscape at Actuncan, Belize. *Research Reports in Belizean Archaeology* 12: 193–204.

Sinclair, Paul J. J., Gullög Nordquist, Frands Herschend, and Christian Isendahl, editors, 2010. *The Urban Mind: Cultural and Environmental Dynamics*. Uppsala University, Uppsala, Sweden.

Smith, Jennifer R., 1998. Geology and Carbonate Hydrogeochemistry of the Lower Mopan and Macal River Valleys, Belize. Unpublished master's thesis, University of Pennsylvania, Philadelphia.

Smith, Michael E., 2007. Form and meaning in the earliest cities: A new approach to ancient urban planning. *Journal of Planning History* 6 (1): 3–47. https://doi.org/1 0.1177/1538513206293713

Smith, Michael E., 2010. The archaeological study of neighborhoods and districts in ancient cities. *Journal of Anthropological Archaeology* 29 (2): 137–154. https:// doi.org/10.1016/j.jaa.2010.01.001

Soja, Edward W., 1989. *Postmodern Geographies: The Reassertion of Space in Critical Social Theory.* Verso, London.

Steadman, Sharon R., 2016. *Archaeology of Domestic Architecture and the Human Use of Space.* Routledge, London.

Stone, Glenn Davis, Robert McC Netting, and M Priscilla Stone, 1990. Seasonally, labor scheduling, and agricultural intensification in the Nigerian Savanna. *American Anthropologist* 92 (1): 7–23. https://doi.org/10.1525/aa.1990.92.1.02a00010

Stuart, David, 1998. "The Fire Enters His House": Architecture and ritual in classic Maya texts. In *Function and Meaning in Classic Maya Architecture,* edited by Stephen D. Houston, pp. 373–426. Dumbarton Oaks Research Library and Collection, Washington, D.C.

Taube, Karl A., William A. Saturno, David Stuart, and Heather Hurst, 2010. *The Murals of San Bartolo, El Petén, Guatemala Part 2: The West Wall.* Ancient America 10. Boundary End Archaeology Research Center, Barnardsville.

Thompson, Amy E., and Keith M. Prufer, 2021. Household inequality, community formation, and land tenure in Classic Period Lowland Maya Society. *Journal of Archaeological Method and Theory,* January. https://doi.org/10.1007/s10816-020-09505-3

Ucko, Peter, and Robert Layton, editors,1999. *The Archaeology and Anthropology of Landscape: Shaping Your Landscape.* Routledge, London.

Watkins, Trevor, 2004. Architecture and "Theatres of Memor" in the Neolithic of Southwest Asia. In *Rethinking Materiality: The Engagement of Mind with the Material World,* edited by Elizabeth DeMarrais, Chris Gosden, and Colin Renfrew, pp. 97–106. McDonald Institute for Archaeological Research, Cambridge.

Wells, E. Christian, and Patricia Ann McAnany, editors, 2008. *Dimensions of Ritual Economy.* Emerald Group Publishing, Bingley, UK.

Wilk, Richard R., and William L. Rathje, 1982. Household archaeology. *American Behavioral Scientist* 25 (6): 617–639.

Wilson, Peter J., 1988. *The Domestication of the Human Species.* Yale University Press, New Haven, Connecticut.

Yaeger, Jason, 2000a. Changing Patterns of Maya Community Structure and Organization: The End of the Classic Period at San Lorenzo, Cayo District, Belize. Unpublished doctoral dissertation, University of Pennsylvania, Philadelphia.

Yaeger, Jason, 2000b. The social construction of communities in the Classic Maya Countryside: Strategies of affiliation in Western Belize. In *The Archaeology of Communities: A New World Perspective,* edited by Marcello A. Canuto and Jason Yaeger, pp. 123–142. Routledge, London.

Yaeger, Jason, 2003a. Small settlements in the Upper Belize River Valley: Local complexity, household strategies of affiliation, and the changing organization. In *Perspectives on Ancient Maya Rural Complexity,* edited by Gyles Iannone and Samuel V. Connell, pp. 42–58. Monograph 49. The Cotsen Institute of Archaeology, Los Angeles.

Yaeger, Jason, 2003b. Untangling the ties that bind: The city, the countryside, and the nature of Maya Urbanism at Xunantunich, Belize. In *The Social Construction of Ancient Cities,* edited by Monica L. Smith, pp. 121–155. Smithsonian Institution Press, Washington, D.C.

8 Holocene sea-level change and evolution of prehistoric settlements around the Yangtze Delta region

Yijie Zhuang and Shenglun Du

The evolution of prehistoric occupation on the Yangtze Delta and the surrounding region is inherently related to Holocene sea level change. As a matter of practicality, changes in sea level created hydrological fluctuations and geomorphological changes that constrained where people could live in inhabitable lands. Additionally, change in sea level had affected the hydrological and pedological regimes in and around the Yangtze Delta region, directly influencing the beginning and subsequent development of prehistoric rice farming as one of the most significant events in history of food production. The seasonality and temporality of prehistoric settlement growth and rice farming are therefore closely linked with a tempo of sea level rise and resultant geomorphological and palaeo-ecological changes. For instance, two chronological hiatuses occurred in the region's well established Neolithic sequence, dated first around 8500–7500 years BP and then again around 7500–7000 years BP (Table 8.1). The causes responsible for these "gaps" have been subjected to heated scholarly debate (see He 2018).

This chapter reconstructs environments and landscapes from around 10,000–9000 years BP until around 4000 years BP, which coincided first with the beginning of the colonisation of the Yangtze Delta region and then with the post-Liangzhu period, which witnessed a dramatic fall of a once flourishing rice-farming-based civilisation. The presentation draws on previously published results and some first-hand palaeo-environmental data derived from our own systematic geoarchaeological surveys. We then discuss technological innovations and prehistoric adaptations to Holocene sea-level change on and surrounding the Yangtze Delta region.

Reconstructing Holocene sea-level change on and around the Yangtze Delta region

Compared to the late Pleistocene, when the sea level was at its lowest, the Holocene sea level experienced a dramatic rise. However, debate persists regarding the processes of Holocene sea-level change, especially during the middle to late Holocene. In particular, disagreement continues to exist on

DOI: 10.4324/9781003139553-8

Table 8.1 Chronological sequence of prehistoric cultures within and surrounding the Yangtze Delta region.

Years (BP)	Period	Taihu lake area	Ningshao Plain	Puyang River Basin
Ca. 10,000–8500	Early Neolithic			Shangshan Culture
Chronological hiatus				
8000–7500	Middle Neolithic		*Jingtoushan site	Kuahuqiao Culture
		Chronological hiatus		
7000–6300		Early-Middle Majiabang Culture	Early Hemudu Culture	
6300–6000		Late Majiabang Culture	Middle Hemudu Culture	Loujiaqiao Type
6000–5300	Late Neolithic	Songze Culture	Late Hemudu Culture	
5300–4300		Liangzhu Culture		
4200–4000	Final Neolithic	Qianshanyang-Guangfulin Culture		
4000–3200	Bronze Age	Maqiao Culture		

whether or not the middle Holocene sea level involved a highstand, plus furthermore if such a highstand occurred in many different areas.

At least three models have been proposed about the sea level history in the study region, based on different lines of evidence and interpretations.

1. In his recent study, He (2018) provides a good summary of these scholarly debates on Holocene sea-level change on and surrounding the Yangtze Delta. Zhao et al. (1979) and Xie and Yuan (2012) consider that the Holocene sea-level underwent cyclic fluctuations, with several periods of high sea level (2–3 m higher than the present level) during the middle Holocene. These events of sea level highstands were punctuated by intervals of low sea level.
2. Zong (2004) and others (Chen et al. 2008; Liu et al. 2018; Zheng et al. 2018) advocate a continuous sea level rise since the early Holocene and even by 7000–6000 years BP, suggesting that when the sea level was at its highest, it was still 2–4 m lower than the present level. The rate of sea level rise, however, was significantly reduced after about 7000 years BP, following the preceding period of rapid increase during the early Holocene.
3. A third possible viewpoint is similar to the aforementioned two options, suggesting that a middle Holocene sea-level highstand occurred in possibly two instances, directly after two other episodes of rapid sea level rise around 9000–7000 years BP and 7000–6500 years BP, respectively, but it

continued to fall after these two events to the present level (Liu 1996; Saito 1998; Song et al. 2013).

Apart from the fact that these reconstructions were based on different environmental proxies with various data resolutions and precisions, the discrepancy of these different viewpoints partly arises from a loose definition of sea level change within the scholarly circle. In a strict sense, sea level change refers to the relative high difference between seawater surface and land surface, both of which are controlled by a series of factors such as tectonism, coastal geomorphologies, and so forth. Hence the reconstruction of sea level fluctuations involves examining multi-scalar evidence, including global glacial and interglacial cycles, regional tectonic activities, and localised landform change, as well as necessary intra-regional and inter-regional comparison to clarify the possible causes of sea level change across neighbouring localities and regions. The latter part is, unfortunately, rarely addressed systematically (but see Zong 2004).

Synthesising the reconstructions of Holocene sea-level change in published studies, with additional information collected from our geoarchaeological surveys and archaeological excavations, we adopt the following chronological framework for a broad-brush reconstruction of Holocene sea-level change on and surrounding the Yangtze Delta region.

1. 10,000–8000 years BP: rapid sea level rise in the entire Yangtze Delta and the surrounding region;
2. 8000–6500 years BP: stabilising rate of sea level rise, reaching the highest level about 6500 years BP;
3. 6500–4500/4400 years BP: stable sea level with a slight decline in some areas;
4. 4500/4400–3500 years BP: small-magnitude sea level rise, possibly beginning from the coastal areas of the Ningshao Plain at first, then inundating areas along the Hangzhou Bay and the Hangjiahu Plain; some high-ground areas that emerged from the rapid land-forming process, however, were not affected by this particular sea level rise; and
5. 3500 years BP–present: sea level decline overall, followed by recent signs of sea level rise.

Land, water, and soils in and around the Yangtze Delta region

The Pleistocene basal landform on and surrounding the Yangtze Delta region can be divided into several characteristic zones (Figure 8.1).

*   On the delta, the late Pleistocene alluvial incision created river terraces on even the most eastern edge of the present delta and deep erosion gullies up to 30–50 m in depth on red clayey sediments which formed during the Last Glacial Maximum (Wang et al. 2012). These reddish

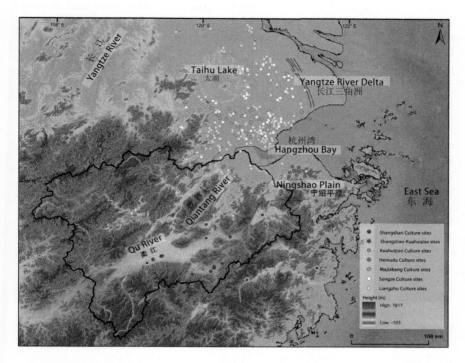

Figure 8.1 Geographic locations of different studied areas and distributions of prehistoric settlements in these areas, modified after Xu et al. (2020).

clayey tablelands became the natural space that accommodated incoming Holocene marine and alluvial sedimentation.

- Moving farther west around the present Taihu Lake, the pre-lacustrine landscape was dominated by hard brownish clayey sediments that were deposited during the Late Pleistocene. The land surface was descending from west to east, and it was further incised into gullies and low hills (Zhang 2009).
- Towards the south of the Yangtze Delta were the Hangjiahu Plain, the Hangzhou Bay, and the Ningshao Plain, mantled by similar late-Pleistocene clayey deposits.
- The north and south parts of the western Palaeo Taihu Lake region were connected to the Palaeo Yangtze and Qiantang Rivers through some palaeo-channels. These were the channels through which seawater inundated the Taihu Lake region during Holocene sea-level rise (Zhang 2009).

Early to middle Holocene sea-level rise resulted in the formation of chenier ridges along shell middens on the Yangtze Delta and the Taihu Lake. First, with the stabilising Holocene sea-level rise from about 7000 years BP, coastal chenier ridges started to form along the shoreline, running south-north and turning westward along Hangzhou Bay area (Zong et al. 2012). These sandy ridges with rich shell deposits continued to develop during the following millennia, and they became natural barriers that protected the westward regions from marine inundation, leading to the formation of massive "enclosed wetlands".

The marine transgression played a vital role in transforming the coastal plain around the present Taihu Lake region into an enormous lake. Between 7000 and 6000 years BP, a dish-like depression was created. In the middle of the depression was a tidal lagoon, surrounded by chenier ridges and other slightly higher altitude places (Hong 1991; Zhang 2009). The late Pleistocene hard clayey sediment became the surface of this newly formed shallow lake that was fed by seawater through previously incised channels in the southern and eastern parts (Zhang 2009).

With important implications, the Palaeo Tiahu Lake initially was divided into west and east lakes with complicated hydrological conditions. The west lake was fed by brackish water, whilst the water regime in the east lake was dominated by freshwater. The two lakes joined and became a freshwater lake only at a later time, around 5500–5000 years BP, when the rate of sea level rise slowed and when the formation of chenier ridges accelerated (Hong 1991; Liu 1996; Sun and Wu 1987).

The ancient setting was complicated by the blockage of some of the river channels due to rapid sedimentation rate with poor drainage, whilst land had started to emerge in other places such as the remnant hills or high ridges from the Late Pleistocene reddish clayey landscape. Together these geomorphological processes and hydrological changes created a typical wetland landscape scattered with a mosaic of smaller-sized lakes. Sites of the Majiabang culture (ca.7000–6000 years BP) started to appear on the raised grounds around the Taihu Lake region.

Different from these geomorphological processes on the central part of the Yangtze Delta and the Taihu Lake region, areas along Hangzhou Bay and the Ningshao Plain region to its south were dominated by marine inundation due to the relatively low-lying terrain. From a series of sediment cores located to the south of the Liangzhu period hydraulic system (dated about 5000 years BP), a deep layer beneath and pre-dating the Liangzhu layer was found to consist of dark greyish clayey sediments with soft structure, typical of marine sediments in the region. Such deposits have been noticed in other places around the Yangtze Delta, clearly indicative of large-scale marine inundation.

While the marine transgression played a crucial role in regional geomorphological processes, the terrestrial sedimentation contributed significantly to the rapid aggradation of the region. The terrestrial input was ascertained by our geoarchaeological data (Zhuang et al. in preparation). At the Tangxi location, for instance, the early Holocene deposits were characterised by thick light yellowish silty sediments, embedded with numerous fine sediment beddings. Most likely, prolonged marine inundation and frequent terrestrial

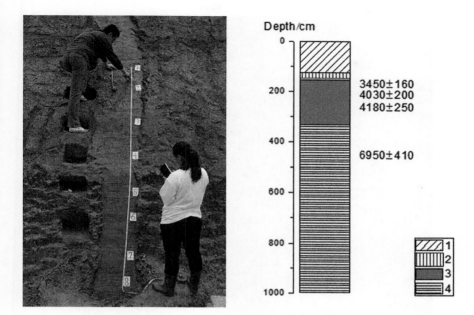

Figure 8.2 Stratigraphy at the Tangxi location (no. ZK3). 1: topsoil; 2: light greyish clayey silt; 3: yellow-brownish clayey silt; 4: light yellowish clayey silt. Dates are OSL dates before present, modified after Jin (2018).

floods both contributed to the Holocene land-forming process. Dates by optically stimulated luminescence (OSL) suggest that the thick sediments (ca. 7 m in thickness) were deposited between 8000 and 4000 years BP, with a particularly rapid sedimentation rate around 8000–6000 years BP (Jin 2018) (Figure 8.2). Sediments of similar characteristics (e.g., with the presence of fine sediment laminae) have been found at the Yujiashan and Dazemiao sites, where they were overlain by cultural deposits of the Liangzhu period around 5300–4300 years BP (Figure 8.3). The Holocene freshwater-dominant hydrological regime experienced a pronounced shift to brackish water by the late Holocene when the sea level started to rise again (Jin et al. 2019; Weisskopf et al. 2015).

In addition to the land-forming processes and changing hydrological regimes during the Holocene, distinctive soil types of early-to-middle Holocene sediments were present in at least two major areas.

1.   In much of the areas on the western part of the Yangtze Delta and on the Hangjiahu Plain, the hard clayey or reworked loess-like deposits are often found directly overlying late Pleistocene loess or red clayey sediments. Such a stratigraphic relationship can be commonly found

Figure 8.3 Light yellowish clayey silty layer at the Dazemiao site. This layer is directly overlain by Songze and Liangzhu culture remains. Note that metal tubes are where OSL samples were collected.

at many Majiabang culture period sites. At the Chuodun site, for instance, the Majiabang culture period rice paddy fields were built directly over loess or reworked loess sediments (Zhuang 2018, 94), and the same overlaying context was observed at the newly excavated paddy field site of Caoxieshan (Rao Zongyue, personal communication).

Figure 8.4 The organic-rich peat-like early to middle Holocene deposits at the Tianluoshan, on top of which rice fields were built. Photography courtesy of Professor Guoping Sun.

2. The early to middle Holocene sediments in the particularly low-lying areas of the Ningshao Plain are dominated by clayey marine sediments or organic-rich alluvium, as well as peat-like sediments (Figure 8.4). On slightly higher grounds, though, the contemporary sediments were characterised by their coarser texture (silt-sized) and fine sediment laminae. These different soil conditions were potentially of great significance to the development of prehistoric rice farming, which we have discussed elsewhere (Zhuang 2018).

Evolution of regional occupation, 10,000–4300 years BP

Shangshan culture period (ca. 10,000–8500 years BP)

The chronological sequence of prehistoric occupation on and surrounding the Yangtze Delta region were summarised in Table 8.1. The earliest Neolithic culture – Shangshan culture – was located in the Jing-Qu Basin and Puyang River catchment. After their detailed analysis of geomorphological and environmental contexts of the 18 Shangshan culture sites discovered to date, Xu et al. (2020) found that most of them were situated on hillslopes with gentle gradients (3°) with average altitudes around 40–50 m above sea level. They were close to tributaries (0.2–1 km), yet they were farther from main channel streams (3–5 km). Their average distance to mountains was 2–8 km. Such environmental locations would have been optimal for access to water and other natural resources whilst avoiding floods and landslides from the mountains. The Shangshan occupation as a whole did not show any strong relationship with sea level change.

First chronological hiatus (8500–8000 years BP)

Although the new radiocarbon dates from the Jingtoushan site have helped to push back the *terminus post quem* of this newly discovered site to around 8300 years BP (see below) and the *terminus ante quem* of the Shangshan culture has undergone some revision (Jiang 2013), a chronological hiatus of a few centuries still exists between the earliest Neolithic culture of Shangshan and the Kuahuqiao culture in the region. Given the geographic overlap between the two cultures, they may have been related to each other, for instance, when noting that the Kuahuqiao culture layer had superimposed the Shangshan culture layer at some sites (Jiang 2013). However, a future investigation would need to ascertain whether or not the Kuahuqiao culture developed directly from the Shangshan culture or perhaps as a later succession with some influence.

Kuahuqiao culture period (8000–7500 Years BP)

The Kuahuqiao culture period at 8000–7500 years BP marked a significant transition in the regional occupation sequence. The Kuahuqiao site was located on coastal wetlands on Hangzhou Bay to the south of the Yangtze Delta. It represented the earliest human adventure to this entirely new territory. People started trial rice farming on this saltmarsh wetland environment with saline water, while they exploited a wide range of resources here. The material culture created by the Kuahuqiao people was of an unprecedented level, including well preserved wooden canoe, the world's earliest lacquerware and many other elaborate artefacts (ZPICRA and Xiaoshan Museum 2004).

The recent excavation of the Jingtoushan site (8300–7800 years BP) reveals another example of an early venture into the littoral environment on the Ningshao Plain at the time of rising sea level. Being covered by thick middle to late Holocene deposits (up to 6–10 m in thickness), the shell middens at Jingtoushan represented the earliest effort to systematically explore marine resources which sustained a rich material culture. The latter include different kinds of wooden artefacts and a developed industry of pottery production. The shell middens at Jingtoushan were, however, probably occupied only seasonally. More permanent residence might have been established on the low hillslopes nearby, which is yet to be excavated.

This pioneering venture into the sea was abruptly interrupted by another episode of marine transgression as directly evidenced by the presence of abundant plankton and brackish diatoms in the upper cultural deposits of Jingtoushan, in contrast to their absence in the lower cultural layers (Wang 2008). It was clear that the rising sea level radically altered the local hydrological regime and other ecological parameters, which eventually forced the abandonment of the Kuahuqiao and Jingtoushan sites.

Second chronological hiatus (7500–7000 years BP)

The chronological frameworks of the Kuahuqiao, Majiabang, and Hemudu cultures are under revision, yet generally, they point to a time gap or hiatus in occupation at approximately 7500–7000 years BP. For instance, the *terminus ante quem* of the Jingtoushan site might be pushed to around 7600 years BP (He 2018: 152–154). However, it is clear that a chronological vacuum of some centuries exists between the Kuahuqiao culture and the succeeding Majiabang and Hemudu cultures. Although a bias in archaeological investigations might be partly responsible for this "gap", most likely, the changing Holocene sea level and environment played a crucial role in causing this cultural hiatus when settlement occupation and food production were abruptly stopped. This hiatus is directly manifested by the thick marine sediments that covered the Kuahuqiao culture remains at both the Kuahuqiao and Xiasun sites (Jiang 2014; ZPICRA and Xiaoshan Museum 2014), but the process of this marine transgression remains to demonstrated (cf. He 2018: 152).

Majiabang and Hemudu culture period (7000–6000 years BP)

After the second chronological hiatus, the Majiabang culture emerged around the Taihu Lake area, while the Hemudu culture emerged on the Ningshao Plain. The region began to witness parallel developments of two independent Neolithic cultural contexts during the next several millennia. The geographic locations and local environments of the Majiabang and Hemudu cultures differed evidently from each other in the southern part of the Yangtze Delta.

Distribution of the Majiabang culture sites demonstrated an eastward expansion. Most sites were situated in the vast areas along the Middle to Lower Puyangjiang River, between the Tianmu Mountains and the south side of Hangzhou Bay. The density of settlement distribution of the Majiabang culture sites in this region is notably lower than that of the Hemudu culture sites.

In addition to the Hemudu site itself, numerous other Hemudu culture sites were located in alluvial settings of the Ningshao Plain (Wu et al. 2012). Noticeably, Hemudu culture sites tend to be concentrated as clusters (Wang and Liu 2005).

Both the Majiabang and Hemudu groups were successful opportunity seekers who began to settle in a wide range of ecological habitats when the regional hydrological and geomorphological conditions continued to be affected by unstable marine processes. Most likely, these developments were enabled by advanced food production and technological innovation.

Both cultures are considered pioneers in intensifying rice production. At the early Majiabang culture site of Luojiajiao (about 7300–6900 years BP), abundant artefacts, animal bones, and rice remains (several hundreds of carbonised rice grains) were recovered, the latter including a large number of cultivated varieties (Zheng et al. 2007; Zhu 2004). Similarly, the Hemudu site is well known for containing abundant rice remains, although the economic and palaeo-ecological implications of these rice remains to the early Hemudu culture subsistence strategies are debatable.

The two cultures started to invest substantial labour force and resources in transforming the wetland environments. These transformations were reflected in the increasing scale of the sites, as well as in the construction of wooden architecture used for accommodation, storage, or defence purposes (ZPICRA Zhejiang Province Institute of Cultural Relics and Archaeology, 2003; ZPICRA Zhejiang Province Institute of Cultural Relics and Archaeology and CCAPKU Center for Chinese Archaeology of Peking University, 2011). The Luojiajiao site of the Majiabang culture measured 12 hectares, while the Hemudu site measured 4 hectares. By the middle to late Majiabang and contemporary middle Hemudu culture periods, regional cultural development began to accelerate, marked by the spread of settlements across the southern Yangtze Delta region. The number of late Majiabang culture sites increased from around 10 to more than 30, whilst the number of Hemudu culture sites increased from around 10 to at least 40 and distributed across a wider area, with a notable southward expansion to the sea (Wang and Liu 2005).

Accompanying these changes were the evident development of rice farming and other structural transformations of the settlements. Perhaps most revealing of these developments, rice farming underwent a decisive expansion during the late Majiabang and middle Hemudu culture periods, with the construction of increasingly larger scales rice fields (Zhuang 2018). Moreover, those larger rice fields coincided with markedly high proportions of domesticated rice spikelet bases in the analysed remains (Fuller et al. 2009).

Songze and Late Hemudu culture period (6000–5300 Years BP)

Building on the aforementioned developments, the archaeological assemblages at 6000–5300 years BP involved new characteristics that in total defined different cultural contexts or associations, known as the Songze culture sites around the Taihu Lake and the Late Hemudu culture sites on the Ningshao Plain. The number of settlements continued to grow, as evidenced by the increased number of the Songze culture sites to about 90 (Zhaobing Zhong, personal communication). Moreover, the Songze sites included large settlement centres.

The Songze culture sites displayed an interesting distribution pattern, with a concentration of sites in the east compared to a more scattered distribution of sites in the west on the Yangtze Delta and surrounding region. Although the geographic area of the Songze culture sites overlapped substantially with that of the previous Majiabang culture sites, a development towards the sea was observable (see Figure 8.1). In this case, access to the sea formed an important part of ancient life here.

The Late Hemudu archaeological context revealed a wave of un-precedented expansion or influence. According to the findings of Late Hemudu artefacts in various places, the cultural influence had reached as far east as the Zhoushan archipelagos and as far south as Fujian province (e.g., the Keqiutou site) (see Figure 8.1). Additionally, long-distance interaction occurred across Changdao Island off the eastern coast (Wang and Liu 2005). Such a significant development benefited from a stable sea level and marine condition.

During the later centuries of the Songze culture period, social stratification started to take shape. Status differential could be noted in the richness of graves, especially noting the graves with large numbers of elaborate jade and other items (Nanjing Museum et al. 2016). Similar developments were hinted at in the Late Hemudu culture sites, but the effects were less dramatic than as seen in the later Songze sites.

Liangzhu culture period (5300–4300 years BP)

During the period of 5300–4300 years BP, large numbers of Liangzhu culture settlements occupied the Yangtze Delta and the surrounding region, defined primarily by the forms and styles of pottery and other artefacts that characterised the cultural layers at Liangzhu and at other sites of the same age. Overall, the material assemblages of the Liangzhu culture sites resembled a unification of the previously diverse cultural contexts of the region. Many of these sites likely had been occupied continuously from the later Songze period and into the Liangzhu period, although the pottery assemblages clearly were distinctive in the two different contexts, such as at the Xiaodouli site (ZPICRA Zhejiang Province Institute of Cultural Relics and Archaeology and Haining Museum 2015).

People established the Liangzhu-associated sites across an unprecedented geographic extent, including settlements in many places that previously were unoccupied, such as the western Taihu Lake region, the Yaojiang region, and the coastal areas (see Figure 8.1). Equally impressive, the number of sites increased to about 300 in the core area, and the total was about 600 for the wider Liangzhu region (Liu et al. 2020). This dramatic change corresponded with the emergence of Liangzhu City, as well as a set of regional centres and sub-centres, indicative of a complex hierarchy that could be interpreted as a "state" form of a political or economic system (Renfrew and Liu 2018).

Liangzhu City commanded an unparalleled position amongst its contemporaries in and around the entire Yangtze Delta. We have discussed elsewhere the scale of the city's construction and its economic, social, and political implications (Liu et al. 2017). Through nearly a thousand years of development, the people of Liangzhu City and the larger Liangzhu-associated culture region enacted a successful adaptation and transformation with the wetland environment, evident in all aspects of the society. This context continued until an abrupt end, apparently caused by external forces, such as the rise of Holocene sea level starting around 4500–4400 years BP, around the same time of influence by the Dawenkou cultural groups from the north.

Chronological summary

The overall chronology of Holocene settlement distribution involved a gradual expansion of people inhabiting newly available or stabilised lands whenever the local sea level had fluctuated and thereby created different configurations of landforms and habitats. In a large-scale view, the Holocene trend revealed a shift of settlements from an older pattern in mountainous valleys to a later pattern in low-lying coastal plains. In a closer examination, more details can be noticed in the evolution of the landscape, including in the aspects of the existence of inhabitable landforms and the responses by people. Some of the effects in the landforms were direct results of higher or lower sea levels, and other factors involved the formation of beach ridges, in combination with hillslope erosion and redeposition regimes.

The relationship between ancient settlement distribution and Holocene sea-level change was not necessarily a simple linear correlation. Despite the far-reaching impact brought by the rapidly rising sea level between 9000 and 7000 years BP, the archaeological sites of this period exhibited at least some degree of marine-oriented activities (see the discussion below). The emerging land from 7000 years BP onwards provided desirable habitats for settlements, yet marine foods continued as large portions of the subsistence food supplies, with implications for the larger cultural setting and range of activities. While coastal and marine resources clearly were important for people, some of the sites in low-lying coastal positions became vulnerable to the changing climate and environmental conditions (e.g., Zheng et al. 2018).

Discussion: innovation, adaptation, and resilience of prehistoric economies to Holocene sea-level change

From marine economies to rice-farming economies

The development of prehistoric economies in the region overall involved emphasis on coastal and marine resources, although rice farming gained more prominence during later periods and especially in the inland or elevated landforms. Even while rice farming became widely practiced, many coastal communities continued to focus on marine foods and on general marine-oriented activities. Most of this evidence has been preserved in the food remains of different sites, notably revealing the amounts of marine shells and animal bones compared with any rice remains.

Evidence about ancient diet has been found even among the sites as old as the Shangshan contexts, around 10,000–8500 years BP. In addition to the pottery artefacts, stone tools of those sites included groundstone objects, flakes, and other objects that may have been used for the processing of foods. Liu et al.'s (2019) starch grain analysis indicated that the Shangshan people may have started exploiting a wide range of plant foods such as acorns, Job's tears, water chestnuts, and other tuber species. Analysis of food residues on stone tools and artefacts suggested that they were used to harvest or work with rice, reed, bamboo, or even leather. Rice remains, in the form of husk impressions, have been found on pottery vessels, together with the discovery of carbonised rice grains. The morphometric characteristics of the rice remains lead some scholars to consider an initial stage of rice cultivation (Zheng and Jiang 2007).

The combined effect of changing climate and rising sea level from 9000 years BP onwards triggered profound changes in human adaptation, to a large extent involving adaption with the marine transgression during the early Holocene. One of the direct consequences of the marine transgression was that the formerly low-lying coastal lands were inundated or drowned beneath the sea, and therefore the new position of the coastline was brought closer to the mountainous valleys that previously had been in inland positions. The populations in those mountains and valleys, therefore, were brought into closer proximity with coastal and marine resources and lifestyles, and many groups started to occupy the newly reconfigured coastal zones (Jiang 2012).

An especially informative adaptation to the early Holocene coastal transformation was documented at the Kuahuqiao site, occupied about 8000–7500 years BP. The Kuahuqiao people relied mostly on hunting-gathering and the possibly limited amount of rice cultivation for food subsistence, while the presence of a small amount of marine animals such as dolphins and turtles indicated the role of coastal and marine foods. The Kuahuqiao people consumed at least some amount of rice, possibly of the early japonica variety (Zheng et al. 2004). The form and extent of rice cultivation may yet be studied in more detail, for example, when considering the local saline water condition in the coastal zone at that time (Zong et al. 2007).

In addition to the evidence of a growing subsistence food base, the Kuahuqiao sites have shown investment in specialised craft production, for instance, in earth-working tools, textile production, a wide range of elaborate objects, and the earliest known domesticated pig in South China (ZPICRA and Xiaoshan Museum 2004). At the early shell midden site of Jingtoushan, roughly contemporary with the Kuahuqiao context, a similarly complex material culture assemblage was supported by a significant contribution from marine food exploitation, as evidenced by the numerous shell remains and related tools. Additionally, rice cultivation was confirmed by the discovery of abundant rice spikelet bases, although further analysis would be necessary in order to distinguish domesticated versus non-domesticated varieties of rice.

During the Hemudu culture period of 7000–6000 years BP, the overall trend continued of relying on both marine resources and rice agriculture. More diverse fish bones and marine resources were discarded at these sites, including the remains of turtle, whale, shark, sea bream, tuna, and crab. Crushed shells were used as tempering material in pottery production (Wu et al., 2012). The contribution of rice agriculture to the economies increased steadily, which is best illustrated by the systematic archaeobotanical research at the Tianluoshan site. The percentage of wild gathered plant foods, for instance, gradually decreased through time at the site, while rice gained more importance in the assemblage (Fuller et al. 2009). Parallel with this development was the slowly increasing percentage of domesticated pigs in the faunal records, although people continued to hunt animals such as deer as sources of animal protein (Zheng, 2016). Recent isotopic studies of human remains from Tianluoshan, however, have questioned the importance of marine food in the prehistoric diet at the site (e.g., Minagawa et al. 2011). These questions may yet be addressed by quantifying the contributions of marine resources in local diets in comparison with the seasonality of acquiring those particular foods.

By the Liangzhu culture period of 5300–4300 years BP, a clear division had appeared between the rice agriculture zone and the marine-oriented economies of the region. Liangzhu City and most of the larger communities were founded by intensified rice farming (Qin and Fuller 2019), yet the coastal residents practised both agriculture and marine-oriented economies. Meanwhile, their offshore island neighbours specialised in marine resources (Jiao 2007).

Technologies and adaptations

The adaptations to fluctuating marine environments benefited from technological innovations, especially in relation to water transportation, architectural constructions, and water management. Some of the relevant evidence has been found in well preserved wooden artefacts at water-logged sites. Other evidence has been found in land-altering creations of dams, levees, and ditches.

Excavations at both Kuahuqiao and Jingtoushan have yielded objects that were used for water transportation, fated generally within the range of 8000–7500 years BP. At Kuahuqiao, a wooden canoe and several wooden oars were dated directly to around 8000 years BP (ZPICRA and Xiaoshan Museum 2004). At Jingtoushan, an exceptionally well preserved wooden paddle was unearthed.

Along with knowing about these material objects from Kuahuqiao and Jingtoushan, the slightly later Hemudu-associated sites of 7000–6000 years BP showed a rapid spread of communities along the coast, thus prompting new ways of thinking about the transportation routes between those sites (Wang and Liu, 2005). Many of these sites of 7000–6000 years BP contained tools for water transportation, including eight wooden oars from the Hemudu Site (ZPICRA Zhejiang Province Institute of Cultural Relics and Archaeology, 2003). These tools were made in diverse designs and shapes, and further analysis and comparisons may yet ascertain their specific usage.

In terms of architectural technologies, a major innovation was accomplished with post-raised houses in the low-lying wetlands along the coast, dated at least as early as the Kuahuqiao contexts of 8000–7500 years BP. Previously, sites of the Shangshan period of 10,000–8500 years BP were semi-subterranean or above-ground houses, supported by a series of posts that could be traced through patterns in post moulds, surrounded by storage pits and burials (Xu et al. 2020). Following this earlier tradition, the Kuahuqiao people, after 8000 years BP, began to build *ganlan* style stilt houses with raised floors, supported by posts or stilts that elevated the living floors above the ground. The excavations of *ganlan* houses have shown patterns of the post moulds, and they often were surrounded by bridges, roads paved with gravels, and other residential facilities (Jiang 2012). The scale and patterning of housing construction were evidenced by the alignments of post moulds, with occasional preservation of wooden planks from the formerly raised floors or related features.

These wooden constructions required advanced wood-working technologies and the ability to acquire large volumes of timber. By 7000–6000 years BP, the Hemudu people in particular already knew how to apply tenon-and-mortise wood-working technology for improving the mechanical stability of houses and other creations, sometimes at large scales. Additionally, at the Hemudu sites, people had used stones or wooden planks as bracing around the edges of posts or stilts before driving pointed-based posts into the

Figure 8.5 Wooden posts discovered from the Tianluoshan site. Photography
courtesy of Professor Guoping Sun.

foundational ground (Huang 2004; Wang et al. 2001; ZPICRA Zhejiang
Province Institute of Cultural Relics and Archaeology 2003) (Figure 8.5).
Outside the Humudu and other related sites, wooden posts were aligned in
multiple rows, forming arc-shaped patterns that may have functioned as
fences for demarcating different residential areas and activity zones at these
sites. Between those fences and the inundated coastal wetlands, rice fields
have been reported (Huang 2004).

Contemporary with the *ganlan* style houses of the Hemudu culture settlements, the Majiabang culture settlements of the same age at 7000–6000
years BP were dominated by ground-floor architectures (Liu and ZPICRA
Zhejiang Province Institute of Cultural Relics and Archaeology 2019). They
were made mostly in rectangular shapes. Their floors or living surfaces often
were carefully built.

Figure 8.6 A wooden feature built in the middle of the rice field at the Tianluoshan site, possibly used as a road in the field. Photography courtesy of Professor Guoping Sun.

By the time of the Liangzhu contexts of 5300–4300 years BP, the scale of construction activities increased dramatically, wherein Liangzhu City was known not only for its large-sized residential areas but also for its earthen walls, palatial buildings, ceremonial monuments, and hydraulic enterprise. At Liangzhu City alone, people had converted about 229 hectares of the wetland into residential areas, where artificially stacked mounds supported numerous houses, and facilities, and infrastructures (Liu et al. 2017) (Figure 8.6). Underpinning the adaptation to wetlands was the sophisticated engineering planning and labour organisation by the Liangzhu society (Liu et al. 2017), including systematic coordination among different working groups or units to acquire the raw materials and to enact the construction activities (Wang et al. 2020).

The Liangzhu hydraulic enterprise, after 5300 years BP, had inherited some of the pre-existing knowledge about water management in the region (Zhuang et al. 2014). Already by the Hemudu contexts of 7000–6000 years BP, people had dug wells inside the residential areas on high grounds, and they created wooden or bamboo structures to protect their walls. These wells were vital for people to obtain fresh water in the coastal region that otherwise was dominated by brackish water with volatile hydrological regimes (Xudun Archaeological Team of Jiangsu Province (XATJP) 1995; Zhang et al. 1987; Lu and Zhu 1984). The water management technologies developed in new directions with rice farming, wherein the Hemudu and Majiabang farmers of 7000–6000 years BP built irrigation and drainage systems, engaged in micro-management of field topographies, and worked in increasingly larger sizes of rice fields (Zhuang, 2018).

All these successful water management practices culminated with the construction of the Liangzhu hydraulic enterprise during 5300–4300 years BP. This enterprise consisted of 11 high and low dams and double-diked levees with a complicated structure. The total volume of water storage and possible functions of this hydraulic enterprise have been discussed elsewhere (Liu et al. 2017).

Concluding remarks

We have synthesised the palaeo-environmental, palaeo-ecological, and archaeological evidence about the long-term chronology of human adaptations with their changing living environments in and around the Yangtze Delta region, profoundly influenced by Holocene sea level fluctuations. The overall growth of populations was steady through time, with a few pronounced periods of expansion and intensification, as well as two noted periods of occupational hiatus. The major trends involved the expansion of communities into newly available lands and habitats, followed by intensification of land-use activities within those new areas. These trends can be traced back to the onset of sedentism and plant food production during the Shangshan contexts of 10,000–8500 years BP, and they continued through the impressive works of the Liangzhu sites of 5300–4300 years BP with the extensive transformation of wetland environments for intensified rice-based economies.

The overall trends in the regional chronology should not be mistaken as mirroring a simple linear progression, and in fact, several fluctuations and nuances occurred through time and across the region. Most notably, at least two major periods of occupational hiatus (at 8500–8000 year BP and again at 7500–7000 years BP) would argue against a continuous linear progression model. Several short-term interruptions occurred in other time periods, as well, often in isolated areas that were affected by episodes of marine transgression that resulted in the deposition of typical greyish clayey sediments.

In the places where people re-inhabited the same spots or farmed the same fields repeatedly, despite periods of interrupting marine deposits or floods, arguably these records could suggest the long-term resilience of communities in living through dramatic environmental change. Such a case has been documented at the Shi'ao Site on the Ningshao Plain, where three separate layers of rice fields (approximately 7000–6000 years BP, 6000–5300 years BP, and 5300–4300 years BP) were interrupted by layers of the typical greyish clayey sediments of marine transgression deposits. The precise timing of each interruption could yet be refined, and the lengths of time between each cultural layer potentially could have lasted for a few centuries.

Further studies could consider the long-term stability of rice farming and its relationship with the changing environmental and ecological setting of the region. In this respect, Qin and Fuller (2019) proposed that rice farming was a "pull factor" to attract people to invest their labour for the benefit of potentially high yields, in many cases leading to local demographic growth and social complexities. If ever those crops should fail to support the increased local populations that depended on this food base, however, then intense periods of stress and possible conflict would arise. Equally important is to examine intra-regional and inter-regional differences in the subsistence economies of the rice farmers, coastal dwellers, and other groups whose subsistence traditions involved diverse environmental and ecological variables.

References

Chen, Zhongyuan, Yongqiang Zong, Zhanghua Wang, Hui Wang, and Jing Chen, 2008. Migration patterns of Neolithic settlements on the abandoned yellow and Yangtze River Deltas of China. *Quaternary Research* 70(2): 301–314.

Fuller, Dorian Q, Ling Qin, Yunfei Zheng, Zhijun Zhao, Xugao Chen, Leo Aoi Hosoya, and Guoping Sun, 2009. The domestication process and domestication rate in rice: Spikelet bases from the Lower Yangtze. *Science* 323(5921): 1607–1610.

He, Keyang, 2018. Micro-Fossil Evidence of the Relationship between the Human-environment Interaction, Evolution of Rice Farming and the Climate and Sea-level Fluctuations in the Lower Yangtze River during the Mid-Holocene. [in Chinese.] Unpublished doctoral dissertation, University of Chinese Academy of Sciences, Beijing.

Hong, Xueqing, 1991. The formation and evolution processes of Taihu Lake. [In Chinese.] *Marine Geology and Quaternary Geology* 4: 87–99.

Huang, Weijin, 2004. A study of wood structures on Hemudu Site. [In Chinese.] *Prehistoric Study*: 264–276.

Jiang, Leping, 2012. Outline of prehistoric cultures in Qiantang River Basin. [In Chinese.] *Relics from South* 81(2): 86–97.

Jiang, Leping, 2013. A study of the early Neolithic and cultural sequence in the Qiantang River catchment. [In Chinese.] *Relics from South* (6): 44–53.

Jiang, Leping, 2014. 跨湖桥文化研究 *[Study of the Kuahuqiao Culture]*. Science Press, Beijing.

Jiao, Tianlong, 2007. *The Neolithic of Southeast China: Cultural Transformation and Regional Interaction on the Coast*. Cambria Press, New York.

Jin, Yuxiang, 2018. *Study on the Relationship between Holocene Environment and Cultural Evolution in the Hang-Jia-Hu Area*. [In Chinese with English Abstract]. Unpublished doctoral dissertation. Peking University, Beijing.

Jin, Yuxiang, Duowen Mo, Yiyin Li, Pin Ding, Yongqiang Zong, and Yijie Zhuang. 2019. Ecology and hydrology of early rice farming: Geoarchaeological and palaeo-ecological evidence from the Late Holocene Paddy Field Site at Maoshan, the Lower Yangtze. *Archaeological and Anthropological Sciences* 11(5): 1851–1863.

Li, Liu, Neil A. Duncan, Xingcan Chen, and Jianxin Cui, 2019. Exploitation of job's tears in Paleolithic and Neolithic China: Methodological problems and solutions. *Quaternary International* 529: 25–37.

Liu, Jinling, 1996. 11,000 years of the formation and evolution of Tiahu Lai. [In Chinese.] *Acta Palaeontologica Sinica* 2: 129–135.

Liu, Bin and ZPICRA (Zhejiang Province Institute of Cultural Relics and Archaeology), 2019. 良渚古城综合研究报告 *[Synthetic Research Report of Liangzhu City]*. Cultural Relics Press, Beijing.

Liu, Rui, Jungan Qin, and Xi Mei, 2014. Sedimentary environment changes of the Ningshao Plain since the later stage of the Late Pleistocene: Evidence from palynology and stable organic carbon isotopes. *Quaternary International* 333: 188–197.

Liu, Bin, Ling Qin, and Yijie Zhuang, 2020. Situating the Liangzhu Culture in Late Neolithic China: An introduction. In *Liangzhu Culture: Society, Belief and Art in Neolithic China*, edited by Bin Liu, Ling Qin, and Yijie Zhuang, pp. 1–17. Routledge, Oxford.

Liu, Bin, Ningyuan Wang, Minghui Chen, Xiaohong Wu, Duowen Mo, Jianguo Liu, Shijin Xu, and Yijie Zhuang, 2017. Earliest hydraulic enterprise in China, 5,100 years ago. *Proceedings of the National Academy of Sciences of the United States of America* 114(52): 13637–13642.

Liu, Yan, Qianli Sun, Daidu Fan, Bin Dai, Fuwei Ma, Lichen Xu, Jing Chen, and Zhongyuan Chen, 2018. Early to Middle Holocene sea level fluctuation, coastal progradation and the Neolithic Occupation in the Yaojiang Valley of Southern Hangzhou Bay, Eastern China. *Quaternary Science Reviews* 189: 91–104.

Lu, Yaohua and Ruiming Zhu, 1984. A wood bucket-shape well of Liangzhu Culture found in Xingang, Zhejiang. [In Chinese.] *Relics* 2: 94–95.

Minagawa, Masao, Sho Matsui, Shin'ichi Nakamura, and Guoping Sun, 2011. Analysis of C\N Isotope composition on human and faunal remains from Tianluoshan Site and inferred food resources and use of livestock in Hemudu Culture. In 田螺山遗址自然遗存综合研究 *[A Synthetic Study on Natural Remines of Tianluoshan Site]*, edited by ZPICRA (Zhejiang Province Institute of Cultural Relics and Archaeology) and CCAPKU (Center for Chinese Archaeology of Peking University), pp. 262–269. Cultural Relics Press, Beijing.

Nanjing Museum, Zhangjiagang Museum Zhangjiagang Cultural Heritage Administration, 2016. 东山村 *[Dongshancun: Excavation Report]*. Cultural Relics Press, Beijing.

Qin, Ling and Dorian Q. Fuller, 2019. Why rice farmers don't sail: Coastal subsistence traditions and maritime trends in Early China. In *Prehistoric Maritime Cultures and Seafaring in East Asia*, edited by Chunming Wu and Barry V. Rolett, pp. 159–191. Springer Singapore, Singapore.

Renfrew, Colin, and Bin Liu, 2018. The emergence of complex society in China: the case of Liangzhu. *Antiquity* 92(364): 975–990. https://doi.org/10.15184/aqy.201 8.60.15184/aqy.2018.60

Saito, Yoshiki, 1998. Sea levels of the Last Glacial in the East China Sea Continental Shelf. [In Japanese with English Abstract.] *The Quaternary Research (Daiyonki-Kenkyu)* 37: 235–242.

Song, Bing, Zhen Li, Yoshiki Saito, Junichi Okuno, Zhen Li, Anqing Lu, Di Hua, Jie Li, Yongxiang Li, and Rei Nakashima, 2013. Initiation of the Changjiang (Yangtze) Delta and its response to the Mid-Holocene sea level change. *Palaeogeography, Palaeoclimatology, Palaeoecology* 388: 81–97.

Sun, Shuncai and Yifan Wu, 1987. Formation, evolution and modern sedimentation of Taihu Lake. [In Chinese.] *Scientia Sinica Series B* 12: 1329–1339.

Wang, Hui, 2008. The rise and fall of Neolithic Kuahuqiao Culture Sites in Hangzhou Bay: A response to the Holocene Sea-level Fluctuations. [In Chinese.] Unpublished master's thesis, East China Normal University, Shanghai.

Wang, Haiming and Shuhua Liu, 2005. The dispersal and diffusion of Hemudu culture. [In Chinese.] *Relics from South* 3: 114–118.

Wang, Haiming, Baoquan Cai, and Liqiang Zhong, 2001. A brief excavation report of Zishan Site in Yuyao, Zhejiang. [In Chinese.] *Archaeology* 10, 14–25: 97–98.

Wang, Ningyuan, Chuanwan Dong, Hongen Xu, and Yijie Zhuang, 2020. Letting the stones speak: An interdisciplinary survey of stone collection and construction at Liangzhu City, prehistoric Lower Yangtze River. *Geoarchaeology: An International Journal* 35: 625–643.

Wang, Zhanghua, Chencheng Zhuang, Yoshiki Saito, Jie Chen, Qing Zhan, and Xiaodan Wang. 2012. Early Mid-Holocene sea-level change and coastal environmental response on the Southern Yangtze Delta Plain, China: Implications for the rise of Neolithic Culture. *Quaternary Science Reviews* 35: 51–62.

Weisskopf, Alison, Ling Qin, Jinglong Ding, Pin Ding, Guoping Sun, and Dorian Q Fuller. 2015. Phytoliths and Rice: From wet to dry and back again in the Neolithic Lower Yangtze. *Antiquity* 89(347): 1051–1063.

Wu, Li, Cheng Zhu, Chaogui Zheng, Feng Li, Chunmei Ma, Wei Sun, Suyuan Li, et al., 2012. Responses of prehistoric cultures to environmental changes in Zhejiang since the Holocene. [In Chinese.] *Acta Geographica Sinica* 67(07): 903–916.

Xudun Archaeological Team of Jiangsu Province (XATJP), 1995. The fifth excavation of Xudun Site in Changzhou. [In Chinese.] *Southeast Culture* 4: 69–94.

Xie, Zhiren and Wanglin Yuan, 2012. A brief discussion of the fluctuating sea-level in the Holocene and its environmental significance. [In Chinese.] *Quaternary Sciences (Chinese)* 32(06): 1065–1077.

Xu, Yiting, Zhou Lin, and Leping Jiang, 2020. The correlation between landscape characteristics in Qu River and the site selection of Shangshan culture. [In Chinese with English Abstract.] *Advances in Geosciences* 10(8): 778–791. https://doi.org/1 0.15184/aqy.2018.60.12677/ag.2020.108078

Zhang, Xiugui, 2009. The historical evolution processes of Taihu Lake. [In Chinese.] *Journal of Chinese Historical Geography* 24(01): 5–12.

Zhang, Minghua, Jian Song, and Xiuling You, 1987. Excavation at Songze Site in Qingpu County, Shanghai in 1987. [In Chinese.] *Archaeology* 03: 204–219, 289.

Zhao, Xitao, Xiushan Geng, and Jingwen Zhang, 1979. The sea-level changing in the past 20000 years in Eastern China. [In Chinese.] *Acta Oceanologica Sinica (Chinse)* 1: 269–281.

Zheng, Xiuwen, 2016. Investigation on the Livelihood of Residents Living in the Lower Yangtze Regions from 6000 BCE to 4000 BCE. [In Chinese with English Abstract.] Unpublished master's thesis, Shandong University, Shandong.

Zheng, Yunfei and Leping Jiang, 2007. Ancient rice remains excavated from Shangshan site and its significance. [In Chinese.] *Archaeology* 9: 2, 19–25, 99.

Zheng, Yunfei, Leping Jiang, and Jianming Zheng, 2004. A study of ancient rice remains from Kuahuqiao Site, Zhejiang. [In Chinese.] *Rice Science* 2: 33–38. https://doi.org/10.15184/aqy.2018.60.16819/j.1001-7216.2004.02.007

Zheng, Yunfei, Guoping Sun and Xugao Chen, 2007. Characteristics of rice spikelets discovered at a 7,000-year-old archaeological site. [In Chinese with English Abstract]. *Chinese Science Bulletin* 52: 1037–1041.

Zheng, Hongbo, Yousheng Zhou, Qing Yang, Zhujun Hu, Guangjiu Ling, Juzhong Zhang, Chunguang Gu, et al., 2018. Spatial and temporal distribution of neolithic sites in Coastal China: Sea level changes, geomorphic evolution and human adaption. *Science China Earth Sciences* 61(02): 123–133.

Zhu, Naicheng, 2004. Origins of rice farming in the Lake Taihu and Hangzhou Bay regions. [In Chinese]. *Southeast Culture* (2): 24–31.

Zhuang, Yijie, 2018. Paddy fields, water management and agricultural development in the Prehistoric Taihu lake region and the Ningshao Plain. In *Water Societies and Technologies from the Past and Present*, edited by Yijie Zhuang and Mark Altaweel, pp. 87–108. University College of London Press, Open Access, London.

Zhuang, Yijie, Pin Ding, and Charles French, 2014. Water management and agricultural intensification of rice farming at the late-Neolithic site of Maoshan, Lower Yangtze River, China. *The Holocene*, 24: 531–545.

Zhuang, Yijie, Chengshuangpin Zhao, Yuxiang Jin, Minghui Chen, Ningyuan Wang, Bin Liu, Ye Zhao, Yonglei Wang, Yuhang Lou, Chou Fan, Xiang Ji, and Duowen Mo, In preparation. Environmental foundation for the rise of the Liangzhu Civilization on the Hangjiahu Plain of the Yangtze Delta, China.

Zong, Yongqiang, 2004. Mid-Holocene sea-level highstand along the Southeast Coast of China. *Quaternary International* 117: 55–67.

Zong, Y., Z. Wang, J.B. Innes, and Z. Chen, 2012. Holocene environmental change and neolithic rice agriculture in the Lower Yangtze Region of China: A Review. *The Holocene* 22(06): 623–635. DOI: https://doi.org/10.15184/aqy.2018.60.1177/0959683611409775

Zong, Y., Z. Chen, J. B. Innes, C. Chen, Z. Wang, and H. Wang, 2007. Fire and flood management of coastal swamp enabled first rice paddy cultivation in East China. *Nature* 449(7161): 459–462.

ZPICRA (Zhejiang Province Institute of Cultural Relics and Archaeology), 2003. 河姆渡 *[Hemudu: Excavation Report]*. Cultural Relics Press, Beijing.

ZPICRA (Zhejiang Province Institute of Cultural Relics and Archaeology) and CCAPKU (Center for Chinese Archaeology of Peking University), 2011. 田螺山遗址自然遗存综合研究 *[A Synthetic Study on Natural Remains of Tianluoshan Site]*. Cultural Relics Press, Beijing.

ZPICRA (Zhejiang Province Institute of Cultural Relics and Archaeology) and Xiaoshan Museum, 2014. 跨湖桥 *[Kuahuqiao: Excavation Report]*. Cultural Relics Press, Beijing.

ZPICRA (Zhejiang Province Institute of Cultural Relics and Archaeology) and Haining Museum, 2015. 小兜里*[Xiaodouli: Excavation Report]*. Cultural Relics Press, Beijing.

9 Palaeolandscapes, radiocarbon chronologies, and the human settlement of southern lowland and Island Papua New Guinea

Ben Shaw

Lowland and island ecosystems played a pivotal role in the dispersal of modern humans through New Guinea (northern Sahul) during the Late Pleistocene, with population densities increasing significantly only during the latter half of the Holocene (Birdsell 1977; Bowdler 1977; Allen and O'Connell 2008; Kealy et al. 2016; Leavesley 2006; O'Connor and Chappell 2003; Spriggs 1997; Summerhayes 2007; Swadling 1997). The Massim islands and southern coastline of Papua New Guinea have been central to modelling human settlement of tropical lowlands (Figure 9.1), and although archaeological research over the last decade has greatly expanded our understanding of the antiquity, extent, and nature of human settlement, significant gaps remain (David 2008; Irwin 1991; Shaw 2019; Summerhayes and Allen 2007). Sequences in southern New Guinea notably lack the time depth of sites on the northern coast and in the Bismarck Archipelago, where evidence for human occupation dates from 44,000 to 40,000 cal. years BP (Groube et al. 1986; Leavesley et al. 2002; O'Connor et al. 2011), and perhaps from 61,000 years ago (Chappell 2002; Roberts 1997). In large part, this is due to sea level rise since the Last Glacial Maximum (LGM, 23,000–18,000 years ago) having inundated the relatively shallow continental shelf between southern New Guinea and Australia, effectively placing early coastal habitation sites out of reach of the archaeologist's trowel. However, human settlement is now known in the Massim islands from at least 17,000 cal. years BP and in the Gulf from 13,500 cal. years BP, reaffirming the potential for investigating the longer-term human history in this geomorphologically complex region, about which little is currently known (David et al. 2007; Shaw et al. 2020a).

This chapter aims to contribute to our understanding of how people adapted to changing lowland and island landscapes of southern island and coastal New Guinea since the Late Pleistocene by assessing the current information from archaeological and palaeo-ecological records. It investigates whether gaps in past cultural records can be explained primarily by post-glacial sea level rise, dynamic geomorphological processes, and the foci of archaeological surveys, or whether cultural decisions also played a

DOI: 10.4324/9781003139553-9

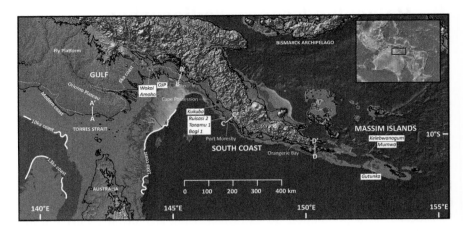

Figure 9.1 Elevation map of Papua New Guinea, showing the geographic areas as mentioned in the text. Coastlines are marked by a thick black line. The black line inland indicates the contour of 100 m above sea level. The map depicts 10x topographic exaggeration. Archaeological sites with Late Pleistocene, Early Holocene, and Middle Holocene dates are shown in boxes. Locations of elevation profiles (shown in Figure 9.4) are marked A, B, C, D. Coastlines are indicated at 6ka (+3 m), 10ka (−30 m), 14ka (−60 m), and 30-18ka (LGM, −120–125 m). Inset = map location in the Asia-Pacific region. Positions of former coastlines during the Late Pleistocene were based on the sea-level history curve and discussion by Lambeck et al. (2014), who investigated global trends in sea and ice volume since the Last Glacial Maximum using far-field studies on plant isotopic proxies, geological data for land movement, and coral records for reef development. Position of the higher mid-Holocene (6ka) coastline was based on work by Chappell (2005: figure 4), who investigated natural landscape evolution using radiocarbon-dated sediment cores. Map was created by the author, using GeoMapApp (www.geomapapp.org) and global multi-resolution topography data compiled by Ryan et al. (2009).

significant role. The south coast was a likely pathway into the mountainous highlands where settlement currently dates from 49,000–44,000 years ago (Summerhayes et al. 2010). It, therefore, looks at the potential for pre-LGM settlement when Australia and New Guinea were connected, and coastal lowlands were relatively stable. It also explores whether the current paucity of Late Pleistocene and Early Holocene settlement indicates very low population densities, or rather, low visibility of archaeological records. Finally, it considers whether increased visibility of human settlement after 5000 years ago can primarily be attributed to the post-glacial stabilisation of coastlines or if social influences were also a significant factor.

Palaeolandscapes and the human settlement of coastal and Island New Guinea

There is now broad consensus amongst researchers that the colonisation of Sahul was a protracted process involving deliberate maritime dispersals, with populations probably entering Sahul along the north coast of New Guinea via Sulawesi and into northern Australia (Arnhem Land) via Timor (Bird et al. 2019; Birdsell 1977; Bradshaw et al. 2019; Kealy et al. 2017). The timing of colonisation is still hotly debated (c.f. Williams et al. 2021), with both long (70,000–60,000 years ago) and short chronologies (51,000–47,000 years ago) proposed (Clarkson et al. 2017; O'Connell et al. 2018). Genetic studies are inconclusive about the timing of colonisation, as estimates vary and span the ranges of these chronologies (Malaspinas et al. 2016; Nagle et al. 2017; Pedro et al. 2020; Tobler et al. 2017). On current evidence, if human groups had penetrated the mountainous interior of northern Sahul by 49,000–44,000 cal. years BP (Vilakuav) and the semi-arid interior of southern Sahul by 53,0000–46,000 cal. years BP (Warratyi), then low-density habitation at least several millennia before 50,000 years ago is likely, and the colonisation process could feasibly have commenced by 60,000 years ago (Figure 9.2).

David et al. (2019b) propose a phased model of human dispersal, hypothesising that people initially inhabited well-watered coastal savannah plains before moving into more diverse landscapes. Bird et al. (2016) argue that colonist populations likely followed permanent water sources along the coast and into interior landscapes, while Bowdler (1977) had argued colonist populations largely kept to the coast. Although Bird and colleagues limit their meta-analysis to Australia, the large water catchments in the mountainous interior of New Guinea means there is a relatively high density of permanently flowing rivers along the south and north coasts. Based on these models, the expansive savannah plains of southern New Guinea would have been an attractive landscape for colonising populations and as a well-watered corridor through which people moved during the Late Pleistocene (see Figure 9.2).

Prior to and during the LGM of 23,000–18,000 years ago, the Torres Strait islands formed a land bridge between Australia and New Guinea. The south Papuan coast would have at this time extended seaward by 2–20 km, and as much as 180 km in the Gulf, making these lowland landscapes potentially more favourable than through much of the Holocene (Ishiwa et al. 2016; Yokoyama et al. 2000). During the 10,000 years following the LGM, global sea level rose by 120–130 m, and around 80,000 km^2 of habitable coastal lowlands in southern Papua New Guinea were submerged, with the landbridge severed by 8000 years ago (Lambeck et al. 2014). Consequently, evidence for Late Pleistocene human presence is likely now preserved only on uplifted coastal landscapes or inland from the former coastline. Sea level subsequently peaked 1.5–3 m above modern level by 7000 years ago

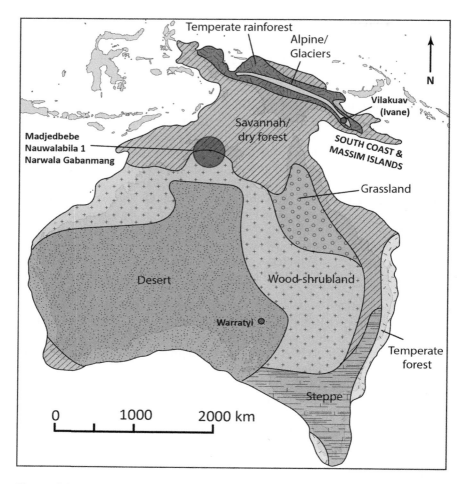

Figure 9.2 Reconstructed vegetation distribution across Sahul during the Last Glacial Maximum, overlaid on modern coastlines of New Guinea, Australia, and Tasmania. Vegetation distribution based on work by Hope et al. (2004: figure 1) who investigated vegetation and habitat change in the Austral-Asian region, using multiproxy records including pollen cores, with alpine/glacial zone added by author based on summarised core data by Hope (2014). Underlying map of the Asia-Pacific region was based on Oceania map from the Perry-Castañeda Library Map Collection, University of Texas.

(mid-Holocene) and remained near this level until ~4000 years BP, falling to +0.5 m from 3500 years BP and reaching near modern levels between 3000 and 2000 years BP (Chappell and Polach 1991; Dougherty et al. 2019; Lewis et al. 2013; Lewis et al. 2008; Sloss et al. 2018). However, progradation and erosion continue to significantly impact coastlines.

Northern New Guinea and the Bismarck Archipelago have steep bathymetries, which limited the horizontal shift of coastlines as sea levels changed. O'Connell and Allen (2012, 2015) argue that colonist populations favoured these steeper coastlines because littoral resources were able to adapt to fluctuations in sea level. Although susceptible to over-predation, shellfish beds could have supported small, highly mobile human groups while they moved along the coast. Such a pattern is supported by pre-LGM archaeological faunal assemblages from Latchitu, Buang Merabak, and Matenkupkum, which also indicate very low population densities (Gosden and Robertson 1991; Leavesley 2004; O'Connor et al. 2011).

The southern coast and the Massim islands would similarly have had steep bathymetries for most of Sahul's human history as the Late Pleistocene coastlines would have followed the edge of the continental shelf (see coastlines in Figure 9.1). The southern New Guinea savannah grasslands were once continuous with those in Arnhem Land, where the earliest habitation sites in Sahul are known at Madjedbebe, Nawarla Gabarnmang, and Nauwalabila 1 (Clarkson et al. 2017; David et al. 2019b; Roberts and Jones 1994) (see Figure 9.2). However, within a few millennia of postglacial sea-level rise commencing, islands in the Massim reduced in size by as much as 90% and vast areas of the shallow continental shelf were inundated with water levels rising by as much as 1 m within a single human lifetime (Shaw et al. 2020a). Social responses to landscape change during the postglacial marine transgression (18,000–8,000 years BP) may therefore have occurred at generational timescales rather than over centuries or millennia.

South Papuan coastal lowlands

The South Papuan coast spans some 1200 km from the eastern tip of New Guinea to where it intersects with the Torres Strait Islands in the west and spans the modern geopolitical provinces of Western, Gulf, Central, National Capital District, and Milne Bay (see Figure 9.1). The lowlands, defined broadly in this context as land less than 100 m above sea level, are bordered to the north by foothills, steep plateaus, and a precipitous mountain range that runs the length of New Guinea (Figure 9.3). Due to the proximity to the equator, there is no major difference in temperature throughout the year, with an annual average of 26–27°C. A monsoonal climate presently characterises these tropical ecosystems, with defined wet (November–April) and dry (May–October) seasons (McAlpine et al. 1983). Prevailing winds blow from the southeast for most of the year, with northwest winds and cyclones prevalent during the wet season. At least 45 languages are spoken by coastal populations along the southern Papuan lowlands. In the Gulf, only Papuan languages are known to belong to the Yam, Pahoturi, and Trans-New Guinea Families (Evans et al. 2018; Wurm 1971). Both Austronesian and

Figure 9.3 Landscapes of the south Papuan coast and the Massim islands. A) Low-lying wetlands near Mai Kussa River, Western Province. B) Yule Island and coastal lowlands near Cape Possession, the boundary between the Gulf and Central Provinces. C) Savannah grasslands overlooking coastal hills and mangrove salt flats in Caution Bay. D) Islands and coral reef in the Louisiade Archipelago, Massim region. Image A: photograph is courtesy of Nick Evans, used by permission. Images B–D: Photographs are by the author, Ben Shaw.

Papuan languages are spoken along the south coast, the former belonging to the Papuan Tip Family and the latter to the Trans-New Guinea Family (Dutton 1975, 2010; Ross 1988). Here, southern Papua New Guinea has been divided into two broad regions of the Papuan Gulf (Western and Gulf Provinces) and the southern coast (Central Province and National Capital District).

The Papuan Gulf

The Gulf lowlands, known broadly as the Fly Platform, consist of extensive alluvial plains, rises ≤55 m above sea level (such as the Oriomo Plateau), river deltas, and swamps mostly below 30 m above sea level that extend almost 400 km inland (Blake and Ollier 1970; Smith 1964) (see Figure 9.3A, Figure 9.4A, 4B). Collectively, the lowlands encompass 220,000 km^2 of southern New Guinea or around one-third of the island's total landmass

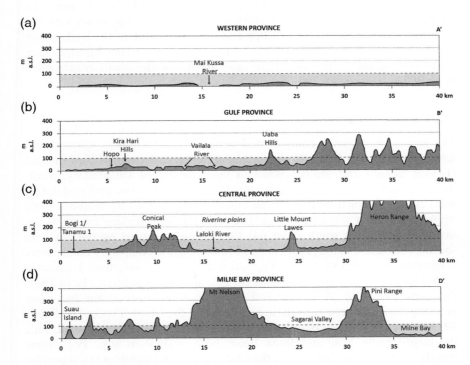

Figure 9.4 Elevation profiles of coastal landscapes in southern Papua New Guinea. The elevation and distance scales are consistent across the profiles. The coast is to the left. Major geographic features and key archaeological sites are labelled. Shaded area indicates coastal lowlands less than 100 m above sea level. Elevation profiles created by author, using path ruler function on Google Earth Pro 7.3.3.7786.

(Löffler 1977: 18). The Fly River – the largest in Oceania by discharge volume – along with two other major rivers (Kikori and Purari), flow across these plains and into the Gulf. Within these river outlets, several large delta islands (e.g. Kiwai, 359 km^2) have formed, probably since the Mid-Holocene, and are currently regressing at a staggering average rate of 42 m per year (Shearman 2010). Grassland, savannah, and tropical forest characterise this landscape which is consistent with the vegetation of northern Australia reflecting the once joint Sahul landmass (Paijmans 1976). Around 10 m of sediment had been deposited on the upper Fly from the mountainous interior within the last 35,000–29,000 cal. years BP[1] (27,100 ± 1100 uncalibrated years BP, sediment), 15 m across the mid-Fly (200 km inland) within the last 8000 years, and 38 m near the river mouths within the last 1600–1000 years (Blake and Ollier 1970, 1971; Chappell 2005; Shearman 2010). During the Mid-Holocene sea level high stand (7000–6000 years ago), land less than 3 m above modern sea level would have been inundated, with the coastline at this time as much as 100 km farther inland (Chappell 2005; Dougherty et al. 2019) (see Figure 9.1). Modern population density (as of 1980) across the region is generally very low (1–8 people/km²), due to the swampy and unstable nature of the land, except for pockets on or near the coast (17–115 people/km²) where banana, yam, sweet potato, and taro rather than sago are the predominant crops (Allen et al. 2002a; Hide et al. 2002b).

South Coast

The geomorphologically complex and heavily faulted southern lowlands extend from Cape Possession (5–30 km inland) to the eastern tip of New Guinea (1–5 km inland) and are characterised by a mosaic of sedimentary, metamorphic, and volcanic plains with alluvial swamps, plateaus, and undulating hills, some of which extend to the coast (Löffler 1977: 16; Mabutt 1965; McAlpine et al. 1983) (see Figure 9.4C and D). The modern extent of the savannah along the south coast has likely been expanded by burning and shifting cultivation in the past (Eden 1974). Drainage is complex, with all rivers originating in highland basins. Vegetation is primarily savannah grasslands with a mosaic of evergreen and drought tolerant Eucalypt forest (Eden 1974; Paijmans 1976: 66). A barrier reef runs discontinuously along the coast marking the former coastline during and prior to the LGM (Davies and Smith 1971; Löffler 1977: 119). Islands and islets dot the coastline, mostly small, with the largest being Yule and Bona Bona, both ~21km² in area. Population density is generally higher than in the Gulf, with greater densities (≥150 people/km²) near the Angabanga River mouth on the Central-Gulf Province border, where the soils are particularly fertile (Allen et al. 2002b). Population densities are lowest (3 people/km²) on the savannah plains near the modern capital, Port Moresby. These savannah plains are some of the driest landscapes in New Guinea, with an average annual rainfall of 1000–1500 mm, significantly lower than the southern Fly (2000–4000 mm) or the Gulf (4000–6000 mm) (Mabutt 1965; McAlpine et al. 1983). The south coast has a particularly marked dry season which can severely impact plant crop growth and traditional subsistence economies (McAlpine et al. 1983: 135; Vasey 1982).

Massim Islands

The Massim Island region encompasses the archipelagos off the eastern tip of New Guinea within the modern province of Milne Bay and is spread over 140,000 km² of ocean, ranging up to 1370 km² in area (Fergusson Island) and 2536 m in elevation (Goodenough Island) (see Figure 9.1). Most of the islands are volcanic, with those in the southern Massim forming a discontinuous extension of the Owen Stanley mountain range that runs the length of New Guinea (Davies and Ives 1965; Smith 1973a, 1973b). Many low-lying coral limestone islands have uplifted on the shallow continental plate (e.g., Trobriand group) and within large barrier reefs (e.g., Panaeati).

The largest reef system (~8000 km^2) encompasses many of the islands of the Louisiade Archipelago, is one of the most biodiverse marine systems in the world, and prior to and during the LGM formed a single large island (Allen et al. 2003; Shaw et al. 2020b). Because of the orographic effects of topography, low-lying islands (<200 m asl) typically receive 1500–2000 mm of rainfall annually and high islands as much as 3000–4000 mm.

The islands can be divided geographically and culturally into northern and southern groups, which are separated by a deep-sea basin. The small islands of Iwa (2 km^2) and Ware (3 km^2), in the regions north and south respectively, have the highest population densities of 296 and 166 people/km^2 (Hide et al. 2002a). Considerably lower population densities (3–7 people/km^2) are recorded on Cape Vogel of the Papuan mainland, where the land is either steep or swampy. A total of 28 languages are spoken across the Massim islands, all of which belong to the Papuan Tip Austronesian Family, except for one Papuan isolate (Yélî Dnye) spoken on Rossel Island, which has tenuous links to West New Britain, Bougainville, or perhaps Highland (Gorokan) languages (Dunn et al. 2008; Levinson 2006; Ross 1988, 2001).

A brief history of archaeological research

Few archaeological excavations were undertaken along the south coast or in the Massim during the first half of the 20th century. However, the handful of published excavations were influential, as they correlated abandoned habitation sites with past changes in settlement patterns and hypothesised social connections between ancestral populations (Austen 1939; Pöch 1907; Williams 1936b). Early archaeological models based on these excavations built upon cultural accounts of local inhabitants reported by colonial missionaries, traders, and administrators (Bevan 1890; Chalmers and Gill 1885; Lindt 1887; MacGillivray 1852; MacGregor 1897; Moresby 1876; Murray 1912) and built on a nascent framework of detailed scholarly research developed by anthropological fieldwork (Haddon 1894, 1900; Malinowski 1915, 1922; Seligman 1910).

Archaeological research commenced in earnest from the middle through the late 1960s, with survey and excavations undertaken along the south coast (Bulmer 1969, 1971; Lampert 1966, 1968; Pretty 1967; Specht 1974; White 1967) and on several Massim islands (Lauer 1971; White and Hamilton 1973: 278–283). Initial investigations started logically at ethnographically 'known' cultural sites connected to traditional groups through oral histories, with a plethora of excavations published through the 1970s and 1980s (Allen 1972a, 1972b, 1978; Bulmer 1978; Egloff 1971; Frankel and Vanderwal 1981; Guise 1985; Irwin 1977, 1991; Rhoads 1980; Sullivan and Sassoon 1987; Swadling 1980; Vanderwal 1973). Armed with developments in radiocarbon dating, these researchers collectively established the first regional cultural sequence that began around 4000 years ago, with an intensive coastal settlement from ~2000 years BP associated with a maritime dispersal of culturally related groups of people and the widespread introduction of pottery (see Summerhayes and Allen 2007). Research also focused on contextualising these cultural remains within a coherent theoretical framework centred around maritime trade, subsistence, and ecology (Allen 1977a, 1977b, 2000; Bulmer 1975, 1979, 1982, 1999; Irwin 1983; Irwin and Holdaway 1996; Macintyre and Allen 1990; Rhoads 1982; Swadling 1981). During the 1990s, the focus shifted to other parts of New Guinea, with the notable exception of Bickler (1998), who presented the first detailed results of archaeological investigations on a Massim island (Woodlark).

Renewed focus on the south Papuan coast and in the Massim began some 15 years ago, which has greatly expanded our understanding of the time depth and nature of human settlement. A terminal Late Pleistocene human presence was identified for the first time in the Gulf, demonstrating the potential for settlement of this antiquity to be preserved in what is generally a geomorphologically unstable environment (David et al. 2007). Excavations on the Massim island of Wari in 2008 was similarly a turning point in our understanding of human history as settlement and pottery here were associated with the Lapita Cultural Complex (Chynoweth et al. 2020; Negishi and Ono 2009). Prior to this discovery, the Lapita settlement had not been identified in Papua New Guinea south of the Bismarck Archipelago "homeland". In 2009–2010, the largest and most detailed excavation program ever undertaken in New Guinea was driven by a natural gas development in Caution Bay, just west of Port Moresby (Richards et al. 2016). Investigations here expanded the time depth for coastal settlement to around 6000 years ago (Rowe et al. 2020), and for the first time, confirmed a Lapita presence on the Papuan mainland (David et al. 2011; McNiven et al. 2011). Late Lapita pottery (~2750–2350 cal. years BP) has since been recovered from excavated contexts in the Gulf Province (Skelley 2014) and elsewhere in the Massim (Shaw et al. 2020b). Most recently, human settlement has been identified on the Massim island of Panaeati from 17,000 cal. years BP at the beginning of the post-glacial marine transgression, demonstrating the potential for locating coastal evidence of Late Pleistocene habitation on a now largely drowned landscape (Shaw et al. 2020a).

Palaeo-ecology

It is likely that the human history of the southern lowlands and islands has a far greater time depth than is currently known. However, comparatively little palaeo-ecological research has been undertaken in lowland New Guinea to frame the environmental context of human settlement and to focus attention on areas with relatively stable landforms (Haberle 2007; Hope and Haberle 2005). Beyond the continental-scale modelling of sea-level change and its impact on coastlines, understanding how local landscapes and vegetation changed is vital for contextualising long-term patterns of adaptive human behaviour (Figure 9.5).

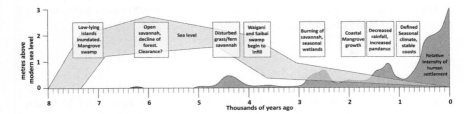

Figure 9.5 Middle to Late Holocene sea levels for the New Guinea region, relative to settlement intensity of the southern Papuan coast and Massim islands with annotated summaries of palaeo-ecological records. Regional sea level data followed Dougherty et al. (2019), Lewis et al. (2008), and Sloss et al. (2018), based on far-field records from the east coast of Australia using aerial photographs, Light Detecting and Ranging (LiDAR), ground penetrating radar (GPR), augers and cores, and radiocarbon dating (n = >120) of coastal beach, chenier ridges, mangrove swamp, shellfish, wave-cut erosional features, and coral deposits. Settlement intensity was based on summarised frequency of dated sites discussed in text and calibrated dates in Tables 9.1, 9.2, and 9.3. Palaeo-ecological record summaries are discussed in the text, except reference to forest clearance at 6 ka was based on the Lake Wamin pollen core record (Garrett-Jones 1979). Each pa-laeo-environmental record was based on sediment pollen analysis. Graphic output was prepared by the author.

South coast

Swampland

Most palaeo-ecological records are concentrated in the Port Moresby region. Coring at Waigani Swamp (~15 m above sea level) inland of Port Moresby provided the first glimpse of lowland savannah ecosystem change over the last 4000 years in southern Papua, particularly in relation to changes in seasonality and rainfall (Osborne et al. 1993). Mangrove forest was present around the swamp from at least 4150–3640 cal, years BP (3580 ± 80 uncalibrated years BP, peat) through to 2770–2360 "cal." years BP (2540 ± 80 uncalibrated years BP, peat), suggesting similar or slightly drier conditions than the present. There was a subsequent transition to seasonally fluctuating (wet versus dry) herbaceous wetland of grass and sedge, with an open water environment after 1830–1410 cal. years BP (1740 ± 90 uncalibrated years BP, sediment) to around 1200 years BP during which time there was an apparent increase in rainfall and flooding of the surrounding riverine plains (see Figure 9.5). Decreased rainfall is suggested after 1060–680 cal. years BP (950 ± 90 uncalibrated years BP, sediment) with a return to a seasonal wetland environment. Osborne et al. (1993) argue that increased rainfall from ~2500 through 1200 years BP would have benefited food crop production in the wider region. Indeed, this period correlates with the large-scale expansion of populations along the south coast and apparent disruptions to settlement patterns from 1200 through 800 years BP is widely referenced as the "ceramic hiccup" (Summerhayes and Allen 2007).

Savannah

A pollen spectrum from the excavated section of the Ruisasi 2 archae-ological site (15–20 m above sea level) in Caution Bay near Port Moresby correlates well with the Waigani core (Rowe et al. 2020). Increased an-thropogenic firing of a grass-fern dominated hinterland savannah occurred from 3000 through 2500 years BP and was associated with the increased prevalence of *Pandanus*, an important economic taxon (see Figure 9.5). Based on coastal topography, Ruisasi 2 likely would have been on the edge of a near-coastal peninsula during the Mid-Holocene when the sea level was higher than today (Figure 9.6). Limited vegetation change between 5800 and 3000 years BP suggests a relatively stable savannah ecosystem with only localised changes. However, on the western margin of Caution Bay in Vaihua Inlet/Galley Reach, relict mudflats 3–5 m above modern sea level and 2–3 km inland indicate lower-lying coastal ecosystems had dramatically transformed during the Mid-Holocene, and perhaps also during the last inter-glacial when sea levels were similarly higher (Pain and Swadling 1980) (see Figure 9.6). In some areas of the inlet, progradation had extended the Late Holocene coastline by up to 4.8 km, with subsequent infilling almost certainly having had an influence on the viability settlement locations in the past (Bulmer 1978; Specht 2012).

Coastal sand dunes

Core records from the mudflats fronting the coastal dune systems in Caution Bay demonstrate the dynamic modification of Late Holocene littoral ecosys-tems (Rowe et al. 2013). Mangrove swamp forest began to proliferate by 2000–1840 cal. years BP (2005 ± 25 uncalibrated years BP, organics), indicating that the inter-tidal coastal fringe had stabilised enough to support vegetation growth as sea level reached near modern level (see Figures 9.5 and 9.6). Prior to this time, discarded shellfish remains in archaeological sites on this dune system suggest much of the coastline likely was exposed with only small pockets of mangrove swamp growth (Faulkner et al. 2020; Thangavelu 2015). A transition to woodland vegetation on the coastal salt flats after 1340–1170 cal. years BP (1363 ± 24 uncalibrated years BP, organics) indicates a drier or highly seasonal climatic regime, consistent with the findings by Osborne et al. (1993) and the timing of the "ceramic hiccup". A prominent shift by 970–820 cal. years BP (1021 ± 24 uncalibrated years BP, organics) to mangrove species adapted to higher levels of salinity suggests the coast had prograded significantly beyond the tidal zone, and within the last 300 years, the savannah was a fire-induced open landscape much like it is today. Rowe et al. (2013) argue that the

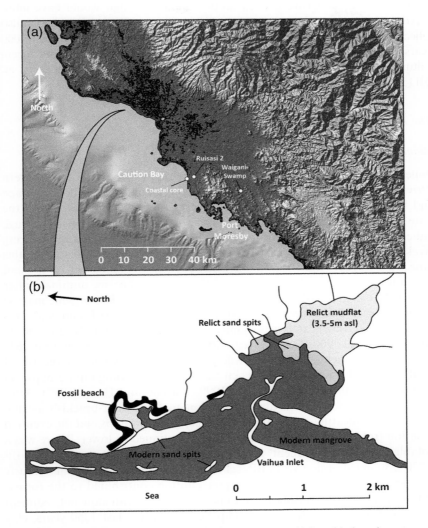

Figure 9.6 Coastal landforms of the Port Moresby area. A) Land below 6 m around
Galley Reach and Vaihua Inlet were susceptible to inundation during the
Mid-Holocene. The contour is based on the modern landform, and it
does not consider past sediment infilling. As such, the low-lying area may
have been more extensive in the past. The palaeo-ecological sample locations are shown. B) Vaihua Inlet, near Caution Bay, showing relict
mudflats and fossil beach at 3.5–5 m above modern sea level, deposited
during the mid-Holocene and perhaps during the last interglacial when
the coast was exposed. Map A was created using GeoMapApp
(www.geomapapp.org) and global multi-resolution topography data.
Map B was based on a base map by Pain and Swadling (1980), derived
from physical survey of the inlet to identify coastal features and archaeological remains.

proliferation of mangrove growth around 2000 years ago would have influenced decisions of settlement location within Caution Bay and probably also reflects sea levels reaching modern levels. Certainly, within the last 1500–1000 years, there was very little sand accumulation on the aeolian dunes where cultural activity since the Mid-Holocene has been recorded (McNiven et al. 2011; Szabo et al. 2020).

The Papuan Gulf

Islands of the former continent

Pollen swamp cores on Saibai Island 3 km off the coast of the Western Province demonstrate the transition from a continental to island landscape and the effects of human-induced transformations of wetlands during the Late Holocene (Barham 1999). Basal clays of assumed Late Pleistocene age indicate that prior to and during the LGM, the island formed a meandering creek-drainage system as part of a more expansive Fly Platform. At this time, vine thicket (e.g. *Argusia argentea*) covered the landscape, consistent with a low-lying coastal plain. Inter-tidal mangroves became dominant after 7560–6790 cal. years BP (6290 ± 170 uncalibrated years BP, organic matrix) when sea levels had reached their Mid-Holocene peak (see Figure 9.5). It is probable that the island was tidally inundated and entirely covered in mangroves until 3900–3270 cal. years BP (3340 ± 120 uncalibrated years BP, humics). Permanent occupation was probably not possible on the island until after 3000–2500 years ago when sea level was receding but perhaps still slightly higher than today (Barham 1999; Barham and Harris 1985). After this time, however, there was a transition to freshwater-brackish swamp sedge taxa, indicating infilling of coastal mangrove swamps and the eventual isolation from marine systems as sea levels continued to lower. There was a notable hiatus in sediment accumulation until 1310–970 cal. years BP (1270 ± 80 uncalibrated years BP, cellulose), and it has been parsimoniously suggested that higher than modern precipitation contributed to the formation of a wetland by raising the groundwater table, with slow net sediment accumulation once infilling was complete. Within the last 1250 years, clay from terrestrial erosion associated with the construction of mound and ditch agricultural systems adjacent to the wetlands was rapidly deposited, indicating an intensified human presence on the island.

Inland riverine

A sediment pollen sample from Epe Amoho, an excavated limestone rock shelter 46 km up the Kikori River in the Gulf Province, further demonstrates a discontinuous Late Holocene sequence of riverine development associated with human presence. During the earliest two phases of sediment deposition (2850 through 2500 cal. years BP and then 1200 through 800 cal. years BP), saline-tolerant mangrove species were present, indicating seawater infiltrated a significant distance up this low-grade river to create a brackish confluence. Pioneer woody species and relatively low taxa diversity in the latter suggests a greater disturbance of the surrounding vegetation, argued to reflect increased human activity in the area. A lack of sediment accumulation between 2500 and 1200 years BP is consistent with the Saibai record and parsimoniously suggests sedimentation processes were, at least partly, affected by regional environmental changes. Within the last 500 years, a transition to freshwater mangrove taxa demonstrates a downriver shift in the brackish confluence, which was likely a result of extensive progradation at the river mouth driven by a slight decrease in sea level. Coastal progradation would have significantly impacted the viability of village settlements on beach ridges in the delta regions, as has been demonstrated by the shifting locations of the ancestral Popo, Otoia, Keveoki, and Meiharo villages within the last 500 years (Barker et al. 2012; Barker et al. 2015; David et al. 2009; Skelly and David 2017; Urwin et al. 2018, 2021). Shifts in upriver ecologies also would have required people to adapt their subsistence strategies as fish populations and vegetation on the flood plains changed.

Submarine cores

Further insight into the landform and vegetation histories of the former coastline in the Gulf has been provided by submarine coring of the once exposed sediments on the continental shelf (Febo et al. 2008). Core sequences spanned the last 38,000 cal. years BP with rates of sediment accumulation peaking between 18,000 and 16,000 cal. years BP, shortly after the LGM when rivers were draining directly off the shelf. The basal clays on Saibai Island likely accumulated at this time. Several layers of volcanic ash were deposited between 28,000 and 18,000 years ago, derived from volcanoes in the surrounding mountains, but it is unclear if these eruptions affected the short-term habitability of the Gulf region. Benthic foraminifera suggest higher seasonality and more defined wet-dry climatic conditions in the post-glacial period compared to earlier millennia. Carbon isotope assays of terrestrial vascular plant remains preserved in the sediments demonstrate a surprising predominance of C3 plants such as mangrove flora and a lack of C4 grasses, which challenges hypotheses that the gulf was primarily savannah grasslands during the LGM. Although some of the river detritus would have derived from highland catchments, a complete absence of C4 grasses is striking and requires coring in lowland locations away from former flood plains to confirm the LGM vegetation structure.

Massim Islands

Unfortunately, no long-term vegetation core records yet have been developed for the Massim. However, a preliminary pollen record from Kelebwanagum, a

limestone cave excavated on Panaeati Island, has provided a discontinuous sequence for canopy forest composition associated with human presence since the
Late Pleistocene. From 17,000 through 12,500 cal. years BP, lower montane
open canopy forest taxa characterised this near sea-level landscape reflecting the
cooler temperatures in the tropics during and shortly after the LGM. It had
been argued that near-surface air temperatures at tropical latitudes did not
significantly change as they did in the New Guinea highlands (Rehfeld et al.
2018). Yet, the presence of lower montane forest suggests it had cooled enough
to alter forest ecology. Prior to and during the LGM, Panaeati was 90% larger
and much closer to Sahul, with lower montane forest likely characteristic of
most of the southern New Guinea lowlands at this time. Lower montane
vegetation is still present at higher elevations on the Massim island of
Goodenough (Brass 1959; Johns et al. 2009). Since 4810–4450 cal. years BP
(4470 ± 30 uncalibrated years BP, shell), changes in pollen composition indicates
the presence of undisturbed canopy taxa characteristic of low-elevation Massim
islands. Since 1350–1290 cal. years BP (1420 ± 20 uncalibrated years BP), the
additional presence of successional scrub taxa represents an anthropogenically
modified landscape similar to today.

The palaeo-ecological records, although patchy in their geographic and
temporal coverage, have collectively indicated:

- Vegetated savannah and island landscapes similar to the modern-day
  had been established by the end of the mid-Holocene, with significant
  transformations to island and coastal landforms prior to this;
- Coastal lowland landscapes were subject to increasingly intensive
  anthropogenic clearance, slope erosion, and coastal progradation over
  the last 3000–2000 years, with significant changes to vegetation and
  landforms since 1300–1200 years BP;
- Inland lowlands were prone to changes in the hydrology of drainage
  basins; and
- Seasonality (wet versus dry seasons) fluctuated in severity which would
  have influenced all modes of subsistence.

Cultural radiocarbon chronologies

Site chronologies

To assess the patterning and relative density of human settlement on the
south Papuan coast and Massim islands in relation to past ecologies,
radiocarbon assays from all excavated and dated archaeological sites have
been collated, totalling 687 dates from 107 sites (Figure 9.7; Tables 9.1, 9.2,
and 9.3). Modern determinations (<1950 AD) are excluded from this total.
Of these, 75 (70%) are open sites, 19 (18%) are caves/rock shelters/niches, 11
(10%) are stone structures, and two sites are artefact find spots (2%) (see
Figure 7d). The stone structures all were excavated on Woodlark (Murua)

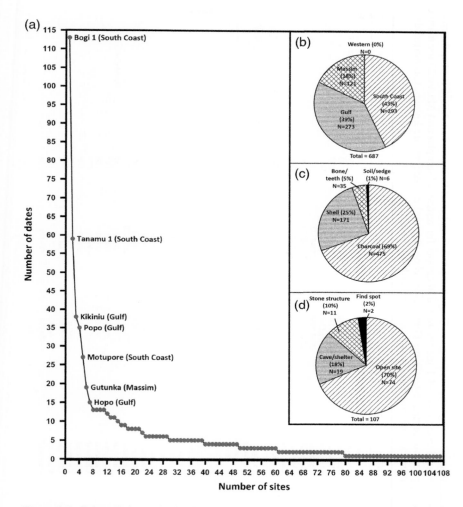

Figure 9.7 Cultural chronologies for the South Papuan coast and Massim islands. A) Number of radiocarbon dates per site. The seven sites with the most dates are labelled. B) Number of radiocarbon dates by region, as discussed in the text. C) Number of radiocarbon dates by sample type. D) Types of sites investigated. See Tables 9.1, 9.2, and 9.3 for the dating details.

Island in the Massim (Bickler 1998, 2006). Geographically, 48 sites (45%) are on offshore islands, and 59 (55%) are on the New Guinea mainland. A total of 77 sites (72%) were situated within 2 km of the coast when they were occupied, with the remaining 30 (28%) sites located farther inland, and of these, 24 (80%) have been recorded upriver in the Gulf (David 2008). There

Table 9.1 Radiocarbon dates from archaeological sites in the Gulf region. Grey shade = Conventional Radiocarbon Age (CRA) with errors of 100 years or more. Dates calibrated using OxCal4.4 and the IntCal20 dataset for terrestrial samples and Marine20 for marine shell samples. No ΔR values applied to marine dates in the Gulf region as no regional corrections are known. Calibrated date ranges rounded to the nearest 10 years, with the highest associated probability bolded.

Site	Lab code	Date	Error	Material	ΔR value	68.3% range	95.4% range	Median	Reference
Hopo	Wk-29539	1425	25	Charcoal		1350-1300	1360-1290	1330	Skelly (2014),
	Wk-31221	212	25	Charcoal		300-150	310-0	190	Skelly et al. (2014)
	Wk-30795	2539	25	Charcoal		2740-2540	2750-2490	2630	
	Wk-31824	1327	25	Charcoal		1300-1170	1300-1170	1260	
	Wk-31227	1561	25	Charcoal		1520-1390	1520-1380	1460	
	Wk-32235	2466	25	Charcoal		2700-2460	2710-2370	2580	
	Wk-32236	2541	25	Charcoal		2740-2540	2750-2500	2630	
	Wk-31228	2513	25	Charcoal		2730-2510	2730-2490	2580	
	Wk-29538	1433	25	Charcoal		1350-1300	1370-1290	1330	
	Wk-31230	1823	25	Sedge		1750-1630	1820-1620	1720	
	Wk-32237	1805	25	Sedge		1740-1630	1780-1610	1700	
	Wk-32238	1863	25	Sedge		1820-1730	1830-1710	1770	
	Wk-31232	2552	25	Charcoal		2750-2540	2750-2510	2710	
	Wk-29540	2522	25	Charcoal		2730-2510	2740-2490	2590	
	Wk-31233	2516	25	Charcoal		2730-2510	2730-2490	2590	
Ihi Kaeke	Wk-18940	84	34	Charcoal		260-30	270-20	110	David(2008)
	Wk-18941	114	39	Charcoal		260-30	280-0	120	
	Wk-18942	1145	38	Charcoal		1180-970	1180-950	1040	
	Wk-18943	1151	33	Charcoal		1180-970	1180-950	1050	
	Wk-19976	1286	44	Charcoal		1280-1170	1300-1070	1220	
Barauni	Wk-18944	97	39	Charcoal		260-30	270-10	110	David (2008)
	Wk-18946	123	33	Charcoal		270-20	280-0	120	
	Wk-18945	148	32	Charcoal		280-0	290-0	140	
	Wk-18947	169	35	Charcoal		290-0	300-0	170	
	Wk-18948	434	34	Charcoal		520-470	530-330	500	
	Wk-18949	1256	36	Charcoal		1270-1120	1290-1070	1210	
Kikiniu	Wk-18884	135	41	Charcoal		270-10	280-0	130	David (2008)
	Wk-18871	146	36	Charcoal		280-0	290-0	140	
	Wk-18896	161	37	Charcoal		290-0	290-0	160	
	Wk-18897	168	35	Charcoal		290-0	300-0	170	
	Wk-18872	190	30	Charcoal		290-0	310-0	180	
	Wk-18883	194	31	Charcoal		290-0	310-0	180	
	Wk-18869	206	32	Charcoal		300-0	310-0	190	
	Wk-18902	209	36	Charcoal		300-0	320-0	190	
	Wk-18881	264	33	Charcoal		430-150	450-150	310	

Site	Lab code	Age	±	Material	Cal range 1	Cal range 2	Median	Reference
	Wk-18900	302	37	Charcoal	440-300	470-290	390	
	Wk-18880	306	31	Charcoal	430-300	460-290	390	
	Wk-18876	401	35	Charcoal	510-330	520-310	470	
	Wk-18903	1145	33	Charcoal	1180-990	1180-950	1040	
	Wk-18868	1162	32	Charcoal	1180-990	1180-970	1070	
	Wk-18908	1187	34	Charcoal	1180-1060	1250-970	1110	
	Wk-20372	1210	35	Charcoal	1170-1070	1270-1000	1130	
	Wk-18870	1212	32	Charcoal	1170-1070	1270-1050	1130	
	Wk-18907	1259	34	Charcoal	1280-1130	1290-1070	1210	
	Wk-18873	1271	31	Charcoal	1280-1180	1290-1120	1220	
	Wk-20191	1307	43	Charcoal	1290-1170	1310-1120	1230	
	Wk-18879	1308	36	Charcoal	1290-1170	1300-1170	1230	
	Wk-18878	1319	31	Charcoal	1290-1170	1300-1170	1250	
	Wk-18901	1349	36	Charcoal	1310-1170	1350-1170	1280	
	Wk-18882	1351	31	Charcoal	1310-1190	1340-1170	1280	
	Wk-18904	1368	41	Charcoal	1350-1190	1350-1170	1290	
	Wk-18895	1373	37	Charcoal	1350-1270	1350-1170	1290	
	Wk-18885	1382	36	Charcoal	1350-1280	1360-1170	1300	
	Wk-18877	1457	32	Charcoal	1360-1300	1360-1300	1340	
	Wk-18899	1554	37	Charcoal	1520-1380	1530-1360	1440	
	Wk-18905	1554	37	Charcoal	1520-1380	1530-1360	1440	
	Wk-18909	1582	35	Charcoal	1520-1410	1540-1380	1460	
	Wk-18906	1597	38	Charcoal	1530-1410	1540-1380	1470	
Kulupuari (Kikiniu)	ANU-1965	280	70	Charcoal	460-150	510-0	340	Rhoads (1980)
	ANU-2173	1390	105	Charcoal	1390-1170	1520-1070	1300	
	ANU-2063	1590	340	Charcoal	1930-1130	2340-800	2345	
	ANU-2064	1210	210	Charcoal	1310-920	1530-720	1120	
	ANU-1962	1360	100	Charcoal	1370-1170	1520-1000	1260	
	ANU-1963	1480	80	Charcoal	1510-1290	1540-1270	1380	
Puriau	Wk-18934	139	34	Charcoal	270-10	280-0	130	David (2008)
	Wk-18929	149	43	Charcoal	280-0	290-0	140	
	Wk-18928	163	34	Charcoal	290-0	290-0	170	
	Wk-18935	186	34	Charcoal	290-0	310-0	180	
	Wk-18937	211	32	Charcoal	300-150	320-0	190	
	Wk-18936	235	34	Charcoal	320-150	430-0	280	
Epe Amoho	Wk-18933	155	34	Charcoal	280-0	290-0	150	David (2008), McNiven et al. (2010)
	Wk-18930	217	33	Charcoal	310-150	310-0	190	
	Wk-18925	223	38	Charcoal	310-150	430-0	200	
	Wk-18931	327	34	Charcoal	450-300	480-300	390	
	Wk-18922	1042	35	Charcoal	980-910	1060-820	950	
	Wk-19971	1065	44	Charcoal	1060-920	1180-830	970	
	Wk-19969	1103	31	Charcoal	1060-950	1070-930	1000	
	Wk-18923	1103	35	Charcoal	1060-950	1180-920	1010	
	Wk-19972	1119	44	Charcoal	1060-960	1180-930	1020	
	Wk-19974	2560	49	Charcoal	2760-2510	2770-2490	2650	

(Continued)

Table 9.1 (Continued)

Site	Lab code	Date	Error	Material	ΔR value	68.3% range	95.4% range	Median	Reference
Porokoiu	Wk-18924	2568	36	Charcoal		2760-2540	2760-2490	2720	
	Wk-19970	2572	33	Charcoal		2760-2620	2760-2510	2720	
	Wk-19973	2616	50	Charcoal		2790-2710	2850-2510	2740	
KG125	Wk-18938	164	35	Charcoal		290-0	290-0	170	David (2008)
	Wk-18939	223	35	Charcoal		310-150	430-0	200	David (2008)
	Wk-18912	177	37	Charcoal		290-0	300-0	180	David (2008)
KG124	Wk-18893	195	43	Charcoal		300-0	310-0	180	David (2008)
	Wk-18887	192	37	Charcoal		290-0	310-0	180	
Wokoi	Wk-18886	208	35	Charcoal		300-0	320-0	190	David (2008)
	Wk-18927	214	38	Charcoal		310-0	430-0	190	
Amoho	Wk-18926	900	34	Charcoal		910-730	920-720	800	David (2008)
	Wk-19968	7365	46	Charcoal		8290-8030	8330-8030	8170	
KG141	Wk-18911	225	33	Charcoal		310-150	430-0	200	David (2008)
	Wk-19982	468	42	Charcoal		530-490	560-330	510	
	Wk-18910	1239	38	Charcoal		1270-1070	1270-1060	1160	
	Wk-19981	1965	46	Charcoal		1980-1820	2000-1740	1890	
	Wk-18890	2754	43	Charcoal		2880-2770	2950-2760	2840	
Leipo	Wk-18874	236	31	Charcoal		320-150	430-0	280	David (2008)
	Wk-18875	442	34	Charcoal		530-480	540-330	500	
KG143	Wk-18920	238	33	Charcoal		320-150	430-0	280	David (2008)
	Wk-18891	263	38	Charcoal		430-150	460-0	310	
	Wk-18921	354	34	Charcoal		480-320	500-310	400	
OJP	Wk-18962	287	37	Human tooth	?	430-290	460-150	380	David (2008)
	Wk-20963	434	37	Human tooth		520-470	540-330	500	
	Wk-18984	708	43	Charcoal		690-560	730-550	660	
	Wk-19983	1068	44	Fish bone	?	680-560	710-540	630	
	Wk-18918	9526	66	Nut		11070-10690	11140-10580	10860	
	Wk-19986	9587	97	Charcoal		11110-10760	11200-10600	10930	
	Wk-19985	9781	98	Charcoal		11320-10880	11610-10770	11200	
	Wk-18919	11,602	70	Nut		13580-13360	13600-13310	13460	
KG140	Wk-19978	333	41	Charcoal		460-310	490-300	390	David (2008)
	Wk-18889	1298	35	Charcoal		1290-1170	1300-1150	1230	
	Wk-18888	1303	39	Charcoal		1290-1170	1300-1130	1230	
	Wk-19979	1307	44	Charcoal		1290-1170	1310-1120	1230	
	Wk-18892	1309	35	Charcoal		1290-1170	1300-1170	1230	
Rupo	Wk-19977	409	42	Charcoal		520-330	530-310	470	David (2008)
	Wk-18916	1330	37	Charcoal		1300-1170	1310-1170	1250	
	Wk-18917	1960	43	Charcoal		1940-1820	2000-1740	1890	
	ANU-2232	910	170	*Batissa violacea*	?	530-190	640-10	360	Rhoads (1980)
	ANU-2233	1160	170	*Batissa violacea*		750-420	930-250	580	
	ANU-2234	2160	180	*Batissa violacea*		1790-1360	2040-1180	1590	

Site	Lab code	Age BP	±	Material	Cal range 1	Cal range 2		Reference
Eremare	Wk-18915	444	47	Charcoal	530-460	550-320	500	David (2008)
	Wk-18913	549	38	Charcoal	630-520	650- 510	550	
	Wk-18914	660	34	Charcoal	670-560	680-550	620	
Keveoki 1	Wk-22221	421	30	Charcoal	520-470	530-330	490	David et al. (2009)
	Wk-22222	434	30	Charcoal	520-480	530-330	500	
	Wk-22743	581	33	Charcoal	630-540	650-530	600	Skelly (2014)
	Wk-22742	550	30	Charcoal	630-520	640-510	600	
	Wk-22741	376	32	Charcoal	500-330	510-310	440	
	Wk-22740	400	33	Charcoal	510-330	520-320	470	
	Wk-22744	306	33	Charcoal	440-300	470-290	390	
EMO/Samoa	Wk-22745	355	33	Charcoal	480-320	500-310	400	David et al. (2010)
	Wk-23052	706	33	Charcoal	680-570	690-560	660	
	Wk-23048	671	30	Charcoal	670-560	680-550	640	
	Wk-23049	662	30	Charcoal	670-560	680-550	620	
	Wk-23050	1574	33	Charcoal	1520-1400	1530-1380	1460	
	Wk-23053	1564	33	Charcoal	1520-1390	1530-1370	1460	
	Wk-23054	1644	43	Charcoal	1580-1410	1690-1400	1520	
	Wk-23055	1646	43	Charcoal	1590-1410	1690-1400	1530	
	Wk-23051	1647	30	Charcoal	1570-1410	1690-1410	1530	
	Wk-23056	1864	33	Charcoal	1830-1730	1870-1700	1770	
	Wk-23057	1860	30	Charcoal	1820-1720	1870-1700	1770	
	I-6153	1850	95	*Shell*	1880-1620	1990-1540	1760	Rhoads (1983)
	ANU-2061b	1220	180		?	Paired sample, ANU-2061a accepted		
	ANU-2061a	2430	370		2930-1990	3440-1610	2500	
Waredaru	Wk-37605	214	20	Charcoal	300-150	310-0	180	David et al. (2015b)
	Wk-37606	212	20	Charcoal	300-150	310-0	180	
	Wk-37607	183	20	Charcoal	290-0	290-0	190	
	Wk-25302	227	30	Charcoal	310-150	420-0	210	
	Wk-37608	169	20	Charcoal	280-0	290-0	180	
	Wk-25303	249	30	Charcoal	420-150	430-0	300	
	Wk-37611	258	21	Charcoal	320-280	430-150	300	
	Wk-37609	1331	20	Charcoal	1300-1190	1300-1170	1270	
	Wk-37610	1334	21	Charcoal	1300-1190	1300-1170	1280	
	Wk-37612	2851	20	Charcoal	3000-2880	3060-2870	2960	
Kumukumu 1	Wk-25291	572	30	Charcoal	630-540	650-520	600	David et al. (2015a)
	Wk-25292	583	30	Charcoal	630-540	650-530	600	
	Wk-25293	495	30	Charcoal	540-510	550-490	520	
	Wk-25295	538	30	Charcoal	630-520	630-510	540	
	Wk-25296	632	30	Charcoal	660-550	660-550	600	
	Wk-25298	531	30	Charcoal	560-510	630-510	540	
	Wk-25299	572	30	Charcoal	630-540	630-520	600	
	Wk-25300	578	30	Charcoal	630-540	650-520	600	
	Wk-25290	604	30	Charcoal	640-550	650- 540	600	David et al. (2012)
	Wk-28652	134	30	Charcoal	270-10	280-0	120	

(Continued)

Table 9.1 (Continued)

Site	Lab code	Date	Error	Material	ΔR value	68.3% range	95.4% range	Median	Reference
Poromoi	Wk-28653	104	30	Charcoal		260-30	270-10	110	
Tamu	Wk-28654	114	30	Charcoal		260-30	270-10	110	
	Wk-25289	562	30	Charcoal		630-530	640-520	590	
	Wk-28655	69	30	Charcoal		260-40	260-30	120	
	Wk-28657	239	30	Charcoal		320-150	430-0	290	Barker et al. (2012)
Otoia 1	Wk-23998	163	30	Charcoal		290-0	290-0	170	
	Wk-23058	101	33	Charcoal		260-30	270-10	110	
	Wk-23059	98	43	Charcoal		260-30	280-10	120	
	Wk-24000	159	30	Charcoal		290-0	290-0	160	
	Wk-23060	442	33	Charcoal		530-480	540-330	500	
	Wk-23999	214	30	Charcoal		310-150	310-0	190	
	Wk-25465	199	32	Charcoal		300-0	310-0	180	
Old Dopima	Wk-27927	144	30	Charcoal		280-0	290-0	140	Barker et al. (2015)
Aiedio	Wk-27918	139	30	Charcoal		270-0	280-0	140	Barker et al. (2015)
	Wk-27919	161	30	Charcoal		290-0	290-0	170	
Kinomere	SUA-1879	410	80	Charcoal		530-320	560-290	440	Frankel et al (1994)
Baikaboria	Wk-35158	146	25	Charcoal		280-0	290-0	140	Barker et al. (2016)
	Wk-38123	629	23	*Bone*	?	650-550	660-550	600	
Meiharo 1	Wk-22750	381	30	Charcoal		500- 330	510-310	450	Skelly et al. (2010)
	Wk-22749	424	32	Charcoal		520-470	530-330	490	
	Wk-22748	97	30	Charcoal		260-30	270-20	110	
	Wk-22747	426	30	Charcoal		520-470	530-330	490	
	Wk-22746	569	30	Charcoal		630-540	650-520	600	
Old Healau	Wk-32240	155	25	Charcoal		280-0	290-0	150	Skelly (2014)
	Wk-33955	117	25	Charcoal		260-30	270-10	110	
	Wk-32241	146	25	Charcoal		280-0	290-0	140	
	Wk-33956	141	25	Charcoal		270-10	280-0	120	
	Wk-33957	142	25	Charcoal		270-10	280-0	130	
Lui Ova	Wk-31825	639	25	Charcoal		660-560	670-550	590	Skelly (2014)
	Wk-31826	583	25	Charcoal		630-540	650-530	600	
	Wk-29546	683	25	Charcoal		670-570	680-560	650	
	Wk-31827	638	25	Charcoal		660-560	670-550	590	
Hivo	Wk-33263	180	25	Charcoal		290-0	300-0	180	Skelly (2014)
	Wk-33954	154	25	Charcoal		280-0	290-0	150	
	Wk-32242	2796	25	Charcoal		2940-2860	2970-2790	2900	
	Wk-33265	153	25	Charcoal		280-0	290-0	150	
Kaveharo	Wk-33266	194	25	Charcoal		290-0	300-0	180	Skelly (2014)
	Wk-33958	1516	28	Charcoal		1410-1350	1520-1310	1380	
	Wk-31831	600	25	Charcoal		640-550	650-540	600	
	Wk-31832	614	25	Charcoal		650-550	650-550	600	

Site	Lab code	Age	±	Material				Reference
	Wk-33959	2453	30	Charcoal	2700–2370	2710–2360	2530	
	Wk-29541	2457	25	Charcoal	2700–2430	2710–2360	2560	
	Wk-29542	572	25	Charcoal	630–540	640–520	600	
	Wk-33961	2465	32	Charcoal	2710–2460	2710–2360	2570	
	Wk-29543	133	25	Charcoal	270–20	280–0	110	
	Wk-33963	2326	30	Charcoal	2370–2330	2470–2180	2350	
	Wk-29544	1536	25	Charcoal	1510–1360	1520–1350	1400	
	Wk-33964	2431	31	Charcoal	2670–2360	2700–2350	2470	
Hohelavi	Wk-29535	183	25	Charcoal	290–0	300–0	180	Skelly (2014)
	Wk-29536	177	25	Charcoal	290–0	300–0	180	
	Wk-31828	2169	25	Charcoal	2300–2110	2310–2050	2180	
	Wk-31829	2449	25	Charcoal	2700–2370	2700–2360	2510	
	Wk-31830	2047	25	Charcoal	2050–1940	2110–1920	1990	
Ira Kahu	Wk-29547	664	25	Charcoal	670–560	670–550	630	Skelly (2014)
	Wk-34212	1720	26	Charcoal	1700–1560	1700–1540	1600	
	Wk-34213	1714	26	Charcoal	1690–1540	1700–1530	1600	
	Wk-33966	1305	28	Charcoal	1290–1170	1300–1170	1230	
Oheo Yopo	Wk-29545	1388	25	Charcoal	1340–1280	1350–1280	1300	Skelly (2014)
	Wk-32243	1592	25	Charcoal	1520–1410	1530–1400	1470	
	Wk-33267	1544	25	Charcoal	1510–1370	1530–1350	1410	
	Wk-33268	1494	25	Charcoal	1390–1340	1410–1310	1370	
Mampaiu	ANU-1832	310	250	*Batissa violacea*	530–0	680–0	340	Rhoads (1980)
Ibira	ANU-2181	410	80	*Batissa violacea*	530–320	560–290	440	Rhoads (1980)
Ouloubomoto	ANU-2235	720	180	*Batissa violacea*	300–0	490–0	210	Rhoads (1980)
	ANU-2236	1050	170	?	640–310	820–120	480	
	ANU-2237	400	170		150–0	300–0	100	
Maero	SUA-1798	310	100	Charcoal	500–150	530–0	360	Frankel et al. (1994)
Mirimua	OZV363	125	25	Charcoal	270–20	270–10	110	Urwin et al. (2021)
	OZV364	110	25	Charcoal	260–30	270–20	110	
Mapoe	OZV365	260	25	Charcoal	430–150	430–150	300	
Popo	ANU-2181	410	80	Charcoal	530–320	560–290	440	Rhoads (1994)
	OZV341	170	25	Charcoal	290–0	290–0	180	
	OZV338	140	25	Charcoal	270–10	280–0	120	
	OZV339	170	25	Charcoal	290–0	290–0	180	
	OZV340	465	25	Charcoal	530–500	540–490	510	
	Wk-41608	131	20	Charcoal	270–20	270–10	110	Urwin et al. (2018),
	OZU282	205	20	Charcoal	300–150	300–0	180	Urwin et al. (2021)
	OZU283	330	25	Charcoal	450–310	470–310	390	
	Wk-41609	305	21	Charcoal	430–300	450–300	390	
	OZU284	330	20	Charcoal	450–310	460–310	380	
	OZU285	415	20	Charcoal	510–470	520–340	490	
	OZV292	260	30	Charcoal	430–150	440–150	310	
	OZV293	240	30	Charcoal	320–150	430–0	290	

(Continued)

Table 9.1 (Continued)

Site	Lab code	Date	Error	Material	ΔR value	68.3% range	95.4% range	Median	Reference
	OZV822	270	35	Charcoal		430-150	460-150	330	
	OZV294	345	25	Charcoal		470-320	480-310	390	
	OZV454	155	25	Charcoal		280-0	290-0	150	
	OZV295	310	25	Charcoal		430-300	460-300	390	
	OZV354	265	20	Charcoal		420-280	430-150	310	
	OZV355	310	20	Charcoal		430-310	450-300	390	
	OZV356	570	25	Charcoal		630-540	640-520	600	
	OZV357	645	20	Charcoal		660-560	670-550	590	
	OZV358	690	25	Charcoal		670-650	680-560	660	
	OZV359	650	25	Charcoal		660-560	670-550	590	
	OZV360	655	25	Charcoal		660-560	670-550	590	
	OZV361	590	25	Charcoal		630-540	650-540	600	
	OZW433	370	20	Charcoal		490-330	500-320	440	
	OZW434	555	25	Charcoal		630-530	630-520	550	
	OZV343	3180	160	Charcoal		3580-3170	3830-2960	3390	
	OZV344	385	30	Charcoal		500-330	510-310	450	
	OZV345	650	25	Charcoal		660-560	670-550	590	
	OZV346	495	25	Charcoal		540-510	550-500	520	
	OZV347	650	25	Charcoal		660-560	670-550	590	
	OZV348	505	25	Charcoal		540-510	550-500	530	
	OZV349	540	30	Charcoal		630-520	630-510	540	
	OZV350	555	25	Charcoal		630-530	630-520	550	
Murua	SUA-1726	700	120	soil		730-550	910-500	660	Frankel and Kewibu (2000)
Upihoi	Wk-22225	110	36	wood hull		260-30	280-0	110	David et al. (2008)
	Wk-22224	171	35	wood hull		290-0	300-0	170	
	Wk-22223	125	35	wood hull		270-20	280-0	120	

Grey shade = Radiocarbon dates with errors of 100 years or more. Dates calibrated using OxCal4.4 and the IntCal20 dataset for terrestrial samples and Marine20 for marine shell samples. No ΔR values applied to marine dates in the Gulf region as no regional corrections are known. Calibrated date ranges rounded to the nearest 10 years. Radiocarbon dates are noted as before present (BP), with BP being AD 1950.

Table 9.2 Radiocarbon dates from archaeological sites in the South Coast region. Dark grey shade = Conventional Radiocarbon Age (CRA) with errors of 100 years or more. Dates calibrated using OxCal4.4 and the IntCal20 dataset for terrestrial samples and Marine13 for marine shell samples. Light grey shade = ΔR values determined by original authors as unreliable due to the feeding behaviour of the dated shellfish. Calibrated date ranges with the highest associated probability are bolded. Calibrated date ranges rounded to the nearest 10 years, with the highest associated probability bolded.

Site	Lab code	Date	Error	Material	ΔR value	68.3% range	95.4% range	Median	Reference
Oposisi	ANU-425	1890	305	charcoal		2300–1420	2710–1280	1860	Vanderwal (1973)
	ANU-729	1530	160	charcoal		1690–1290	1820–1070	1440	
	ANU-728	1600	210	charcoal		1720–1290	2000–1070	1520	
	ANU-727	1920	180	charcoal		2100–1610	2330–1410	1860	
	ANU-726	940	180	charcoal		1060–680	1270–550	870	
	ANU-725	1180	200	charcoal		1300–920	1510–680	1090	
	WK-21610	1573	33	charcoal		1520–1400	1530–1380	1460	Allen et al. (2011)
	WK-21611	1607	33	charcoal		1540–1410	1540–1400	1470	
	WK-21612	1639	30	charcoal		1550–1410	1590–1410	1520	
	WK-21613	1834	30	charcoal		1820–1700	1830–1630	1740	
	WK-21614	2004	30	charcoal		2000–1890	2010–1830	1940	
	WK-21615	2041	30	charcoal		2050–1930	2100–1890	1980	
	WK-21616	2022	30	charcoal		2000–1920	2050–1840	1960	
Abe	ANU-731	1560	85	charcoal		1530–1360	1690–1300	1450	Vanderwal (1973)
Urourinia	ANU-730	720	105	charcoal		740–550	910–520	670	Vanderwal (1973)
Kukuba cave	ANU-395a	3980	105	charcoal		Original dates combined in OxCal			Vanderwal (1973)
	ANU-395b	3920	90	charcoal					Vanderwal (1973)
	ANU-395 combined	3946	69	charcoal		4520–4250	4580–4150	4390	
Nebira 4	ANU-732	1250	85	charcoal		1280–1070	1310–970	1160	Allen (1972)
	Gak-2667	880	250	charcoal		1060–560	1340–340	840	
	Gak-2990	3340	160	charcoal		3830–3390	4070–3180	3590	
	I-5796	1760	90	charcoal		1740–1540	1880–1410	1650	
Nebira 2	GaK-2672	280	80	charcoal		470–150	510–0	330	Bulmer (1978)
	GaK-2675	380	120	*bone*	?	520–300	650–0	400	
	GaK-2673	660	150	charcoal		730–520	920–330	630	
	GaK-2346	720	80	charcoal rich soil		730–560	790–540	660	
	GaK-2345	390	90	ashy soil		510–310	630–150	420	
Eriama	GaK-2671	210	70	*bone*	?	300–0	430–0	180	Bulmer (1978)
	GaK-2668	380	120	charcoal		520–300	650–0	400	
	GX-3334	600	125	charcoal		670–510	790–320	590	

(*Continued*)

Table 9.2 (Continued)

Site	Lab code	Date	Error	Material	ΔR value	68.3% range	95.4% range	Median	Reference
Taurama	GaK-2670	1930	230	charcoal		2150-1580	2430-1340	1880	Bulmer (1978)
	I-6862	560	85	charcoal		650-510	680-460	580	
	I-6887B	775	85	charcoal		790-650	910-550	710	
	I-6863	865	140	charcoal		920-680	1070-550	800	
Papa Salt Pan	SUA 1524	1280	170	charcoal		1350-670	1530-800	1180	Swadling and Kaiku (1980)
Ava Garau	SUA 515	1220	95	charcoal		1280-1050	1300-950	1140	Swadling (1981)
Motupore	I-5901	380	90	charcoal		510-310	630-150	410	Allen (2017)
	I-5902	715	90	charcoal		730-560	900-520	660	
	I-5903	740	105	charcoal		780-550	910-530	690	
	ANU-1512	280	40	charcoal		430-280	470-150	370	
	ANU-1511	550	70	charcoal		640-510	670-490	570	
	ANU-1177	360	50	charcoal		490-320	510-310	400	
	ANU-1508	400	70	charcoal		520-320	540-300	440	
	ANU-1219	1010	80	charcoal		1050-790	1180-730	910	
	ANU-1178	580	50	charcoal		640-540	660-520	600	
	ANU-1510	810	150	charcoal		910-650	1060-520	760	
	ANU-1210	590	70	charcoal		650-540	670-510	590	
	ANU-1209	600	70	charcoal		650-540	670-510	600	
	ANU-1163	740	80	charcoal		740-560	900-540	680	
	ANU-1211	830	80	charcoal		900-670	920-650	750	
	ANU-1509	480	60	charcoal		560-470	650-320	520	
	ANU-1218	310	60	charcoal		460-300	510-150	380	
	ANU-1217	370	80	charcoal		500-310	540-150	410	
	ANU-1212	390	70	charcoal		510-320	530-300	430	
	ANU-1179	550	50	charcoal		630-520	650-500	560	
	ANU-3373	240	60	bone apatite		250-10	270-0	100	
	ANU-3374	310	220	bone collagen	?	400-0	530-0	230	
	ANU-3375	190	70	bone apatite		240-0	270-0	90	
	ANU-3371	350	100	charcoal		500-310	560-0	390	
	ANU-3370	290	70	charcoal		460-150	510-0	360	
	ANU-3369	450	70	charcoal		550-330	630-310	490	
	ANU-3368	440	70	charcoal		540-330	630-310	480	
	ANU-4444	250	70	charcoal		440-0	490-0	290	
Loloata	ANU-4808	2300	100	shell	?	2050-1770	2180-1610	1910	Sullivan and Sassoon (1987)
O1	ANU-1541	780	230	charcoal		930-540	1250-310	750	Irwin (1985)

Site	Lab code	Age	±	Material	δ13C				Reference
Mailu 3	ANU-1661	2030	100	*shell*	?	1700–1430	1820–1330	1570	
	ANU-1662	2120	80	*shell*		1780–1560	1890–1460	1670	
	ANU-1230	1690	110	charcoal		1710–1410	1830–1350	1580	Irwin (1985)
	ANU-1229	1900	70	charcoal		1890–1720	2000–1620	1820	
	ANU-1432	220	75	charcoal		430–0	450–0	210	
	ANU-1433	410	70	charcoal		520–320	540–310	450	
	ANU-1434	410	70	charcoal		520–320	540–310	450	
	ANU-1435	520	70	charcoal		630–500	660–330	540	
	ANU-1436	710	80	charcoal		730–560	780–540	660	
Selai	ANU-1316	1790	70	charcoal		1750–1570	1870–1530	1680	Irwin (1985)
	ANU-1317	1770	70	charcoal		1740–1570	1840–1520	1680	
Agila	Wk-42519	743	20	*anadara shell*	−141±27	430–280	490–200	350	Skelly et al. (2018)
	Wk-42520	713	20	*anadara shell*		410–250	470–150	320	
	Wk-42521	713	20	*anadara shell*		410–250	470–150	320	
	Wk-42522	704	20	*anadara shell*		460–140	460–140	310	
	Wk-42523	379	23	charcoal		500–330	500–320	450	
	Wk-42524	386	23	charcoal		500–330	510–320	460	

CAUTION BAY

Site	Lab code	Age	±	Material	δ13C				Reference
Tanamu 1	Wk-29957	117	30	Charcoal		260–30	270–10	110	Thangavelu (2015)
	Wk-29966	123	30	Charcoal		270–20	280–0	110	
	Wk-32532	593	25	*Anadara antiquata*	−129±30	270–90	310–0	180	
	Wk-32533	575	25	*Anadara antiquata*		250–80	290–0	160	
	Wk-27504	193	30	Charcoal		290–0	310–0	180	
	Wk-32534	538	25	*Anadara antiquata*	−129±30	200–20	260–0	120	
	Wk-29967	117	30	Charcoal		260–30	270–10	110	
	Wk-29968	769	30	Charcoal		730–670	730–660	690	
	Wk-32535	2971	30	*Anadara antiquata*	−129±30	2820–2660	2900–2540	2730	
	Wk-29958	66	33	Charcoal		260–40	260–30	120	
	Wk-27505	826	30	Charcoal		740–680	790–680	720	
	Wk-29959	158	30	Charcoal		290–0	290–0	160	
	Wk-32536	3042	26	*Gafrarium tumidum*	−60±30	2780–2710	2830–2690	2750	
	Wk-32537	3053	28	*Anadara antiquata*	−129±30	2890–2730	2990–2680	2820	
	Wk-32538	3080	31	*Anadara antiquata*	−60±30	2920–2750	3030–2700	2850	
	Wk-32540	2990	31	*Gafrarium tumidum*		2750–2680	2780–2600	2710	
	Wk-32539	2993	31	*Anadara antiquata*	−129±30	2840–2690	2940–2580	2760	David et al. (2011)
	Wk-32541	3000	27	*Anadara antiquata*		2840–2690	2940–2600	2770	Thangavelu (2015)
	Wk-32542	3078	26	*Anadara antiquata*		2920–2760	3020–2700	2850	
	Wk-32543	3024	26	*Anadara antiquata*		2860–2710	2960–2630	2790	
	Wk-42506	2842	30	Charcoal		3000–2880	3060–2860	2950	
	Wk-32544	3055	27	*Anadara antiquata*	−129±30	2890–2740	2990–2680	2820	

(Continued)

Table 9.2 (Continued)

Site	Lab code	Date	Error	Material	ΔR value	68.3% range	95.4% range	Median	Reference
	Wk-32545	3035	28	Anadara antiquata		2870-2720	2970-2660	2800	
	Wk-32546	3024	29	Anadara antiquata		2860-2710	2960-2640	2790	
	Wk-32547	3350	26	Anadara antiquata		3290-3100	3360-3000	3190	
	Wk-32548	4076	27	Anadara antiquata		4210-3990	4300-3900	4100	
	Wk-27508	4032	29	Anadara antiquata		4150-3940	4240-3850	4040	
	Wk-29961	3734	30	Anadara antiquata		3750-3550	3830-3470	3650	
	Wk-29962	3715	30	Charcoal		4150-3980	4160-3970	4050	
	Wk-29969	3829	32	Charcoal		4290-4150	4410-4090	4220	
	Wk-29963	3858	32	Charcoal		4400-4160	4410-4150	4280	
	Wk-27714	3864	30	Charcoal		4410-4230	4410-4150	4290	
	Wk-27643	3919	30	Charcoal		4420-4290	4430-4240	4350	
	Wk-29970	3895	31	Charcoal		4410-4290	4420-4230	4330	
	Wk-27644	3968	30	Charcoal		4520-4410	4530-4290	4440	
	Wk-29341	3941	39	Charcoal		4510-4290	4520-4250	4380	
	Wk-29340	3968	30	Charcoal		4520-4350	4530-4290	4440	
	Wk-28805	4053	33	Charcoal		4580-4440	4790-4420	4520	
		4021	30	Charcoal		4530-4420	4580-4410	4480	
	Wk-31008	4268	25	Anadara antiquata	-129±30	4440-4240	4540-4150	4360	
	Wk-31007	4285	25	Gafrarium tumidum	-60±30	4390-4270	4540-4210	4320	
	Wk-31009	4313	25	Anadara antiquata	-129±30	4520-4320	4610-4220	4440	
	Wk-27645	4042	30	Charcoal		4570-4440	4780-4410	4500	
	Wk-27646	4012	30	Charcoal		4520-4420	4570-4410	4480	
	Wk-27647	4037	30	Charcoal		4570-4430	4620-4410	4490	
	Wk-27971	3969	32	Charcoal		4520-4410	4530-4290	4440	
	Wk-29977	3949	30	Charcoal		4520-4300	4520-4300	4410	
	Wk-29972	3978	31	Charcoal		4520-4410	4530-4300	4460	
	Wk-29978	3965	32	Charcoal		4520-4400	4530-4290	4430	
	Wk-32550	4318	37	Gafrarium tumidum	-60±30	4420-4290	4490-4230	4360	
	Wk-28604	4071	30	Charcoal		4790-4440	4800-4430	4560	
	Wk-32551	4029	27	Anadara antiquata	-129±30	4140-3930	4230-3850	4040	
	Wk-29974	3949	30	Charcoal		4520-4300	4520-4290	4410	
	Wk-29984	4154	27	Charcoal		4820-4620	4830-4570	4700	
	Wk-29964	3971	30	Charcoal		4520-4410	4530-4290	4440	
	Wk-29965	4093	30	Charcoal		4800-4520	4810-4440	4600	
	Wk-29212	4091	35	Charcoal		4800-4520	4820-4440	4600	
	Wk-32552	4766	30	Anadara antiquata	-129±30	5070-4860	5220-4810	4990	
	Wk-32553	4727	30	Gafrarium tumidum	-60±30	4920-4820	4980-4800	4880	

Site	Lab code			Material	δ				Reference
Bogi 1	Wk-28414	2384	30	*Anadara granosa*	**-201±30**	2150-2030	2240-1980	2100	David et al. (2012)
	Wk-28266	1603	30	Charcoal	13±31	1530-1410	1540-1400	1470	McNiven et al. (2011)
	Wk-31010	2529	25	*Conomurex*	156±16	2270-2130	2310-2070	2190	Petchey et al. (2012)
	Wk-32565	2713	27	*laevistrombus*	31±37	2300-2180	2320-2130	2230	
	Wk-32571	2511	27	*gibberulus*	70±42	2240-2050	2290-2000	2140	
	Wk-32577	2461	27	*Euprotomus*	71±53	2100-1950	2150-1870	2020	
	Wk-32591	2557	26	*Lambis*		2260-2070	2300-1980	2150	Petchey et al. (2013)
	Wk-30326	2507	25	*Gafrarium tumidum*	**-60±30**	2130-2030	2190-1970	2080	
	Wk-32554	2524	27	*G.pectinatum*	53±16	2180-2050	2270-2010	2120	
	Wk-30327	2397	25	*Anadara granosa*	**-201±30**	2170-2050	2260-2010	2120	
	Wk-30328	2427	25	*Anadara antiquata*	**-129±30**	2150-1950	2260-1870	2050	
	Wk-28267	1599	30	Charcoal		1530-1410	1540-1400	1470	McNiven et al. (2011)
	Wk-31011	2390	25	*Conomurex*	13±31	2070-1940	2120-1890	2000	Petchey et al. (2012)
	Wk-32566	2510	27	*laevistrombus*	156±16	2030-1920	2090-1870	1970	
	Wk-32572	2546	27	*gibberulus*	31±37	2270-2120	2310-2050	2190	
	Wk-32578	2590	27	*Euprotomus*	70±42	2280-2130	2320-2050	2190	
	Wk-32586	2579	25	*Canarium shell*	55±34	2280-2140	2310-2070	2200	
	Wk-30329	2501	25	*Gafrarium tumidum*	**-60±30**	2130-2020	2170-1960	2070	Petchey et al. (2013)
	Wk-32555	2473	27	*G.pectinatum*	53±16	2110-2000	2150-1950	2060	
	Wk-30330	2378	25	*Anadara granosa*	**-201±30**	2140-2030	2210-1980	2090	
	Wk-30331	2358	25	*Anadara granosa*		2120-2020	2160-1960	2070	
	Wk-30332	2460	25	*Anadara antiquata*	**-129±30**	2180-1990	2290-1930	2100	
	Wk-31037	2439	25	*Echinoida*	11±17	2120-2010	2160-1960	2060	
	Wk-31012	2485	25	*Conomurex*	13±31	2200-2040	2280-2000	2130	Petchey et al. (2012)
	Wk-32582	2491	30	*Canarium shell*	63±20	2120-2000	2180-1940	2070	
	Wk-32567	2687	27	*laevistrombus*	156±16	2270-2150	2310-2110	2210	
	Wk-32573	2460	27	*gibberulus*	31±37	2130-1990	2240-1920	2070	
	Wk-32587	2492	26	*Canarium shell*	55±34	2140-2000	2240-1940	2080	
	Wk-32592	2435	25	*Lambis*	71±53	2070-1910	2150-1840	1990	
	Wk-33810	2459	25	*cerithidea*	-59±14	2260-2140	2300-2100	2190	
	Wk-33811	2662	25	*cerithidea*		2460-2350	2550-2320	2420	
	Wk-32533	2521	25	*Gafrarium tumidum*	**-60±30**	2150-2040	2240-1980	2100	Petchey et al. (2013)
	Wk-32556	2512	26	*G.pectinatum*	53±16	2150-2040	2240-1990	2110	
	Wk-30334	2326	25	*Anadara granosa*	**-201±30**	2090-1980	2130-1930	2030	
	Wk-30335	2434	25	*Anadara antiquata*	**-129±30**	2150-1970	2270-1890	2060	
	Wk-32560	2299	29	*P.erosa*	-154±23	2150-2030	2250-1980	2100	
	Wk-32563	2219	27	*B.violacea*	-207±28	2120-2000	2180-1930	2060	

(Continued)

Table 9.2 (Continued)

Site	Lab code	Date	Error	Material	ΔR value	68.3% range	95.4% range	Median	Reference
	Wk-31038	2451	25	*Echinoida*	11±17	2140-2030	2190-1970	2080	McNiven et al. (2011)
	Wk-28268	1537	30	*Charcoal*		1510-1360	1520-1350	1410	Petchey et al. (2012)
	Wk-31013	2440	25	*Conomurex*	13±31	2120-2000	2190-1930	2060	
	Wk-32568	2570	27	*laevistrombus gibberulus*	156±16	2100-1990	2150-1940	2050	
	Wk-32574	2474	27	*gibberulus*	31±37	2150-2000	2260-1950	2090	
	Wk-32579	2501	27	*Euprotomus*	70±42	2140-1990	2250-1930	2070	
	Wk-32583	2462	27	*Canarium shell*	63±20	2090-1980	2130-1920	2030	
	Wk-32588	2489	25	*Canarium shell*	55±34	2130-2000	2240-1920	2070	
	Wk-32593	2492	25	*Lambis*	71±53	2140-1970	2260-1900	2060	
	Wk-33812	2152	25	*cerithidea*	-59±14	1870-1770	1890-1710	1810	
	Wk-33813	2471	25	*cerithidea*		2270-2150	2300-2110	2210	
	Wk-30336	2537	25	*Gafrarium tumidum*	**-60±30**	2180-2050	2270-2010	2120	Petchey et al. (2013)
	Wk-32557	2530	26	*G.pectinatum*	53±16	2190-2060	2280-2020	2130	
	Wk-30337	2373	25	*Anadara granosa*	**-201±30**	2140-2030	2200-1970	2080	
	Wk-30338	2448	25	*Anadara antiquata*	**-129±30**	2170-1980	2290-1910	2080	
	Wk-32561	2295	27	*P.erosa*	-154±23	2150-2030	2240-1980	2090	
	Wk-32564	2253	27	*B.violacea*	-207±28	2170-2030	2260-1990	2110	
	Wk-31039	2459	25	*Echinoida*	11±17	2140-2030	2210-1970	2090	
	Wk-33814	2209	25	*cerithidea*	-59±14	1920-1830	1970-1790	1880	Petchey et al. (2012)
	Wk-33815	2355	25	*cerithidea*		2100-2000	2140-1950	2050	
	Wk-31047	2097	27	*Charcoal (carb. Fruit)*		2110-2000	2150-1950	2060	
	Wk-31048	2110	28	*Charcoal*		2120-2000	2290-1990	2070	Petchey et al. (2013)
	Wk-31014	2447	25	*Conomurex*	13±31	2130-2000	2220-1940	2070	Petchey et al. (2012)
	Wk-32584	2531	27	*Canarium shell*	63±20	2180-2040	2270-2000	2120	
	Wk-32569	2558	27	*laevistrombus*	156±16	2090-1980	2140-1930	2030	
	Wk-32575	2395	27	*gibberulus*	31±37	2060-1920	2120-1860	1990	
	Wk-32580	2496	31	*Euprotomus*	70±42	2140-1980	2250-1910	2060	
	Wk-32589	2446	25	*Canarium shell*	55±34	2090-1960	2140-1900	2020	
	Wk-32594	2570	26	*Lambis*	71±53	2270-2090	2310-2000	2170	
	Wk-33816	2309	25	*cerithidea*	-59±14	2040-1940	2100-1890	1990	
	Wk-33817	2452	25	*cerithidea*		2250-2130	2300-2090	2190	
	Wk-30340	2498	30	*Gafrarium tumidum*	**-60±30**	2120-2010	2180-1950	2070	Petchey et al. (2013)
	Wk-32558	2467	26	*G.pectinatum*	53±16	2100-1990	2140-1940	2050	
	Wk-30341	2425	29	*Anadara granosa*	**-201±30**	2240-2100	2290-2050	2160	

Wk-30342	2433	29	*Anadara antiquata*	**-129±30**	2150-1960	2270-1880	2060	
Wk-32562	2273	27	*P. erosa*	-154±23	2120-2000	2180-1940	2060	
Wk-31040	2467	25	*Echinoida*	11±17	2150-2040	2240-1990	2100	McNiven et al. (2011)
Wk-30458	2090	25	Charcoal		2100-2000	2130-1950	2050	Petchey et al. (2012)
Wk-31049	2114	27	Charcoal (nut)		2120-2000	2290-1990	2080	
Wk-31050	2101	27	Charcoal (monocot culm)		2110-2000	2150-1990	2060	
Wk-31051	2192	25	Charcoal		2310-2140	2310-2120	2240	Petchey et al. (2013)
Wk-31015	2444	25	*Conomurex*	13±31	2130-2000	2210-1930	2070	Petchey et al. (2012)
Wk-32570	2558	27	*laevistrombus*	156±16	2090-1980	2140-1930	2030	
Wk-32576	2456	27	*gibberulus*	31±37	2130-1990	2220-1910	2060	
Wk-32585	2539	28	*Canarium shell*	63±20	2200-2050	2280-2020	2130	
Wk-32590	2485	26	*Canarium shell*	55±34	2130-2000	2220-1930	2070	
Wk-30343	2495	29	*Gafrarium tumidum*	**-60±30**	2120-2010	2170-1950	2060	Petchey et al. (2013)
Wk-32559	2470	27	*G. pectinatum*	53±16	2110-2000	2150-1940	2050	
Wk-30344	2346	30	*Anadara granosa*	**-201±30**	2110-1990	2150-1940	2050	
Wk-30345	2453	29	*Anadara antiquata*	**-129±30**	2180-1980	2290-1920	2090	
Wk-31041	2452	25	*Echinoida*	11±17	2140-2030	2190-1970	2080	David et al. (2012)
Wk-27154	2215	30	Charcoal		2310-2150	2330-2140	2230	McNiven et al. (2011)
Wk-29210	2148	32	Charcoal		2300-2050	2310-2000	**2130**	David et al. (2012)
Wk-25748	2134	30	Charcoal		2290-2050	2300-1990	2100	McNiven et al. (2011)
Wk-28270	2147	30	Charcoal		2300-2060	2310-2000	2130	David et al. (2012)
Wk-25749	2140	30	Charcoal		2290-2050	2300-2000	2120	McNiven et al. (2011)
Wk-28271	2159	30	Charcoal		2300-2070	2310-2000	2150	
Wk-27500	2223	30	Charcoal		2320-2150	2340-2140	2230	David et al. (2012)
Wk-25750	2130	30	Charcoal		2310-2000	2300-1990	2100	
Wk-25751	2180	30	Charcoal		2150-2120	2320-2060	2220	
Wk-29953	2448	25	Charcoal		2700-2370	2700-2360	2500	McNiven et al. (2011)
Wk-25752	2229	30	Charcoal		2320-2150	2340-2140	2230	David et al. (2012)
Wk-29954	2508	25	Charcoal		2720-2510	2730-2490	2580	McNiven et al. (2011)
Wk-27713	2538	30	Charcoal		2740-2530	2750-2490	2620	
Wk-27712	2537	30	Charcoal		2740-2520	2750-2490	2620	
Wk-27711	2775	30	Charcoal		2930-2790	2960-2780	2870	
Wk-27707	2783	30	Charcoal		2940-2840	2960-2780	2880	
Wk-30465	3491	25	Charcoal		3830-3710	3840-3690	3760	
Wk-27710	2533	30	Charcoal		2740-2520	2750-2490	2620	
Wk-29956	3667	25	Charcoal		4080-3920	4090-3900	4000	
Wk-27708	3826	30	Charcoal		4290-4150	4410-4090	4220	
Wk-28272	3824	30	Charcoal		4290-4150	4400-4090	4210	
Wk-28273	3811	30	Charcoal		4250-4140	4360-4090	4200	
Wk-28274	3833	30	Charcoal		4300-4150	4410-4090	4230	

(Continued)

Table 9.2 (Continued)

Site	Lab code	Date	Error	Material	ΔR value	68.3% range	95.4% range	Median	Reference
Edubu 1	Wk-28275	3816	30	Charcoal		4250–4140	4390–4090	4200	McNiven et al. (2012)
	Wk-28278	4053	30	Charcoal		4580–4440	4790–4420	4520	
	Wk-27302	2339	30	Charcoal		2370–2330	2470–2310	2350	
	Wk-27301	2440	30	Charcoal		2690–2360	2700–2350	2480	
	Wk-27510	2531	30	Charcoal		2730–2510	2750–2490	2610	
	Wk-27511	2537	30	Charcoal		2740–2520	2750–2490	2620	
	Wk-27512	2520	30	Charcoal		2730–2510	2740–2490	2590	
	Wk-27514	2502	30	Charcoal		2720–2500	2730–2470	2590	
	Wk-27515	2514	30	Charcoal		2730–2510	2740–2490	2590	
	Wk-27516	2546	30	Charcoal		2750–2540	2750–2490	2630	
Moiapu 1	Wk-27501	2470	30	Charcoal		2710–2470	2720–2370	2580	David et al. (2011)
	Wk-30042	2551	26	Charcoal		2750–2540	2750–2510	2710	
	Wk-27632	2515	30	Charcoal		2730–2510	2740–2490	2590	
	Wk-27633	2529	30	Charcoal		2730–2510	2750–2490	2600	
Moiapu 3	Wk-36369	2814	25	Anadara antiquata	−129±30	2670–2460	2710–2360	2550	David et al. (2019)
	Wk-36371	2714	25	Anadara granosa	−201±30	2620–2450	2680–2390	2530	
	Wk-36372	2778	25	Anadara antiquata	−129±30	2610–2390	2690–2330	2500	
	Wk-36373	2689	25	Anadara granosa	−201±30	2570–2390	2650–2350	2480	
	Wk-36374	2674	25	Anadara granosa		2510–2360	2610–2340	2450	
	Wk-36375	2812	25	Anadara antiquata	−129±30	2660–2460	2710–2350	2540	
Ruisasi 1	Wk-29342	2031	31	Gafrarium sp.	60±11	1590–1480	1620–1400	1530	David et al. (2016)
	Wk-29343	1969	31	Gafrarium tumidum	−60±30	1510–1400	1540–1350	1450	
	Wk-29344	1807	33	Gafrarium tumidum		1330–1250	1380–1220	1290	
	Wk-29345	2016	25	Gafrarium sp.	60±11	1560–1460	1590–1410	1510	
	Wk-29346	2068	27	Gafrarium tumidum	−60±30	1610–1510	1680–1480	1560	
Ruisasi 2	Wk-29347	1599	30	Charcoal		1530–1410	1540–1400	1470	Rowe et al. (2020)
	Wk-29348	1601	30	Charcoal		1530–1410	1540–1400	1470	
	Wk-29349	2588	30	Charcoal		2750–2720	2770–2540	2740	
	Wk-38680	578	24	Conus	11±17	260–140	290–0	200	
	Wk-38681	580	23	Conomurex	13±31	260–140	290–110	200	
	Wk-38683	613	25	Conomurex	13±31	290–150	310–130	250	
	Wk-38684	2766	26	Ostreidae	−71±15	2690–2540	2710–2480	2610	
	Wk-30017	5432	30	charcoal		6290–6200	6300–6190	6240	
	Wk-30018	5085	30	charcoal		5910–5750	5920–5740	5810	

	Lab code			Species	ΔR			Thangavelu (2015)
JA24	WK-31109	2421	28	*A.?*	-39±22	2160-2030	2260-1990	2110
	WK-31110	2397	26	*Anadara granosa*	**-201±30**	2170-2050	2260-2010	2120
	WK-31111	2441	29	*Anadara granosa*		2260-2130	2300-2080	2190
	WK-27498	2311	35	*Anadara granosa*		2070-1940	2120-1890	2010
	WK-31112	1911	29	*Gafrarium tumidum*	**-60±30**	1430-1320	1490-1300	1380
	WK-31113	1869	28	*Gafrarium tumidum*		1380-1290	1430-1260	1340
	WK-31114	2607	27	*G.?*	60±11	2290-2170	2310-2130	2220
	WK-31115	2596	31	*Anadara antiquata*	**-129±30**	2340-2160	2450-2080	2260
	WK-27499	2284	38	*Anadara granosa*	**-201±30**	2040-1910	2100-1860	1980

Grey shade = Radiocarbon dates with errors of 100 years or more. Dates calibrated using OxCal4.4 and the IntCal20 dataset for terrestrial samples. Marine samples with bold text were calibrated with ΔR correction for Marine20 curve. Other marine shell samples were calibrated with Marine13. Calibrated date ranges rounded to the nearest 10 years. Radiocarbon dates are noted as before present (BP), with BP being AD 1950.

Table 9.3 Radiocarbon dates from archaeological sites in the Massim region. Grey shade = Conventional Radiocarbon Age (CRA) with errors of 100 years or more. Dates calibrated using OxCal4.4 and the IntCal20 dataset for terrestrial samples and Marine20 for marine shell samples. Calibrated date ranges rounded to the nearest 10 years, with the highest associated probability bolded. ^20% ΔR correction applied to bone samples to account for probable marine dietary input.

Site	Lab code	Date	Error	Material	ΔR value	68.3% range	95.4% range	Median	Reference
Kelebwana-gum	Beta-479384	14480	40	Barbatia sp.	-135±20	17000-16750	17090-16600	16870	Shaw et al. (2020a)
	Beta-502620	13250	40	Acanthople-ura sp.		15400-15150	15530-15050	15280	
	OZX-907	12430	40	Turbo sp. opercula		14080-13860	14190-13770	13980	
	Beta-502621	11290	40	Turbo sp. opercula		12830-12680	12920-12610	12760	
	OZX-908	11010	35	Turbo sp. opercula		12610-12450	12680-12330	12520	
	OZX-905	4,470	30	Donax sp. shell		4730-4520	4810-4450	4630	
	Beta-502622	3,720	30	Asaphis violascens		3730-3540	3820-3470	3640	
	OZX-906	2,355	25	Turbo sp. opercula		2050-1880	2140-1810	1970	
	OZX-904	1,420	20	Charcoal		1350-1290	1350-1290	1340	
	OZX-903	885	25	Charcoal		800-730	910-720	770	
	OZX-902	220	20	Charcoal		300-150	310-0	190	
Kasasinab-wana	Wk-25603	2463	38	Tridacna sp.	-135±20	2200-1990	2300-1930	2110	Negishi and Ono (2009)
	Wk-25604	2921	36	Tridacna sp.		2770-2600	2840-2490	2680	
	Wk-25605	2726	38	Tridacna sp.		2530-2330	2660-2280	2440	
Malakai	Beta-479380	2720	30	Anadara antiquita	-135±20	2520-2330	2650-2280	2430	Shaw and Dickinson (2017); Shaw et al. (2020b)
	Beta-479375	2450	30	Mytilidae		2180-1990	2290-1930	2090	
	Beta-479379	2320	30	Tridacna crocea		2010-1830	2100-1750	1920	

Site	Lab code			Material					Reference
	Beta-479378	2170	30	*Tridacna crocea*		1830-1660	1910-1580	1740	
	ANU-32536	1405	20	Charcoal		1340-1290	1350-1290	1310	
	Beta-487396	420	30	Charcoal		520-470	530-330	490	
	Beta-487398	420	30	Charcoal		520-470	530-330	490	
	Beta-479374	420	30	Charcoal		520-470	530-330	490	
	Beta-479377	360	30	Charcoal		480-320	500-310	400	
	Beta-487397	280	30	Charcoal		430-290	450-150	380	
	ANU-33532	325	35	Human bone	-135±20	320-150	430-0	280	
	ANU-33529	230	35	Charcoal		310-150	430-0	210	
	Beta-479376	140	30	Charcoal		270-10	290-0	130	
Mt Yeme	ANU-25133/25136	2398	22	Charcoal		2460-2350	2490-2340	2410	Shaw (2015)
Bwapawe	ANU-25132	1705	30	Charcoal		1690-1540	1700-1530	1590	Shaw (2015)
Ghakpo	ANU 32606	1315	25	Charcoal		1290-1170	1300 -1170	1250	Shaw (2015)
	ANU 32539	895	20	Charcoal		900-740	910-730	780	
	ANU 33519	255	30	Charcoal		430-150	430-150	300	
	ANU 32605	700	30	Charcoal		680-570	690-560	660	
	ANU 33520	245	30	Charcoal		320-150	430-0	290	
	ANU-33530/33523	703	25	Charcoal		680-650	690-560	660	
Pambwa	ANU 32534	330	50	Charcoal		460-310	500-300	390	Shaw (2015)
	ANU 33526	400	35	Charcoal		510-330	520-310	460	
	ANU 32535	605	45	Charcoal		650-550	660-530	600	
	ANU 33527	370	35	Charcoal		500-320	510-310	430	

(Continued)

Table 9.3 (Continued)

Site	Lab code	Date	Error	Material	ΔR value	68.3% range	95.4% range	Median	Reference
Wule	ANU 32537	440	30	Charcoal		520-480	540-340	500	Shaw (2015)
	ANU 33525	400	35	Charcoal		510-330	520-310	460	
	ANU-32538/32607	286	22	Charcoal		430-290	440-290	390	
Ndapa	ANU 32531	350	60	Charcoal		480-310	510-300	400	Shaw (2015)
Low:a	ANU 32609	230	30	Charcoal		310-150	430-0	220	Shaw (2015)
Keyvu	ANU 25130	155	30	Charcoal		280-0	290-0	150	Shaw (2015)
	ANU 25131	165	30	Charcoal		290-10	290-0	170	
Morpa	ANU 25133	130	45	Charcoal		270-20	280- 0	130	Shaw (2015)
Njaru	ANU 25126	240	30	Charcoal		320-150	430-0	290	Shaw (2015)
Puti	ANU 25129	140	30	Charcoal		270-10	280- 0	130	Shaw (2015)
Oilobogwa	Ua-15989	105	70	Charcoal		270-20	290- 0	130	Burenhult (2002)
Odubekoya	Ua-15467	930	80	Human bone	-135±20^	900-680	930-660	770	Burenhult (2002)
	Ua-15468	1100	70	Human bone		1050-800	1080-740	930	Burenhult (2002)
	Ua-15986	755	70	Human bone		680-550	730- 530	630	
	Ua-15987	1045	80	Human bone		960-780	1060-690	860	
Selai	Ua-15487	315	55	Human bone	-135±20^	420-0	440-0	230	Burenhult (2002)
Obuwaga	Ua-15985	445	75	Human bone	-135±20^	480-320	530-150	400	Burenhult (2002)
Bwara Tudava	Ua-15990	200	85	Human bone	-135±20^	260-0	290-0	140	Burenhult (2002)
BMY	AA25106	1500	45	Human bone	-135±20	1110-930	12200-870	1030	Bickler (1998)

Site	Lab								Reference
BNA	AA25107	1355	45	Human bone	-135±20	950-780	1040-710	870	Bickler (1998)
MUY-131 (Undalai)	AA25112	465	40	Human bone	-135±20	120-0	240-0	80	Bickler (1998)
	AA25113	475	40	Human bone		130-0	240-0	80	
BND (Bunmu-yuw)	AA25108	1260	50	Human bone	-135±20^	1180-1000	1250-960	1100	Bickler (1998)
	AA25109	395	45	Human bone		430-300	480-280	380	
BPJ (Munob-wag)	AA25110	865	45	Human bone	-135±20^	730-670	780-650	710	Bickler (1998)
	AA25111	745	45	Human bone		670-560	680-550	620	
MUY-192 (Diblimai)	AA25116	530	45	Human bone	-135±20^	530-460	550-320	490	Bickler (1998)
	AA25117	465	65	Human bone		500-320	520-300	420	
MUY-205	AA25118	650	55	Human bone	-135±20^	630-520	650-500	580	Bickler (1998)
	AA25119	995	55	Human bone		910-750	930-710	830	
	AA25120	685	55	Human bone		640-540	670-520	600	
	AA25121	780	55	Human bone		690-560	720-550	650	
MUY-206 (Kilibwai)	AA25122	820	55	Human bone	-135±20^	730-640	780-550	680	Bickler (1998)
	AA25123	1065	55	Human bone		960- 790	1050-740	870	

(Continued)

Table 9.3 (Continued)

Site	Lab code	Date	Error	Material	ΔR value	68.3% range	95.4% range	Median	Reference
MUY-213 (Uselola)	AA25124	685	60	Human bone	-135±20^	650-540	670-510	590	Bickler (1998)
	AA25125	675	55	Human bone		640-530	660-510	590	
MUY-214	AA25126	870	55	Human bone	-135±20^	750-660	900-570	710	Bickler (1998)
MUY-221	AA25127	930	60	Human bone	-135±20^	800-680	910-670	760	Bickler (1998)
	AA25128	975	60	Human bone		900-730	920-690	810	
MUY-227	AA25129	645	55	Human bone	-135±20^	630-510	650-500	570	Bickler (1998)
Rainu	ANU-371A	1240	145			Dates combined in Oxcal				Egloff (1979)
	ANU-371B	810	100						
	R_combine	956	83	Charred material		930-770	1060-680	850	
	ANU-419	560	80			650-510	680-480	580	
	ANU-369A	770	90			Dates combined in Oxcal			
	ANU-369B	670	235						
	R_combine	757	85	Charcoal		780-560	910-550	700	
	ANU-370	1040	90	Charcoal		1060-800	1180-730	950	
	ANU-361	880	60	Charcoal		910-720	920-680	790	
	ANU-362	810	95	Charcoal		900-660	930-560	740	
	ANU-363	920	85	Charcoal		920-740	970-670	830	
	ANU-416	600	100	Charcoal		660-530	730-460	590	
	ANU-418	650	120	Charcoal		690-530	910-460	620	
	ANU-417	510	120	Charcoal		660-340	680-300	530	
Guntuka	Beta-535172	3860	30	Charcoal		4400-4180	4410-4150	4280	Shaw, unpublished
	UNSW-37	3796	13	Charcoal		4240-4140	4240-4090	4190	

Site	Lab code	Age	±	Material	δ13C				Reference
	OZZ294	3630	35	Charcoal		4060-3880	4090-3840	3940	
	UNSW-225	2975	12	Charcoal		3210-3070	3210-3070	3150	
	UNSW-36	3028	13	Charcoal		3320-3180	3330-3160	3230	
	OZZ292	3280	35	Charcoal		3560-3450	3580-3400	3500	
	Beta-515726	2910	30	Charcoal		3140-2960	3160-2960	3050	
	UNSW-35	2436	12	Charcoal		2670-2370	2690-2360	2460	
	UNSW-224	2543	11	Charcoal		2740-2620	2740-2540	2720	
	UNSW-34	2522	12	Charcoal		2730-2540	2730-2500	2620	
	UNSW-33	3041	12	Charcoal		3330-3210	3340-3170	3250	
	OZZ291/ UNSW-32 combine	2536	12	Charcoal		2730-2540	2740-2520	2710	
	UNSW-31	2500	13	Charcoal		2710-2510	2720-2490	2580	
	Beta-515724	2510	30	Charcoal		2730-2510	2740-2490	2590	
	UNSW-29/30 (R_combine)	2493	9	Charcoal		2710-2500	2720-2490	2590	
	UNSW-28	2234	12	Charcoal		2320-2160	2330-2150	2220	
	UNSW-27	357	11	Charcoal		470-330	480-320	400	
	Beta-482841	2380	30	Charcoal		2460-2340	2670-2340	2400	
	Beta-479381	1780	60	Charcoal		1740-1580	1830-1540	1660	
Mumwa	Beta-482842	3,920	30	Charcoal		4420-4290	4430-4240	4350	Shaw et al. (2020a)
	Beta-502623	820	30	Charcoal		760-690	790-680	730	
	Beta-515728	420	30	Charcoal		520-470	530-330	490	Shaw, unpublished
Bwenabwe-nama	ANU 5131	2440	80	*Shell*	-135±20	2210-1950	2320-1830	2080	(Irwin et al. 2019)
	ANU 5133	1630	80	*Shell*	-135±20	1270-1060	1350-940	1160	
Conus	Wk-25782	1209	30	*Conus shell*		790-660	900-610	730	Ambrose et al. (2012)
	Wk-25781	1251	30	*Conus shell*		840-680	910-640	770	
	Wk-31234	944	31	*Conus shell*		580-450	650-380	510	
	AA-25130	1205	45	*Conus shell*	-135±20	800-650	910-590	730	
Lagisuna	NZ5187	890	28	*Strombus luhuanus*		540-400	600-310	470	Shaw et al. (2021)

(Continued)

Table 9.3 (Continued)

Site	Lab code	Date	Error	Material	ΔR value	68.3% range	95.4% range	Median	Reference
	NZ5188	758	55	*Strombus luhuanus*		440-270	510-170	350	
Tupwana	NZ5190	776	55	*Strombus luhuanus*	-135±20	450-290	520-190	370	Shaw et al. (2021)
Tupwalaul-au	NZ5189	476	76	*Strombus luhuanus*	-135±20	150-0	270-0	100	Shaw et al. (2021)

Grey shade = Radiocarbon dates with errors of 100 years or more. Dates calibrated using OxCal4.4 and the IntCal20 dataset for terrestrial samples and Marine20 for marine shell samples. Calibrated date ranges rounded to the nearest 10 years. Radiocarbon dates are noted as before present (BP), with BP being AD 1950. ^20% Marine20 input and ΔR correction applied to bone samples to account for probable marine dietary input.

has therefore been a predominant focus on coastal open settlements, with less attention on inland locations and caves.

There are clear biases in the resolution of some site chronologies. More dates are associated with archaeological sites excavated in recent decades due to the per-sample cost for dating having become less prohibitive and with more laboratories having radiocarbon capacity. This is most apparent with the Caution Bay sites, which were excavated as part of a large-scale commercial salvage project (Richards et al. 2016). Notably, seven sites stand out as having a larger number of associated radiocarbon dates than most (Bogi 1, Tanamu 1, Kikinui, Popo, Motupore, Gutunka, and Hopo) (see Figure 9.7A). Of these, Bogi 1 (≤4790–4420 cal. years BP) and Tanamu 1 (≤5200–4880 cal. years BP) in Caution Bay have by far the most robust chronologies with 113 and 59 dates respectively, together representing 25% of the total number of dated samples. McNiven (personal communication) states in Thangavelu (2015), there are 142 dates from Bogi 1, although again, not all are currently available. Of those dates available at Bogi 1, 60% (n = 68) were obtained on the shell from a well defined 26-cm-thick midden deposit for the purpose of developing species-specific Delta-R corrections (Petchey et al. 2013; Petchey et al. 2012). Although the comparability between these dated samples is excellent, they represent only a small part of the temporal sequence.

Regional chronologies

By region, the South Coast is best represented and comprises 43% of the total radiocarbon sample but only 22% of the site total, equating to 293 dates from 24 sites (see Figure 9.7B and Table 9.2). A large number of sites in Caution Bay have yet to be reported in full, and only generalised age ranges have so far been provided and are therefore not discussed here (e.g. Mialanes et al. 2016). David et al. (2012a: 75) have stated that ~1000 AMS dates from 122 excavated sites in Caution Bay are to be published in forthcoming monographs, which are eagerly awaited and will expand on the meta-analytic model presented here. Other than the Caution Bay sites, Motupore (≤1180–730 cal. years BP) has the most robust chronology with 27 dates (Allen 1978, 2017). The Gulf region is similarly well represented with 273 dates (39% of the total sample) from 44 sites (41% of site total) (see Table 9.1). Here, Kikinui (≤1540–1380 cal. years BP), Popo (≤680–560 cal. years BP) and Hopo (≤2750–2510 cal. years BP) have the most robust chronologies (David 2008; Rhoads 1980; Skelly et al. 2014; Urwin et al. 2021).

The Massim Islands represent 18% (n = 121) of the total dated sample from 40 sites (37% site total) (Table 9.3). The lower ratio between radiocarbon dates and excavated sites in the Massim reflects the exploratory nature of archaeological investigations in the region thus far, with single dates obtained from 17 sites as part of preliminary test excavations, compared to only six sites in the Gulf and five along the south coast.

Nonetheless, the number of radiocarbon dates for the Massim has more than doubled within the last decade because of renewed research activity in the region, with Gutunka (≤4410–4150 calBP) having the most robust chronology (Unpublished, but see Shaw 2019).

A radiocarbon chronology for the Western province (within the Gulf region here) is entirely lacking, despite a considerable suite of ethnographic information on the peoples who inhabit these mosaic landscapes (Chalmers 1903; Ely 1988; Hitchcock 2004; Landtman 1927; Ohtsuka 1983; Williams 1936a). Across this vast region, preliminary excavations have been undertaken only in a limestone cave on the Oriomo Plateau and ditch and mound agricultural systems at Waidoro (Padad Kao). However, the cultural deposits have not been dated or analysed in detail (Harris & Laba 1982; Hitchcock 2010; Lampert 1966). A lack of investigations elsewhere in the Western Province is largely because cultural sites with extended antiquity will likely be poorly preserved or deeply buried on the alluvial plains. Researchers have therefore focused efforts and finite funding elsewhere.

Dated material

Of the dated samples, 69% (n = 475) were on wood charcoal (see Figure 9.7c). With few exceptions, charcoal fragments were from unidentified taxa, which can, in some circumstances, have in-built errors of up to several hundred years (Allen and Wallace 2007; Gavin 2001). While this issue cannot be avoided when using published radiocarbon dates, the effects of inbuilt age can be minimised by assessing single dates within regional cultural chronologies and, in appropriate cases, by using Bayesian statistics to identify clear outliers (Ramsey 2009). Marine shell dates make up 25% (n = 171) of the total dataset. Marine shells often can be unreliable on account of variation in the local marine carbon reservoir for which location-specific corrections are required (Heaton et al. 2020). Species-specific Delta-R corrections have been developed for Caution Bay spanning the last 3000 years (Petchey et al. 2013; Petchey et al. 2012), and regional corrections are available for the Massim (Ambrose et al. 2012; Petchey and Ulm 2012). However, the applicability of these corrections to each island remains uncertain and offset errors must be assumed until higher resolution reservoir corrections can be developed.

A smaller proportion of dates were obtained from human and animal bone, and teeth fragments (5%, n = 35) which are equally as problematic for dating as calibration requires additional carbon, nitrogen and sulphur isotope datasets to estimate the dietary input from marine foods (Petchey and Green 2005; Petchey et al. 2011). With the exception of the human burial at the Malakai site on Nimowa Island, the data are not available for human bone samples (Shaw 2015: 151). A 20% marine input, therefore, has been conservatively estimated for human bone samples obtained from the Massim islands. Fishbone from marine environments, if identified, can be calibrated in the same manner as shell samples. Finally, three-quarters of dates (n = 520, 76%) have uncalibrated errors of ±40 or less, which is typical for Holocene determinations using modern accelerator mass spectrometry (AMS) techniques. A total of 43 dates (6%) have uncalibrated errors equal to or greater than ±100 years, all of which were obtained before the advent of AMS. Most of these dates are less than 1000 years in age and therefore have limited interpretive value.

Late Pleistocene and Early Holocene settlement (17,000–8000 years BP)

When the radiometric dataset across these regions is considered, it is apparent that the last 5000 years are now relatively well documented, whereas earlier contexts are represented by only 12 dates from four sites (Figure 9.8). The three earliest sites dating to the Late Pleistocene (Kelebwanagum, OJP) and Early Holocene (Wokoi Amoho) are cave deposits. It is not surprising that caves have yielded the only Late Pleistocene evidence as sedimentary cave deposits are often not exposed to the same dynamic erosional processes as open site contexts.

Kelebwanagum cave

Kelebwanagum cave on the uplifted limestone island of Panaeati in the Massim region is the oldest site in southern New Guinea and the smallest island in the Australasian region with evidence for Late Pleistocene human habitation (Shaw et al. 2020a). The cave is located 160 m inland and ~10 m above modern sea level, on the edge of an uplifted limestone plateau (Figure 9.9A). Excavation of a 70 cm deposit revealed the cave had been utilised intermittently starting 17,090–16,600 cal. years BP (14,480 ± 40 uncalibrated years BP, shell) and continuing until 12,680–12,330 cal. years BP (11,010 ± 35 uncalibrated years BP, shell), after which time there was no sediment deposition for nearly 8000 years (Figure 9.10A). When the cave first was used, the island was 280 km² in area, but it had reduced to 30 km² by the time it had been abandoned, equating to ~90% reduction in land area. West Fergusson obsidian was transported across distances of 200 km to the cave during the earliest phase of occupation, indicating the continued use of maritime transport as well as social connections between island populations. Late Pleistocene sites are likely to be found on Fergusson Island, which is located just offshore from Sahul but was never connected to it. Faunal remains indicate the early inhabitants harvested chiton (*Acanthopleura* sp.) and crab (Decapoda) from the exposed limestone shores near the cave and opportunistically caught turtle and fish. *Turbo* sp. shells were used for the production of net weights and were perhaps used for fishing. It is argued that people on Panaeati began to move away from the coast as sea level started to rise and subsequently abandoned the island in favour of larger "refugia" islands when the shallow platform was rapidly inundated (see Figure 9.9).

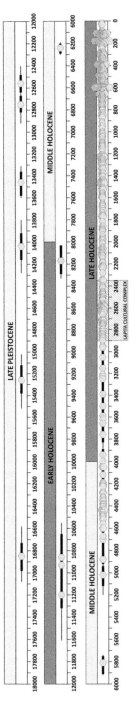

Figure 9.8 Meta-chronology of radiocarbon dates (N = 609) from the South Papuan lowlands and Massim Islands. The shell dates for Bogi 1 (N = 78), reported by Petchey et al. (2012, 2013), were not included. See Tables 9.1, 9.2, and 9.3 for the dating details.

(a)

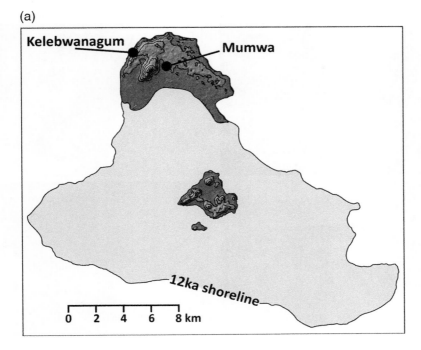

(b)

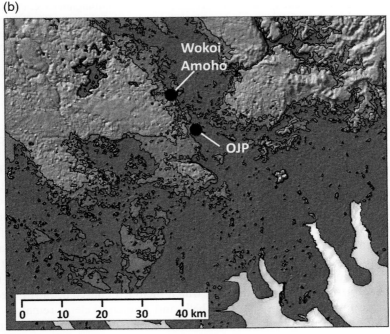

Figure 9.9 Topography of landscapes where Late Pleistocene and Early Holocene sites have been excavated. A) Panaeati Island, showing 30 m contours. The edge of the lagoon is shown that marks the former shoreline prior to 12,000 years ago. B) Kikori River basin, showing 30 m and 60 m contours. Maps were created using GeoMapApp (www.geomapapp.org) and global multi-resolution topography data compiled by Ryan et al. (2009).

Figure 9.10 Environmental and cultural contexts. A) Kelebwanagum cave, Panaeati Island, a 70cm deposit with 17ka cultural evidence. B) Gutunka, Brooker Island, a 280cm deposit with the earliest cultural evidence (4410-4150 calBP) below the water table. C) Bogi 1, Caution Bay, a 350cm deposit beginning 4790-4420 calBP. Note the beachrock near the base of the excavation indicating the former inter-tidal zone. D) Photos A-D: Ben Shaw.

OJP cave

OJP is the only other Late Pleistocene site known in the southern Papua New Guinea lowlands. It is a small limestone cave 40 km upstream of the Kikori River and 10 m above the flood-prone river plains in the Gulf Province (David et al. 2007). Intermittent human presence was identified during the Late Pleistocene-Early Holocene transition, starting about 13,600–13,300 cal. years BP (11,602 ± 70 uncalibrated years BP, nut) until 11,150–10,550 cal. years BP (9526 ± 66 uncalibrated years BP, nut), after which time it was not used again until within the last 1000 years. Unfortunately, the shallow deposit (~26 cm) had undergone substantive post-depositional mixing, and there was no clear association between the dated samples and cultural material, which included charcoal, burnt seeds, vertebrate faunal remains, marine shell, and human skeletal remains. At least two of the four Late Pleistocene radiocarbon dates were obtained from the top 10 cm of the deposit in close association with skeletal and dental remains from secondary burials directly dated within the last 500 years (David 2008). David et al. (2007) argued that the presence of charcoal of this antiquity nonetheless confirms a Late Pleistocene human presence in what was probably a lowland rainforest ecosystem with high annual rainfall where natural fires are virtually unknown.

Wokoi Amoho cave

Besides OJP, the only site with evidence for human presence in the southern lowlands or Massim islands during the Early Holocene is Wokoi Amoho, a rock shelter 10.6 km farther inland from OJP (David 2008). The rock shelter was visited since 8350–8000 cal. years BP (7365 ± 46 uncalibrated years BP, charcoal), at the Early-Middle Holocene transition. The cave similarly was not reused again until within the last 1000 years. Unfortunately, no details of these dates are provided in relation to their stratigraphy or their association with cultural material, and the depth of the deposit was not reported. The site has been interpreted as a hunting camp which is supported at least in more recent contexts by oral traditions (McNiven et al. 2010: 44). Of particular importance is that Wokoi Amoho was utilised at a time when much of the coastal lowlands were flooding, and the coastline would have begun to recede from its maximum extent. The cave, therefore, potentially demonstrates that people lived or moved through the interior lowlands while the coastline was relatively unstable.

Post-glacial population dynamics in southern Papua New Guinea

Settlement during the Late Pleistocene and Early Holocene appears to have been infrequent and involved small and relatively mobile groups of people, as was the case in the Bismarck Archipelago and New Guinea Highlands at this time (Allen et al. 1989; Leavesley 2006; Summerhayes et al. 2017). Population density was very low during the Late Pleistocene and probably remained low until the mid-late Holocene. The presence of Late Pleistocene sites on both island and inland riverine landscapes suggests populations had adapted to living across a broad spectrum of lowland ecosystems. However, cave sites provide relatively limited insights into human behaviour, as they typically were not used for long-term settlement, and comparison with open sites is needed to expand on the scope of existing models.

The shallow depth of the sedimentary deposits suggests that these sheltered localities had not been affected by the significant landscape changes that occurred across the surrounding lowlands throughout the Holocene. Yet, temporal resolution is generally poor, and use of the caves over more recent millennia has had an adverse impact on *in situ* stability of cultural materials. For instance, while the Late Pleistocene chronology from Kelebwanagum cave has no significant date inversions, it nonetheless forms a palimpsest record of human use at this time. The relatively limited representation of Late Pleistocene cultural sites can be attributed, in part, to caves typically not being the focus of survey plans and excavation strategies or a complete lack of suitable caves or rock shelters in surveyed areas.

Highland records (Ivane and Simbai Valleys) suggest that elevations at or above 2000 m above sea level were abandoned during the LGM, with people not returning until several millennia later (Summerhayes et al. 2010; Field, Shaw, and Summerhayes, submitted). It is likely that people moved to lower elevations during this time and perhaps favoured lowland locations bordering montane forests where a range of terrestrial fauna could be hunted, and access to marine and freshwater resources could also be maintained. Currently, no Late Pleistocene archaeological sites are known on mainland Papua New Guinea between 100 m (Watinglo) and 1280 m above sea level (Yuku), forming a substantial blank in our understanding of human-environmental dynamics (Bulmer 1975; O'Connor et al. 2011; Summerhayes et al. 2017). Until this blank is filled, models attempting to explain the processes of New Guinea colonisation and adaptive behaviour will be inherently inadequate. While this blank is certainly due to a lack of targeted survey at these elevations, it raises the possibility that people preferably lived on the coast and only made short term trips to higher elevations during the Late Pleistocene and Early Holocene (Summerhayes et al. 2017). If this were the case, then sites with evidence for Late Pleistocene settlement may be detected on broad ridges, spurs, and saddles in the lowland foothills that form natural pathways into the mountainous interior.

Middle Holocene (8000–4000 years BP)

More regular coastal settlement is evident from the end of the Middle Holocene when sea level was still above modern but had begun to recede (Figure 9.11). Based on palaeo-ecological records, lowland ecosystems were similar to today and could potentially have supported a sizable population, irrespective of whether swidden agriculture or other forms of plant management were practised. Seven sites with Mid-Holocene evidence are now known, including four along the south coast (Ruisasi 2, Tanamu 1, Bogi 1, and Kukuba Cave) and three on Massim islands (Kelebwanagum cave, Mumwa, and Gutunka). There has been a significant increase in the number of mid-Holocene sites reported over the last decade, as only one (Kukuba Cave) was known prior to this. There are currently no mid-Holocene sites in the Gulf, nor are there sites anywhere in the study region with chronologies that span the earlier half of the Mid-Holocene.

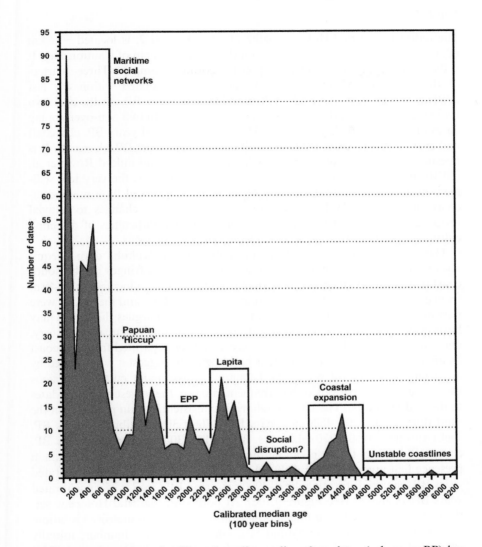

Figure 9.11 Frequency of calibrated median radiocarbon dates (cal. years BP) by 100-year periods (N = 609). Shell dates from Bogi 1 from Petchey et al. (2012, 2013) were excluded. Four notable peaks in date frequency are present at 4600–4000 (mid-Holocene), 2800–2400 (Lapita), 1600–1100 (Papuan "hiccup") and ≤800 cal. years BP (expansion of maritime social networks). See Tables 9.1, 9.2, and 9.3 for the dating details.

South Coast

The oldest of the mid-Holocene suite of sites, Ruisasi 2, is located ~1.6 km inland and ~15–20 m above sea level on the savannah plain of Caution Bay. It is the same locality from which the pollen record was sampled. Three-quarters of the excavated 45 cm sedimentary deposit accumulated within the last 2710–2480 cal. years BP (2766 ± 26 uncalibrated years BP, shell), and two broken chert flakes in the basal clay were associated with two non-overlapping dates of 6300–6190 cal. years BP (5432 ± 30 uncalibrated years BP, charcoal) and 5920–5740 cal. years BP (5085 ± 30 uncalibrated years BP, charcoal). It is questionable whether this represents human activity, and indeed Rowe et al. (2020) suggested the evidence was equivocal. Nonetheless, the very low sediment accumulation rate during the Mid-Holocene followed by a marked increase in the Late Holocene demonstrates substantive changes to alluvial processes on savannah landscapes and highlights the difficulties of detecting cultural deposits of this age in slow-developing clay deposits.

The cultural records from Tanamu 1, Bogi 1 and Kukuba cave unequivocally demonstrate a human presence on the coastal fringes of the south coast from 5000–4000 years ago. Tanamu 1 and Bogi 1 are located only 140 m from each other on an aeolian sand dune in Caution Bay, and they likely were part of a contiguous cultural landscape. Specht (2012) argues these sites may have been on an inshore island for at least part of their occupation history, and while a part of this dune system may have been separated from the coast by an intertidal zone at times, there is no evidence to suggest it was isolated by any significant body of water. At Tanamu 1, three major cultural horizons were separated by periods of less frequent site use, the lowest of which dates to the mid-Holocene. There was a relatively consistent human presence on the sand dune from 5200–4880 cal. years BP (4766 ± 30 uncalibrated years BP, shell) through 4160–3970 cal. years BP (3715 ± 30 uncalibrated years BP, charcoal). At Bogi 1, four major cultural horizons were defined, with a human presence first apparent from 4790–4420 cal. years BP (4053 ± 30 uncalibrated years BP, charcoal) through 4090–3900 cal. years BP (3667 ± 25 uncalibrated years BP, charcoal) (see Figure 9.10c). At both sites, the Mid-Holocene settlement was followed by a millennium of infrequent, low-intensity visitation.

Kukuba cave near Hall Sound is a small limestone chamber, initially excavated in 1964 and more substantively in 1969–1970 (Vanderwal 1973: 44–47; White 1967: 23–24). The cave contained evidence for two distinct periods of cultural activity underlying dense bat guano, with a hearth within the earliest deposit dated to 4580–4150 cal. years BP (3946 ± 69 uncalibrated years BP, charcoal, ANU-395 combined). It is uncertain for how long it was periodically utilised during the mid-Holocene, but later occupation did not occur until around 1200 years ago and was associated with pottery.

Mid-Holocene settlement of the south coast was associated primarily with flaked stone and shell technology. At Kukuba, shell and bone were absent, but 43 low-quality chert artefacts (18 notched stones and 25 utilised flakes) reflected the exploitation of locally available lithic resources during infrequent human visitation to the cave, perhaps while people gardened nearby or hunted in the savannah (Vanderwal 1973: 175–176). The reduction characteristics of the Kukuba chert indicate relatively expedient use of locally available material and is similar in nature to Mid-Holocene technologies from cave sites in highland, lowland, and island contexts elsewhere in Papua New Guinea (Bulmer 1975; Christensen 1975; Fredericksen et al. 1993; Gorecki et al. 1991; White 1972).

Little is yet known of the Mid-Holocene material assemblage from Bogi 1. McNiven et al. (2011) noted, however, the presence of fully ground stone axes dating to ~4200 cal. years BP and a well preserved, complete flexed burial, but the details have yet to be reported in full. The burial is probably the earliest complete interment in the Pacific, and an association with marine shells might suggest a close ritual connection to the marine environment. A lack of disarticulation potentially reflects individual perceptions of personhood. We await the publication of the full studies and findings.

At Tanamu 1, Szabo et al. (2020) reported a range of modified shell artefacts (n = 16) from Mid-Holocene contexts, including a *Conus* sp. adze, *Nautilus* sp. disc bead, and several ring preforms. While shell working is known in Late Pleistocene and Mid-Holocene contexts in the Bismarck Archipelago and Massim (Gosden et al. 1994; Shaw et al. 2020a; Smith and Allen 1999), the presence of forms most often found in Lapita cultural contexts suggests these technologies were either already in use prior to the arrival of Lapita groups, or perhaps during initial contact between Lapita and incumbent populations. In particular, the beads and rings may have been worn as indicators of social status, as has been argued for Lapita groups and would imply some degree of social stratification in these Mid-Holocene cultural groups. Furthermore, analysis of the non-worked shellfish indicates communities at Tanamu 1 and Bogi 1 primarily exploited intertidal sandflats, estuaries, mangroves, and rocky shore habitats (Faulkner et al. 2020; Thangavelu 2015). In line with the palaeo-ecological records, discarded fauna indicate coastal ecosystems similar to the present today were established by the Mid-Holocene.

Massim

Three Mid-Holocene sites are now known in the Massim islands, although the details so far have been published for only one site. Kelebwanagum cave was reused after a long hiatus from 4810–4450 cal. years BP (4520 ± 30 uncalibrated years BP, shell) until at least 3820–3470 cal. years BP (3720 ± 30 uncalibrated years BP, shell). The density of discarded faunal remains increased substantively during the Mid-Holocene relative to both the Late Pleistocene and Late Holocene layers and notably included shellfish from the reef and sandy shore habitats. The harvested shellfish demonstrate that the expansive Deboyne reef-lagoonal system around the island had formed by this time, unlike the exposed coasts which would have characterised the island during the Late Pleistocene. Shell and bone were used to manufacture tools in the absence of good quality lithic resources. Obsidian discard had also increased during the Mid-Holocene, which derived from at least three West Fergusson sub-sources, with an expanded social network between island populations inferred.

The open site of Mumwa, located on a limestone plateau in the centre of Panaeati Island 2.5 km to the southeast of Kelebwanagum and 1.6 km from the coast, appears to have been infrequently inhabited from 4430–4240 cal. years BP (3920 ± 30 uncalibrated years BP, charcoal) (Shaw et al. 2020a). Although full details of this cultural sequence are forthcoming, it can be noted that during the Mid-Holocene, the limestone plateau would have consisted of exposed coral limestone with very little soil development – similar to most locations on southern Panaeati today. Its location, on high ground and adjacent to a lower-lying basin near the base of the volcanic mountain where freshwater naturally accumulated, made it an attractive place to settle rather than nearer to the coast where no such water traps are present (see Figure 9.9). More intensive settlement at Mumwa probably occurred within the last 2000–1500 years associated with "Early Papuan Pottery" (cf. Irwin et al. 2019).

Finally, excavations at the Gutunka site on Brooker Island have yielded a sequence containing multiple periods of cultural activity, with settlement commencing during the Mid-Holocene from 4410–4150 cal. years BP (3860 ± 30 uncalibrated years BP, charcoal) and continuing intermittently until 3160–2960 cal. years BP (2910 ± 30 uncalibrated years BP, charcoal) (see Figure 9.10b). It is uncertain at present if there was a hiatus in settlement at this site prior to the arrival of Lapita cultural groups. Modified turtle bone, flakes, and obsidian was associated with the mid-Holocene settlement, demonstrating a reliance on both local and non-local resources. The full details of this site are forthcoming.

Stabilisation of coastal ecologies and mid-Holocene human settlement

Currently, it appears that population densities in the southern lowlands and on Massim islands prior to 5000 years BP were very low, and groups were relatively transient. A lack of evidence for coastal habitation in the southern Papuan lowlands for much of the Mid-Holocene and Early Holocene is likely to be related to higher than modern sea level having flooded the former coastline and inundated parts of the present coastline. However, changes in sea level cannot explain a current lack of evidence for Mid-Holocene habitation farther inland across the lowland savannah or in the interior of islands. While the interior lowlands along much of the south coast have not been the focus of the survey, it is striking that no Mid-Holocene cultural deposits were detected in the upriver Gulf region where Late Pleistocene and Early Holocene evidence have otherwise been reported.

It raises the possibility that coastal locations and a focus on maritime subsistence strategies were favoured by populations during the Mid-Holocene, as suggested thus far by the substantive assemblages at Bogi 1 and Tanamu 1.

The beach ridge on which Bogi 1 and Tanamu 1 are situated is 4–5 m above modern sea level, with buried evidence for Mid-Holocene settlement 1–3 m above the modern level. When considered against regional sea level records, settlement from 5000 years ago must have occurred when the dune initially began to form and was just above the heightened Mid-Holocene sea level. Indeed, the presence of concreted sand (beach rock) immediately below the basal date of 4053 ± 30 uncalibrated years BP at Bogi 1, which typically forms in intertidal zones, provides supporting evidence for occupation of the dune during the initial stages of formation (Mauz et al. 2015). The sea level reached its modern position between 3000 and 2000 years ago, and by this time, the dune system was well established. Notably, Mid-Holocene evidence at Mumwa and at Ruisasi 2 lacked the temporal control observed in beachfront sites because of the comparatively slow development of sediments away from the coast. At these locations, erosional processes may occur at a similar or greater rate as sediment deposition.

Mid-Holocene occupation at Gutunka similarly must have occurred as soon as a coastal fringe had developed on Brooker Island. Many of the smaller Massim islands, especially neighbouring islands in the Calvados chain, lack the sheltered bays characteristic of Brooker and would not be well suited for long term settlement prior to 5000–4500 years BP. Probably just a few small islands were inhabited at this time, with a focus on larger islands with a more diverse range of ecosystems and natural resources to support a subsistence base. Certainly, it has been demonstrated elsewhere in New Guinea (Arawe and Nissan Island groups) that coastal and island settlement during the Mid-Holocene commenced shortly after habitable fringes had developed (e.g. Gosden et al. 1994; Spriggs 1991). If an earlier settlement were present in the Massim, then it would be preserved only on elevated relict beach deposits or on low-grade spurs, ridges, and plateaus.

Late Holocene (4000 years BP–present)

An enigmatic millennium? (4000–3000 years BP)

It has become increasingly apparent with the development of longer and more robust chronologies, admittedly at relatively few locations, that there was less frequent habitation between 4000 and 3000 years BP (see Figure 9.11). Of the 107 sites and 687 radiocarbon dates included in this meta-analysis, only 11 dates (median) from five sites fall within this millennium, including Tanamu 1 (n = 2), Bogi 1 (n = 1), Popo (n = 1), Kelebwanagum (n = 1), and Gutunka (n = 6). Nebira 4 yielded a date of 3340 ± 160 uncalibrated years BP, but this was a duplicate of a sample that yielded an age of 1760 ± 90 uncalibrated years BP obtained from the aberrant Gakushuin Laboratory and is therefore not considered here (Allen 1972b). The single date from Popo (3180 ± 180 uncalibrated years BP, charcoal) is at odds with the robust chronology indicating human settlement of this rapidly forming landscape no later than 680–560 cal. years BP (690 ± 20 uncalibrated years BP, charcoal), but does raise questions as to how charcoal of this age was deposited in a terrestrial environment that likely would not have existed at this time (Urwin et al. 2021).

Of the remaining sites, it is unlikely that habitation of Kelebwangum ever was permanent, so a pattern of intermittent visitation is not unusual. At Gutunka, it appears that coastal settlement had been intermittent since it was first inhabited around 4400 years ago and was more intensively utilised after 3500 years ago, but it remains uncertain if the island was inhabited when Lapita cultural groups arrived. The tighter temporal control of the Bogi 1 and Tanamu 1 sequences, however, does suggest the dunes were only occasionally utilised between 4000 and 3000 years ago. Sea level had begun to recede by 4000 years ago, coastal dunes had stabilised, and core records indicate that rainfall patterns were similar to today. It is therefore likely that a change in the intensity of coastal settlement, at least in Caution Bay but perhaps across a wider geographic area, was predominantly driven by social factors. Had populations dispersed further into the interior as savannah landscapes became more conducive for sustained settlement, or was there a reorganisation of settlements across lowland localities? We await the results of further survey and excavation to provide clarity on these issues.

Lapita, indigenous populations, and modified landscapes (3000–2300 years BP)

There are now 18 sites dating between 3000 and 2300 years BP in the Gulf (n = 7), along the south coast (n = 7), and in the Massim (n = 4). Of these, nine are associated with Lapita cultural deposits (Tanamu 1, Bogi 1, Edubu 1, Moiapu 1, Moiapu 3, Hopo, Kasasinabwana, Malakai, and Gutunka), of which five are within 6.5 km of each other in Caution Bay. Mialanes et al. (2016) and McNiven et al. (2011) together list an additional 11 sites in Caution Bay associated with decorated Lapita pottery that have not yet been published in any detail (JD17, JD10, JD14, Tanamu 3, ABKL, Ataga 1, AAIT, AAUJ, Nese 1, Moiapu 2, Edubu 2). Many of these sites overlapped in age and were likely part of an integrated series of villages that probably also included indigenous cultural groups (McNiven et al. 2012: 144).

All known Lapita sites in southern Papua New Guinea and the Massim are located within 2 km of the coast. The earliest dates for Lapita occupation have been documented at Tanamu 1, located only 25 m from the modern coast and dating as early as 3060–2860 cal. years BP (2842 ± 30 uncalibrated years BP, charcoal) (David et al. 2011). However, this date is out of sequence, with dates above (n = 8) and below (n = 3) dating no earlier than 2950–2750 cal. years BP. The earliest dates attributed to Lapita at Bogi 1 at 2960–2780 cal. years BP (2783 ± 30 and 2775 ± 30 uncalibrated years BP, charcoal) are from the base of a dense Lapita midden layer with all other dates (n = 4) within this layer started around 2750–2490 cal. years BP. As such, the Lapita settlement of southern Papua New Guinea likely occurred at least a century later than previously published, with most evidence dating since 2750–2500 cal. years BP. Finer temporal control over this chronological period is difficult because of a flattening in the calibration curve at this age (Reimer et al. 2020). In contrast to the coastal location of Tanamu 1, the Hopo site in the Gulf is 4.5 km inland from the modern shoreline, but sediment analyses confirm it was adjacent to the coast when it was occupied since 2750–2510 cal. years BP (2552 ± 25 uncalibrated years BP, charcoal). It is clear that populations with pottery bearing "Lapita" culture predominantly settled on coastal margins but expanded several kilometres inland to exploit terrestrial resources (fauna and lithics) while maintaining a heavy reliance on the marine environment (McNiven et al. 2012). In part, these inland settlements may have been occupied as Lapita groups expanded connections with indigenous populations living in the hinterland and elsewhere along the coast.

Excavations on the savannah hinterland in Caution Bay have demonstrated that 30–90 cm of silty clay had developed within 300–400 years of Lapita settlement, at most, and is a marked increase in depositional rate compared to earlier contexts. On Brooker Island, 140 cm of sediment had accumulated within no more than 150 years of Lapita arrival, in contrast to the underlying 70 cm that accumulated over 1450 years. The lowland coastal landscapes, therefore, had changed significantly within the last 3000–2500 years as a result of increased burning and slope erosion associated with more intensive land use with the arrival of Lapita cultural groups (David et al. 2019a; McNiven et al. 2012). It is probable that the preservation and visibility of earlier cultural evidence were affected negatively by slope erosion that increasingly occurred during the Late Holocene, particularly across relatively shallow clay deposits characteristic of savannah plains.

Of the other nine sites dating at approximately 3000–2300 cal. years BP that do not have any clear affiliation with Lapita cultural groups, six are located in the Gulf (Epe Amoho, KG141, EMO, Waredaru, Kaveharo, Hohelavi), two along the south coast (Ruisasi 1, Ruisasi 2), and one in the Massim (Mt Yeme). A date within this range from the Gulf site of EMO has since been confirmed as erroneous (David et al. 2010). Of the remaining sites, the dates at KG141, Ruisasi 1, and Ruisasi 2 could not be correlated with any cultural material or social group as they derived from palimpsest deposits (David 2008; David et al. 2016; Rowe et al. 2020). At Epe Amoho, human occupation between about 2850 and 2500 cal. years BP was associated with fish and shark remains, consistent with the associated palaeoecological record indicating that a brackish confluence extended 45 km up the Kikori River. Bamboo remains here have been argued to reflect the manufacture of organic tools and utensils. Notably, this deposit did not contain any stone artefacts, pottery, or shellfish remains which were present in later layers of occupation and were likely used as a short-term shelter only at this time (McNiven et al. 2010).

Excavations at the open site of Kaveharo in the Gulf revealed small quantities of red-slipped pottery, some with shell impressed decoration that is regionally consistent with traditions in the Port Moresby region dating to the latter end of this time range (Skelly and David 2017: 253–321, 452). Certainly, by 2500 years BP, regional pottery traditions had begun to emerge alongside those that are characteristic of Lapita, although Lapita pottery persisted in simplified forms for another one to two centuries. At Edubu 1, similar shell-impressed pottery was found in contexts alongside dentate stamped Lapita pottery and may represent the development of regional systems of social organisation and cultural identities that also involved trade to the Gulf (McNiven et al. 2012). At Hohelavi, an open site only 140 m from Kaveharo, occupation commenced at 2700–2360 cal. years BP (2449 ± 25 uncalibrated years BP, charcoal) but is associated only with a single chert flake, with pottery deposition only occurring after 2000 years BP (see Figure 9.10d). This location was likely on the periphery of more intensively settled areas until this time.

In the Massim, Mount Yeme Cave on Rossel Island was first visited at 2490–2340 cal. years BP (2398 ± 22 uncalibrated years BP, charcoal). Although no cultural material was recovered from the deposit, sediment phytolith analyses revealed banana in the lower deposit, which was likely transported to the cave with people. Banana phytoliths were notably absent from overlying culturally sterile sediments (Shaw 2015: 138). The cave is on an isolated mountainside on the island and was probably used as a shelter while people gardened on the adjacent slopes or during periods of conflict. It was abandoned after 1700–1530 cal. years BP (1705 ± 30 uncalibrated years BP) and appears to have coincided with the partial collapse of the chamber and entrance. It is now a sacred location where spirits of the deceased enter the underworld. Taboos associated with its sacred designation may have prevented people from using it, at least in recent centuries.

Sites without any clear ties to Lapita cultural groups are therefore known across once-coastal sand dunes, upriver terraces, and islands. On present evidence, it cannot be certain if these sites represent "Lapita without pots" or reflect unrelated cultural groups that lived alongside migrant Lapita communities. To some extent, and as argued earlier by Gorecki (1992), a focus on Lapita in Papua New Guinea has been at the expense of defining in more detail the social existence of indigenous populations at this time. Nevertheless, it is clear that the increased visibility of pottery-producing cultural groups across lowland and island environments within the last 3000 years was coupled with increased anthropogenic modifications of lowland landscapes.

Settlement of a modified landscape (2300–800 years BP)

By 2300–2000 years BP, modern or nearly modern sea level had been reached, and the lowlands had, in parts, already been altered substantively through anthropogenic clearance and burning for at least a millennium. Mangrove forests had begun to proliferate along formerly open coastlines (Rowe et al. 2013; Shaw et al. 2020b). In this context, Barham (1999) suggests that increased habitation in the Gulf and probably elsewhere along the south coast within the last 2500–2000 years may have been an opportunistic response to geomorphological changes following the formation of biodiverse back-mangrove freshwater wetlands and stabilisation of coastal landscapes.

There is the greater resolution of settlement spanning the last 2300–2000 years along the south coast, but the comparable resolution is currently lacking for the Massim islands. A total of 39 sites (36% of site total) have median dates within this range. Pottery bearing sites dating between 2200 and 1600 years BP have been linked with the maritime dispersal of cultural groups who occupied island (Oposisi, Mailu, Moturina), coastal (Bogi 1) and inland (Nebira) locations. These communities produced and imported pottery to villages as far as the Gulf (e.g. Hopo), and probably to the Massim islands (Irwin et al. 2019; Skelly and David 2017; Summerhayes and Allen 2007). Prior to the discovery of Lapita, these pottery-bearing sites had been argued to represent the initial introduction of pottery to the region associated with a rapid maritime dispersal of people along the coast and into the hinterland. While it is no longer the earliest pottery, this argument is sustained as the similarity of these "Early Papuan" pottery forms and decorations suggests close cultural connections between its makers (Allen 1977a, 2010). As with Lapita, the subsistence focus of these populations was derived from both terrestrial and marine resources in the form of gardening, hunting, and fishing.

Between 1300 and 800 years BP, there was an apparent disruption to coastal settlements and established maritime social networks, the so-called "ceramic hiccup" (Irwin 1991). The relative homogeneity of earlier pottery styles became increasingly diverse. It has been argued to have involved increased contact with populations in the Massim either directly through migration or indirectly through trade, identified primarily through the appearance of "Massim" style pottery in sites along the south coast, and particularly in the Port Moresby region (Bulmer 1978; Swadling 1981). However, it is clear that trade of pottery to the Gulf also continued from south coast villages prior to and during this time (Marsaglia et al. 2016).

Sediment and core records indicate substantive landscape modification, both natural and anthropogenic, around the same time, which may have been a contributing factor to the observed social changes. The core records at Waigani swamp suggests a decrease in rainfall around 800 years BP, which would have put pressure on subsistence systems, particularly in the drier climes of the Port Moresby region. If this were the case, then there would likely have been increased competition for naturally irrigated fertile soil beds for swidden gardening. During this time, the inland settlement at the base of Nebira hill was abandoned and resumed shortly after on the naturally defensible hilltop (Allen 1972b; Bulmer 1978). Occupation of the offshore Motupore Island commenced shortly after 1000 cal. years BP (1010 ± 80 uncalibrated years BP, charcoal) and permanently from 800 cal. years BP (810 ± 150 uncalibrated years BP, charcoal) (Allen 2017). Settlement of defensive locations suggests conflict may have been a significant social factor of this observed disruption.

Reliance on maritime trade: The last 800 years

Of the 107 dated sites, 88 (82%) have evidence of human activity within the last 800 years, demonstrating both the high visibility of cultural deposits within the last millennium and the increasingly widespread populations in lowland environments (see Figure 9.11). Vasey (1982) had estimated a pre-European contact population of ~2000 in the hill country surrounding Port Moresby, which was noted as having been extensively gardened 150 years ago when European observers visited the region (Moresby 1876). Although populations densities would not have been consistent along the length and breadth of the south coast, areas around Hood Bay, the Aroma coast, and Amazon Bay would also have supported relatively large populations as they do in modern times (Allen et al. 2002b; Chalmers 1887; Irwin 1985; Saville 1926).

Lowland landscapes by this time were similar to the modern-day, with increasing population densities continuing to impact slope erosion through forest clearance and swiddening, which ultimately altered coastal and lowland ecologies. Core records also suggest the volume and frequency of rainfall fluctuated over this time which influenced the efficacy of gardening in these landscapes (Osborne et al. 1993). From this time, localised pottery traditions appear which have little connection to earlier traditions, and such is the case at Ava Garua (Swadling 1981) and in the southern Massim islands (Irwin et al. 2019), which has been attributed to population movement, the establishment of new villages, and reinforcement of social identities (Allen 2010; Shaw et al. 2020b).

The intensification of several separate but overlapping maritime trade systems such as the *Kula*, *Mailu*, and *Hiri* developed as a risk mitigation strategy employed as larger populations adapted to a drier climate relative to earlier centuries, particularly in the highly seasonal savannah ecologies around Port Moresby and on smaller islands (Dutton 1982; Irwin 1985; Irwin et al. 2019; Skelly and David 2017). Notably, even with drier conditions, it has been convincingly argued that local yields of food would have been enough to sustain a population of 2000–3000 people. Large-scale maritime trade was, therefore, not an inevitable result of environmental degradation or change in climatic conditions but was a multi-generation strategy to maintain social connections between distant communities that involved the exchange of foodstuffs (Allen 1977a; Macintyre and Allen 1990; Vasey 1982).

In the Massim islands, extensive progradation of beach deposits has been identified over the last 300 years (Shaw et al. 2020b). Indeed, many currently inhabited beaches had formed within the last two to three centuries, and in some cases, the pre-contact movement of villages to the coast is recounted in oral histories (Shaw 2015; Skelly and David 2017; Urwin et al. 2018). Allen (2017: 47–49) has also demonstrated that the sand spit on Motupore Island progressively expanded over the last 1000 years with human settlement and had formed on top of beachrock, which developed between 3340 and 2360 cal. years BP (2940 ± 80 uncalibrated years BP, 2530 ± 80 bp, beachrock), suggesting the coast was largely inundated prior to this time. To this end, Shaw et al. (2020b) argued that several small islands (e.g. Dobu, Amphlett) became major hubs in the *Kula* network within the last few centuries because coastal progradation made settlement of these islands and formerly narrow and swampy coastal fringes increasingly conducive. Consequently, there were large-scale movements of populations within and between islands, along coastlines and between regions in response to developing opportunities for trade, territory, subsistence, and social relationships (See also Shaw and Coxe 2021).

Concluding remarks

The island of New Guinea, of which Papua New Guinea encompasses the eastern half, hosts around 1200 languages, or one-sixth of the world's linguistic diversity on less than 1% of the worlds land area and with <0.1% of the world's population. Around 73 Papuan and Austronesian languages are currently spoken along the south coast and in the Massim, attesting to a long and complex human history involving millennia of social interaction and adaptation since the Late Pleistocene. Such remarkable diversity requires explanation, and yet there is less focus on archaeological research in New Guinea than in other parts of the Asia-Pacific region. At present, there is simply not enough resolution on past human records to adequately model the multicausal drivers for this linguistic and cultural diversity. Detailed site records are desperately needed.

The history of human landscape use in the Massim Islands and southern Papuan lowlands is one of increasing use within the last 5000–4000 years, particularly the last 1000 years, and low-density, intermittent use during earlier millennia. Unlike the north coast of New Guinea and the Bismarck Archipelago, these landscapes are fringed by a shallow continental shelf that was inundated in the millennia after the Last Glacial Maximum. Despite early coastal sites now mostly being underwater, the southern lowlands would likely have been a significant corridor through which colonist populations moved and as a possible route to highland valleys where evidence for human use dates from about 49,000–44,000 cal. years BP. It is clear that the initial populations of Sahul were adaptable, resilient, and not limited to coastal ecosystems. The southern lowlands and islands, therefore, have much to contribute to our understanding of the peopling of Sahul and movement between ecozones. Focused attention is now needed in the interior lowlands, where evidence for pre-LGM settlement may be found.

Archaeological records across Sahul have demonstrated low population densities until the Mid-Holocene. Even with the current paucity of Late Pleistocene and Early-Mid Holocene sites across the southern lowlands and islands, archaeological records in these regions are consistent with this broader pattern. The increasing intensification of settlement and landscape use has likely had a negative impact on the visibility of earlier cultural sites, made more difficult by low-level visibility of deposits prior to the introduction of pottery. As Egloff and Kaiku (1983) note, populations with a reliance on organic materials will, even in recent centuries, only leave a faint trace on the landscape. The higher frequency of sites recorded containing pottery sherds can attest to this, and logically, pottery-bearing sites have been the focus of most archaeological surveys from which a regional framework of human settlement was developed. In this regard, the increase in the number of sites dating to the mid-Holocene has been due to the intensive nature of investigations and targeted survey within project areas, only made possible by the foundational research approaches of the past 60 years (McNiven et al. 2011; Richards et al. 2016; Shaw et al. 2020a).

Thus far, all Mid-Holocene sites have been found on the coast, and it is no coincidence that these sites date from or after 5000–4000 years BP as this is when coastal landscapes had stabilised as sea levels had begun to decline. In all cases, Mid-Holocene cultural deposits were deeply buried, and even in shallower savannah substrates, there were no indications of cultural material of this antiquity on the surface. Unlike pottery for which relative chronologies have been developed to aid in determining the approximate age of surface finds, technologies which are known to span the mid-Holocene are few, relatively rare, and most often found in other parts of New Guinea (e.g. formally manufactured pestles, mortars, and stemmed obsidian blades) (Swadling 2016; Swadling et al. 2008; Torrence et al. 2013). The distribution of Mid-Holocene populations is almost certainly more widespread than is currently known, and the next decade should see an increase in the number of sites reported if the current trend of fieldwork intensity in the wider region continues.

These landscapes are fundamentally important to indigenous past narratives, with archaeological and palaeo-ecological research over the last 60 years having been hugely influential in framing future research directions (Allen 1972a). The long-term history of the people who once lived on these landscapes is still shrouded by the fog of time and requires increased input from indigenous scholars and communities than is currently the case. This is their history, and the combination of traditional knowledge and scientific approaches will, over the next decades and centuries, continue to reveal the rich tapestry of human history in these dynamic lowland and island environments.

Acknowledgements

First and foremost, I would like to extend a heartfelt *tenkyu tru* to all the communities in Papua New Guinea who have collaborated on the many archaeological projects that have taken place over the decades and to the past and present staff of the National Museum and Art Gallery of Papua New Guinea who have been the custodians of this material history, and who ensure research is ethical and inclusive. To the pioneering researchers whose work has formed a strong foundation on which to build and whose dedication to the intricacies of the past we aspire to follow. I would also like to thank Mike Carson for the invitation to contribute to this edited volume and to Glenn Summerhayes, Pamela Swadling, and Simon Coxe for providing thoughtful comments on a draft. Any errors are my own.

Note

1 Calibrated 95.4% date ranges are presented as cal. years BP, with the original corrected but uncalibrated determination provided in brackets (uncalibrated years BP) with the dated material. When dates are interpolated by the original authors, these are presented as BP.

References

Allen, Bryant, Robin Hide, Richard Bourke, W. Akus, D. Fritsch, R. Grau, G. Ling, and E. Lowe, 2002a. Agricultural Systems of Papua New Guinea: Western Province. Working Paper Number 4. Australian National University, Canberra.

Allen, Bryant, T. Nen, Richard Bourke, Robin Hide, D. Fritsch, R. Grau, P. Hobsbawn, and S. Lyon, 2002b. Agricultural Systems of Papua New Guinea: Central Province. Working Paper Number 15. Australian National University, Canberra.

Allen, Gerald, Jeff Kinch, Sheila McKenna, and Pamela Seeto, 2003. A Rapid Marine Biodiversity Assessment of Milne Bay Province, Papua New Guinea - Survey II (2000), RAP Bulletin of Biological Assessment. Conservation International, Washington DC.

Allen, Jim, 1972a. The first decade in New Guinea archaeology. *Antiquity* 46: 180–190.

Allen, Jim, 1972b. Nebira 4: An early Austronesian site in Central Papua. *Archaeology and Physical Anthropology in Oceania* 7: 92–124.

Allen, Jim, 1977a. Fishing for wallabies: Trade as a mechanism for social interaction, integration and elaboration on the central Papuan coast, In *The evolution of social systems*, edited by J. Friedman and M. J. Rowlands, pp. 419–455. Gerald Duckworth and co. Ltd, London.

Allen, Jim, 1977b. Sea traffic, trade and expanding horizons. In *Sunda and Sahul: Prehistoric Studies in Southeast Asia, Melanesia and Australia*, edited by J. Allen, J. Golson, and R. Jones, pp. 387–417. Academic Press, New York.

Allen, Jim, 1978. The physical and cultural setting of Motupore Island, Central Province, Papua New Guinea. *Bulletin of the Indo-Pacific Prehistory Association* 1: 47–55.

Allen, Jim, 2000. From beach to beach: The development of maritime economies in prehistoric Melanesia. In *East of Wallace's Line: studies of past and present maritime cultures of the Indo-Pacific region*, edited by S. O'Connor and P. M. Veth, pp. 130–176. Balkema, Rotterdam.

Allen, Jim, 2010. Revisiting Papuan ceramic sequence changes: Another look at old data. *Artefact* 33: 4–15.

Allen, Jim, 2017. Excavations on Motupore Island, Central District, Papua New Guinea. 2 Volumes, Working Papers in Anthropology, Number 4. University of Otago, Dunedin.

Allen, Jim, Chris Gosden, and Peter White, 1989. Human Pleistocene adaptations in the tropical island Pacific: Recent evidence from New Ireland, a Greater Australian outlier. *Antiquity* 63: 548–561.

Allen, Jim, Glenn Summerhayes, Herman Mandui, and Matthew Leavesley, 2011. New data from Oposisi: Implications for the Early Papuan Pottery phase. *Journal of Pacific Archaeology* 2: 69–81.

Allen Jim, and James O'Connell. 2008. Getting from Sunda to Sahul. In *Islands of Inquiry: Colonisation, Seafaring and the Archaeology of Maritime Landscapes*, edited by G. Clark, F. Leach, and S. O'Conner, pp. 31–46. ANU E-Press, Canberra.

Allen, Melinda, and Rod Wallace, 2007. New evidence from the East Polynesian gateway: Substantive and methodological results from Aitutaki, Southern Cook Islands. *Radiocarbon* 49: 1–17.

Ambrose, Wal, Fiona Petchey, Pamela Swadling, Harry Beran, Elizabeth Bonshek, Katherine Szabo, Simon Bickler, and Glenn Summerhayes, 2012. Engraved prehistoric *Conus* shell valuables from southeastern Papua New Guinea: Their antiquity, motifs and distribution. *Archaeology in Oceania* 47: 113–132.

Austen, Leo, 1939. Megalithic structures in the Trobriand Islands. *Oceania* 10: 30–53.

Barham, Anthony, 1999. The local environmental impact of prehistoric populations on Saibai Island, northern Torres Strait, Australia: Enigmatic evidence from Holocene swamp lithostratigraphic records. *Quaternary International* 59: 71–105.

Barham, Anthony, and David Harris, 1985. Relict field systems in the Torres Strait Region. In *Prehistoric intensive agriculture in the tropics*, edited by I. S. Farrington, pp. 247–283.British Archaeological Reports, Oxford.

Barker, Bryce, Lara Lamb, Bruno David, Kenneth Korokai, Alois Kuaso, and Joanne Bowman, 2012. Otoia, ancestral village of the Kerewo: Modelling the historical emergence of Kerewo regional polities on the island of Goaribari, south coast of mainland Papua New Guinea. In *Peopled landscapes: Archaeological and Biogeographic Approaches to Landscapes*, edited by S. Haberle and B. David, pp. 157–176. The Australian National University, Canberra.

Barker, Bryce, Lara Lamb, Bruno David, Robert Skelly, and Kenneth Korokai, 2015. Dating of *in situ* longhouse (*dubu daima*) posts in the Kikori River delta:

Refining chronologies of island village occupation in the lower Kikori River delta, Papua New Guinea. *Quaternary International* 385: 27–38.

Barker, Bryce, Lara Lamb, and Tiina Manne, 2016. Baikaboria Ossuary and the origins of the Kesele Clan, Upper Kikori River, Papua New Guinea. *Journal of Pacific Archaeology* 7: 89–105.

Bevan, Theodore, 1890. *Toil, Travel and Discovery in British New Guinea*. Kegan Paul, Trench, Trubner, London.

Bickler, Simon, 1998. Eating Stone and Dying: Archaeological Survey on Woodlark Island, Milne Bay Province, Papua New Guinea. Eating Stone and Dying: Archaeological Survey on Woodlark Island, Milne Bay Province, Papua New Guinea. Unpublished doctoral thesis, University of Virginia, Charlottesville.

Bickler, Simon. 2006. Prehistoric stone monuments in the northern region of the Kula ring. *Antiquity* 80: 38–51.

Bird, Michael, Scott Condie, Sue O'Conner, Damien O'Grady, Christian Reepmeyer, Sean Ulm, Mojca Zega, Frederik Saltre, and Corey Bradshaw, 2019. Early human settlement of Sahul was not an accident. *Scientific Reports* 9: article 8220.

Bird, Michael, Damien O'Grady, and Sean Ulm, 2016. Humans, water, and the colonization of Australia. *Proceedings of the National Academy of Sciences of the United states of America* 113: 11477–11482.

Birdsell, Joseph, 1977. The recalibration of a paradigm for the first peopling of greater Australia, In *Sunda and Sahul: Prehistoric Studies in Southeast Asia, Melanesia and Australia*, edited by J. Allen, J. Golson, and R. Jones, pp. 111–167. Academic Press, London.

Blake, David, and Cliff Ollier, 1970. Geomorphological evidence of quaternary tectonics in southwestern Papua. *Revue de Geomorphologie Dynamique* 19: 28 32.

Blake, David, and Cliff Ollier, 1971. Alluvial plains of the Fly River, Papua. *Zeitschrift für Geomorphologie* 12: 1–17.

Bowdler, Sandra, 1977. The coastal colonisation of Australia. In *Sunda and Sahul: Prehistoric studies in Southeast Asia, Melanesia, and Australia*, edited byJ. Allen, J. Golson, and R. Jones, pp. 205–245. Academic Press, London.

Bradshaw, Corey, Sean Ulm, Alan Williams, Michael Bird, Richard Roberts, Zenobia Jacobs, Fiona Laviano, Laura Weyrich, Tobias Fridrich, Kasih Norman, et al., 2019. Minimum founding populations for the first peopling of Sahul. *Nature Ecology and Evolution* 3: 1057–1063.

Brass, Leonard, 1959. Results of the Archibold expeditions. Number 79: Summary of the fifth Archibold expedition to New Guinea (1956–1957). *Bulletin of the American Museum of Natural History* 118: 1–70.

Bulmer, Susan, 1969. Recent archaeological discoveries in Central Papua. *Australian Natural History* 16: 229–233.

Bulmer, Susan, 1971. Prehistoric settlement patterns and pottery in the Port Moresby area. *Journal of Papua New Guinea Society* 5: 28–91.

Bulmer, Susan, 1975. Settlement and economy in prehistoric Papua New Guinea: A review of the archaeological evidence. *Journal de la Société des Océanistes* 46: 7–75.

Bulmer, Susan, 1978. Prehistoric culture change in the Port Moresby region. Unpublished doctoral thesis, University of Papua New Guinea, Port Moresby.

Bulmer, Susan. 1979. Prehistoric ecology and economy in the Port Moresby region. *New Zealand Journal of Archaeology* 1: 5–27.

Bulmer, Susan, 1982. Human ecology and cultural variation in prehistoric New Guinea, In *Biogeography and Ecology of New Guinea*, edited by J. L. Gressitt, pp. 169–206. Springer, New York.

Bulmer, Susan, 1999. Revisiting Red Slip: The Laloki style pottery of Southern Papua and its possible relationship to Lapita. In *The Western Pacific 5000-2000 BP: Colonisations and transformations*, edited by J. C. Galipaud and I. Lilley, pp. 543–577. IRD Publishers, Paris.

Burenhult, Goran, 2002. The Archaeology of the Trobriand Islands, Milne Bay Province, Papua New Guinea: Excavation season 1999, British Archaeological Reports. Archaeopress, Oxford.

Chalmers, James, 1887. Pioneering in New Guinea. Religious Tract Society, London.

Chalmers, James, 1903. Notes on the Natives of Kiwai Island, Fly River, British New Guinea. *The Journal of the Anthropological Institute of Great Britain and Ireland* 33: 117–124.

Chalmers, James, and William Gill, 1885. *Work and Adventure in New Guinea, 1877–85*. The Religious Tract Society, London.

Chappell, John, 2002. Sea level changes forced ice breakouts in the Last Glacial cycle: New results from coral terraces. *Quaternary Science Reviews* 21: 1229–1240.

Chappell, John, 2005. Geographic changes of coastal lowlands in the Papuan Past. In *Papuan Pasts: Cultural, Linguistic and Biological Histories of Papuan-speaking Peoples*, edited by A. Pawley, R. Attenborough, J. Golson, and R. Hide, R, pp. 525–539. Australian National University, Canberra.

Chappell, John, and Henry Polach, 1991. Post-glacial sea-level rise from a coral record at Huon Peninsula, Papua New Guinea. *Nature* 349: 147–149.

Christensen, Ole, 1975. Hunters and horticulturalists: A preliminary report of the 1972–74 excavations in the Manim valley, Papua New Guinea. *Mankind* 10: 24–36.

Chynoweth, Merryn, Glenn Summerhayes, Anne Ford, and Yo Negishi, 2020. Lapita on Wari Island: What's the problem? *Asian Perspectives* 59: 100–116.

Clarkson, Chris, Zenobia Jacobs, Ben Marwick, Richard Fullagar, Lynley Wallis, Mike Smith, Richard Roberts, Elspeth Hayes, Kelsey Lowe, Xavier Carah, et al., 2017. Human occupation of northern Australia by 65,000 years ago. *Nature* 547: 306–310.

David, Bruno, 2008. Rethinking cultural chronologies and past landscape engagement in the Kopi region, Gulf Province, Papua New Guinea. *The Holocene* 18: 463–479.

David, Bruno, Andrew Fairbairn, Ken Aplin, Lesley Murepe, Michael Green, John Stanisic, Marshall Weisler, Douglas Simala, Thomas Kokents, John Dop, et al., 2007. OJP, a terminal Pleistocene archaeological site from the Gulf Province lowlands, Papua New Guinea. *Archaeology in Oceania* 42: 31–33.

David, Brun, Nick Araho, Alois Kuaso, Ian Moffat, and Nigel Tapper, 2008. The Upihoi Find: Wrecked Wooden Bevaia (Lagatoi) Hulls of Epemeavo Village, Gulf Province, Papua New Guinea. *Australian Archaeology* 66: 1–14.

David, Bruno, Nick Araho, Bryce Barker, Alois Kuaso, and Ian Moffat, 2009. Keveoki 1: Exploring the *Hiri* ceramics trade at a short-lived village site near the Vailala River Papua New Guinea. *Australian Archaeology* 68: 11–22.

David, Bruno, Jean-Michel Geneste, Ken Aplin, Jean-Jacques Delannoy, Nick Araho, Chris Clarkson, Kate Connell, Simon Haberle, Bryce Barker, Lara Lamb, et al., 2010. The Emo Site (OAC), Gulf Province, Papua New Guinea: Resolving long-standing questions of antiquity and implications for the history of the ancestral Hiri maritime trade. *Australian Archaeology* 70: 39–54.

David, Bruno, Ian McNiven, Thomas Richards, Sean Connaughton, Matthew Leavesley, Bryce Barker, and Cassandra Rowe, 2011. Lapita sites in the Central Province of mainland Papua New Guinea. *World Archaeology* 43: 576–593.

David, Bruno, Lara Lamb, Jean-Jacques Delannoy, Frank Pivoru, Cassandra Rowe, Max Pivoru, Tony Frank, Nick Frank, Andrew Fairbairn, and Ruth Pivoru, 2012a. Poromoi Tamu and the case of the drowning village: History, lost places and the stories we tell. *Journal of Historical Archaeology* 16: 319–345.

David, Bruno, Ian McNiven, Matthew Leavesley, Bryce Barker, Herman Mandui, Thomas Richards, and Robert Skelly, 2012b. A new ceramic assemblage from Caution Bay, South Coast of Mainland PNG: The linear shell edge-impressed tradition from Bogi 1. *Journal of Pacific Archaeology* 3: 73–89.

David, Bruno, Ken Aplin, Fiona Petchey, Robert Skelly, Jerome Mialanes, Holly Jones-Amin, John Stanisic, Bryce Barker, and Lara Lamb, 2015a. Kumukumu 1, a hilltop site in the Aird Hills: Implications for occupational trends and dynamics in the Kikori River delta, south coast of Papua New Guinea. *Quaternary International* 385: 7–26.

David, Bruno, Jerome Mialanes, Fiona Petchey, Ken Aplin, Jean-Michel Geneste, Robert Skelly, and Cassandra Rowe, 2015b. Archaeological investigations at Waredaru and the origins of the Keipte Kuyumen clan estate, upper Kikori River, Papua New Guinea. *PALEO* 26: 33–57.

David, Bruno, Holly Jones-Amin, Thomas Richards, Jerome Mialanes, Brit Asmussen, Fiona Petchey, Ken Aplin, Matthew Leavesley, Ian McNiven, Camille Zetzmann, et al., 2016. Ruisasi 1 and the earliest evidence of mass-produced ceramics in Caution Bay (Port Moresby region), Papua New Guinea. *Journal of Pacific Archaeology* 7: 41–60.

David, Bruno, Ken Aplin, Helene Peck, Robert Skelly, Matthew Leavesley, Jerome Mialanes, Katherine Szabo, Brent Koppel, Fiona Petchey, Thomas Richards, et al., 2019a. Moiapu 3: Settlement on Moiapu Hill at the very end of Lapita, Caution Bay hinterland, In *Debating Lapita: Distribution, Chronology, Society and Subsistence*, edited by S. Bedford and M. Spriggs, pp. 61–88. ANU E Press, Canberra.

David, Bruno, Jean-Jacques Delannoy, Jerome Mialanes, Chris Clarkson, Fiona Petchey, Jean-Michel Geneste, Tiina Manne, Michael Bird, Bryce Barker, Thomas Richards, et al., 2019b. 45, 610–52, 160 years of site and landscape occupation at Nawarla Gabarnmang, Arnhem Land plateau (northern Australia). *Quaternary Science Reviews* 215: 64–85.

Davies, Hugh, and Denis Ives, 1965. The geology of Fergusson and Goodenough Islands, Papua. Bureau of Mineral Resources, Geology and Geophysics, Canberra.

Davies, Hugh, and Ian Smith, 1971. Geology of Eastern Papua. *Geological Society of America Bulletin* 82: 3299–3312.

Dougherty, Amy, Zoe Thomas, Chris Fogwill, Alan Hogg, Jonathon Palmer, Eleanor Rainsley, Alan Williams, Sean Ulm, Kerrylee Rogers, Brian Jones, et al.,

2019. Redating the earliest evidence of the mid-Holocene relative sea-level high-stand in Australia and implications for global sea-level rise. *PLoS ONE* 14, e0218430.

Dunn, Michael, Stephen Levinson, Eva Lindstrom, Ger Reesink, and Angela Terril, 2008. Structural phylogeny in historical linguistics: Methodological explorations applied in Island Melanesia. *Language* 84: 710–759.

Dutton, Tom, 1975. South-Eastern Trans-New Guinea Phylum Languages, In *New Guinea Area Languages and Language Study Vol 1: Papuan Languages and the New Guinea linguistic scene*, edited by S. A. Wurm, pp. 613–664. ANU Press, Canberra.

Dutton, Tom, 1982. *The Hiri in history: Further aspects of long distance Motu trade in Central Papua*. Australian National University, Canberra.

Dutton, Tom, 2010. *Reconstructing Proto Koiarian: The History of a Papuan Language Family*. ANU Press, Canberra.

Eden, Michael, 1974. The origin and status of Savanna and Grassland in Southern Papua. *Transactions of the Institute of British Geographers* 63: 97–110.

Egloff, Brian, and Resonga Kaiku, 1983. Prehistory and paths in the upper Purari River basin, In *The Purari: Tropical Environment of a High Rainfall Basin*, edited by T. Petr, pp. 475–491. Doctor W. Junk Publishers, the Hague.

Egloff, Brian, 1971. Collingwood Bay and the Trobriand Islands in recent prehistory: Settlement and interaction in coastal and island Papua. Unpublished doctoral thesis, Australian National University, Canberra.

Egloff, Brian, 1979. *Recent Prehistory in Southeast Papua*. Terra Australis 4. Australian National University, Canberra.

Ely, Thomas, 1988. *Hunters of the Reefs: The Marine Geography of the Kiwai, Papua New Guinea*. University of California Press, Berkley.

Evans, Nicholas, Wayan Arka, Matthew Carroll, Yun Jung Choi, Christian Dohler, Volker Gast, Eri Kashima, Emil Mittag, Bruno Olsson, Kyla Quinn, et al., 2018. The languages of southern New Guinea, In *The Languages and Linguistics of the New Guinea Area: A Comprehensive Guide*, edited by B. Palmer, pp. 641–774. De Gruyter Mouton, Berlin.

Faulkner, Patrick, Anbarasu Thangavelu, Redbird Ferguson, Samantha Aird, Bruno David, Tanya Drury, Cassandra Rowe, Bryce Barker, Ian McNiven, Thomas Richards, et al., 2020. Middle to Late Holocene near-shore foraging strategies at Caution Bay, Papua New Guinea. *Journal of Archaeological Science: Reports* 34: article 102629.

Febo, Lawrence, Samuel Bentley, John Wrenn, Andre Droxler, Gerald Dickens, Larry Peterson, and Bradley Opdyke, 2008. Late Pleistocene and Holocene sedimentation, organic-carbon delivery, and paleoclimatic inferences on the continental slope of the northern Pandora Trough, Gulf of Papua. *Journal of Geophysical Research* 113: https://doi.org/10.1029/2006JF000677.1029/2006JF000677

Field, Judith, Ben Shaw, and Glenn Summerhayes, Submitted. Pathways to the Interior: Human settlement in the Simbai-Kaironk Valleys of the Madang Province, PNG. Manuscript submitted for review.

Frankel, David, and Ron Vanderwal, 1981. Archaeological investigations near Kerema, Gulf Province, P.N.G. 1980–81 preliminary field report. *Research in Melanesia* 6: 20–32.

Frankel, David, and Vincent Kewibu, 2000. Early Ceramic Period Pottery from

Murua Site (ODR), Gulf Province, Papua New Guinea, In *Australian Archaeologist: Collected Papers in Honour of Jim Allen*, edited by A. Anderson and T. Murray, pp. 279–290. ANU Press, Canberra.

Frankel, David, K Thompson, and Ron Vanderwal. 1994. Kerema and Kinomere. In *Archaeology of a coastal exchange system: Sites and ceramics of the Papuan Gulf*, edited byFrankel D and Rhoads, J (eds.), pp. 1–48. ANU Press, Canberra.

Fredericksen, Clayton, Matthew Spriggs, and Wal Ambrose, 1993. Pamwak rock-shelter: A Pleistocene site on Manus Island, Papua New Guinea. In *Sahul in review: Pleistocene archaeology in Australia, New Guinea and Island Melanesia*, edited by M. A. Smith. M. Spriggs, and B. Fankhauser, pp. 144–152. The Australian National University, Canberra.

Garrett-Jones, Samuel, 1979. Evidence for changes in Holocene vegetation and lake sedimentation in the Markham Valley, Papua New Guinea. Unpublished doctoral thesis, Australian National University, Canberra.

Gavin, Daniel, 2001. Estimation of inbuilt age in radiocarbon ages of soil charcoal for fire history studies. *Radiocarbon* 43: 27–44.

Gorecki, Pawel, 1992. A Lapita smoke screen? In *Poterie Lapita et Peuplement*, edited by J. C. Galipaud, pp. 27–47. ORSTOM, NouméA, New Caledonia.

Gorecki, Pawel, Mark Mabin, and John Campbell, 1991. Archaeology and geomorphology of the Vanimo Coast, Papua New Guinea: Preliminary results. *Archaeology in Oceania* 26: 119–122.

Gosden, Chris, and Nola Robertson, 1991. Models for Matenkupkum: Interpreting a Late Pleistocene site from southern New Ireland, Papua New Guinea, In *Report of the Lapita Homeland Project*, edited by J. Allen and C. Gosden, pp. 20–45. Australian National University, Canberra.

Gosden, Chris, John Webb, Brendan Marshall, and Glenn Summerhayes, 1994. Lolmo Cave: A Mid-to Late Holocene Site, the Arawe Islands, West New Britain Province, Papua New Guinea. *Asian Perspectives* 33: 97–119.

Groube, Les, John Chappell, John Muke, and David Price, 1986. A 40,000-year-old human occupation site at Huon Peninsula, Papua New Guinea. *Nature* 324: 453–455.

Guise, Alu, 1985. *Oral Tradition and Archaeological Sites in the Eastern Central Province*. PNG National Museum Record, Number 9. Papua New Guinea National Museum, Port Moresby.

Haberle, Simon, 2007. Prehistoric human impact on rainforest biodiversity in highland New Guinea. *Philosophical Transactions of the Royal Society B* 362: 219–228.

Haddon, Alfred, 1894. *The Decorative Art of British New Guinea: A Study in Papuan Ethnography*. Cunningham Memoirs. Royal Irish Academy, Dublin.

Haddon, Alfred, 1900. A classification of the stone clubs of British New Guinea. *Journal of the Royal Anthropological Institute of Great Britain and Ireland* 30: 221–250.

Harris, David, and Billai Laba, 1982. The mystery of the Papuan Mound Builders. *The Geographical Magazine* 54: 386–391.

Heaton, Timothy, Peter Köhler, Martin Butzin, Edouard Bard, Ron Reimer, William Austin, Christopher Ramsey, Pieter Grootes, Konrad Hughen, Bernd Kromer, et al., 2020. Marine20 – the marine radiocarbon age calibration curve (0–55,000 cal BP). *Radiocarbon* 62: 1–42.

Hide, Robin, Richard Bourke, Bryant Allen, T. Betitis, D. Fritsch, R. Grau, L. Kurika, E. Lowes, D. Mitchell, S. Rangai, et al., 2002a. *Agricultural Systems of Papua New Guinea: Milne Bay Province*. Working Paper Number 6. Australian National University, Canberra.

Hide, Robin, Richard Bourke, Bryant Allen, N. Fereday, D. Fritsch, R. Grau, E. Lowes, and M. Woruba, 2002b. *Agricultural Systems of Papua New Guinea: Gulf Province*. Working Paper Number 5. Australian National University, Canberra.

Hitchcock, Garrick, 2004. *Wildlife is Our Gold: Political Ecology of the Torassi River Borderland, Southwest Papua New Guinea*. University of Queensland, Brisbane.

Hitchcock, Garrick, 2010. Mound-and-ditch taro gardens of the Bensbach or Torassi River area, southwest Papua New Guinea. *The Artefact* 33: 70–90.

Hope, Geoff, 2014. The sensitivity of the high mountain ecosystems of New Guinea to climatic change and anthropogenic impact. *Arctic, Antarctic, and Alpine Research* 46: 777–786.

Hope, Geoff, and Simon Haberle, 2005. The history of the human landscapes of New Guinea, In *Papuan pasts: Cultural, linguistic and biological histories of Papuan speaking peoples*, edited by A. Pawley, R. Attenborough, J. Golson, and R. Hide, pp. 541–554. Pacific Linguistics, Canberra.

Hope, Geoff, Peter Kershaw, Sander van der Kaars, Sun Xiangjun, Ping-Mei Liew, Linda Heusser, Hikaru Takahara, Matt McGlone, Norio Miyoshi, and Patrick Moss, 2004. History of vegetation and habitat change in the Austral-Asian region. *Quaternary International* 118–11: 103–126.

Irwin, Geoff, 1977. The emergence of Mailu as a central place in the prehistory of Coastal Papua. Unpublished doctoral thesis Australian National University, Canberra.

Irwin, Geoff, 1983. Chieftanship, kula, and trade in Massim prehistory, In *The Kula: New perspectives on Massim exchange*, edited by J. W. Leach and E. Leach, pp. 29–72. Cambridge University Press, Cambridge.

Irwin, Geoff, 1985. *The Emergence of Mailu as a Central Place in Coastal Papuan Prehistory*. Terra Australis 10. Australian National University, Canberra.

Irwin, Geoff, 1991. Themes in the prehistory of Coastal Papua and the Massim, In *Man and a half: Essays in Pacific anthropology and ethnobiology in honour of Ralph Bulmer*, edited by A. Pawley, pp. 503–510. Polynesian Society, Auckland.

Irwin, Geoff, and Simon Holdaway, 1996. Colonization, trade and exchange: From Papua to Lapita, In *Oceanic Culture History: Essays in Honour of Roger Green*, edited by J. M. Davidson, G. Irwin, B. F. Leach, A. Pawley, and D. Brown, pp. 225–235. New Zealand Journal of Archaeology Special Publication, Wellington.

Irwin, Geoff, Ben Shaw, and Andrew McAlister, 2019. The origins of the Kula Ring: Archaeological and maritime perspectives from the southern Massim and Mailu areas of Papua New Guinea. *Archaeology in Oceania* 54: 1–16.

Ishiwa, Takeshige, Yusuke Yokoyama, Yosuke Miyairi, Stephen Obrochta, Takenori Sasaki, Akihisa Kitamura, Atsushi Suzuki, Minoru Ikehara, Ken Ikehara, Katsunori Kimoto, et al., 2016. Reappraisal of sea-level lowstand during the Last Glacial Maximum observed in the Bonaparte Gulf sediments, northwestern Australia. *Quaternary International* 397: 373–379.

Johns, Robert, Osia Gideon, J. Simaga, T. Kuria, and G. Bagoeara, 2009. An introduction to the flora of the Milne Bay Archipelago. *Blumea* 54: 251–254.

Kealy, Shimona, Julien Louys, and Sue O'Connor, 2016. Islands under the sea: A review of early modern human dispersal routes and migration hypotheses through Wallacea. *The Journal of Island and Coastal Archaeology* 11: 364–384.

Kealy, Shimona, Julien Louys, and Sue O'Connor, 2017. Reconstructing Palaeogeography and inter-island visibility in the Wallacean Archipelago during the likely period of Sahul Colonization, 65–45,000 years ago. *Archaeological Prospections* 24: 259–272.

Lambeck, Kurt, Helene Rouby, Anthony Purcell, Yiying Sun, and Malcolm Sambridge, 2014. Sea level and global ice volumes from the Last Glacial Maximum to the Holocene. *Proceedings of the National Academy of Sciences of the United States of America* 111: 15296–15303.

Lampert, Ron, 1966. Archaeological Reconnaissance in Papua & New Guinea: 1966. Unpublished report to Department of Anthropology, Australian National University, Canberra.

Lampert, Ron, 1968. Some archaeological sites of the Motu and Koiari areas. Proceedings of presentations at the History of Melanesia Seminar, Port Moresby, 30 May through 5 June 1968, pp. 411–422. The Australian National University, Canberra.

Landtman, Gunnar, 1927. *The Kiwai Papuans of British New Guinea.* Macmillan and Company London.

Lauer, Peter,1971. Preliminary report on ethno-archaeological research in the northwestern Massim, T. P. N. G. *Asian Perspectives* 14: 69–75.

Leask, Maurice, 1943. A kitchen midden in Papua. *Oceania* 13: 235–242.

Leavesley, Matthew, 2004. Trees to the sky: Prehistoric hunting in New Ireland, Papua New Guinea. Unpublished doctoral thesis, Australian National University, Canberra.

Leavesley, Matthew, 2006. Late Pleistocene Complexities in the Bismarck Archipelago, In *Archaeology of Oceania: Australia and the Pacific Islands*, edited by I. Lilley, pp. 189–204. Blackwell Publishing, London.

Leavesley, Matthew, Michael Bird, Kieth Fifield, P. Hausladen, G. Santos, and M. di Tada, 2002. Buang Merabak: Early evidence for human occupation in the Bismarck Archipelago, Papua New Guinea. *Australian Archaeology* 54: 55–57.

Levinson, Stephen, 2006. The language of space in Yélî Dnye In *Grammars of Space: Explorations in Cognitive Diversity*, edited by S. C. Levinson and D. P. Wilkins, pp. 157–204. Cambridge University Press, Cambridge.

Lewis, Stephen, Craig Sloss, Colin Murray-Wallace, Colin Woodroffe, and Scott Smithers, 2013. Post-glacial sea-level changes around the Australian margin: A review. *Quaternary Science Reviews* 74: 115–138.

Lewis, Stephen, Raphael Wüst, Jody Webster, and Graham Shields, 2008. Mid-Late Holocene sea-level variability in eastern Australia. *Terra Nova* 20: 74–81.

Lindt, John, 1887. *Picturesque New Guinea, with an Historical Introduction and Supplementary Chapters on the Manners and Customs of the Papuans.* Longmans Green and Company, London.

Löffler, Ernest, 1977. *Geomorphology of Papua New Guinea.* Australian National University Press, Canberra.

Mabutt, Jack, 1965. Geomorphology of the Port Moresby-Kairuku area. Commonwealth Scientific and Industrial Research Organization, Land Research Series 14. CSIRO, Melbourne.

MacGillivray, John, 1852. *Narrative of the Voyage of the H. M. S. Rattlesnake.* Facsimile edition. T&W Boone, London.

MacGregor, William, 1897. *British New Guinea: Country and People.* John Murray, London.

Macintyre, Martha, and Jim Allen, 1990. Trading for subsistence: The case from the Southern Massim. In *Pacific production systems: Approaches to economic prehistory*, edited by D. E. Yen and J. M. J. Mummery, pp. 120–136. The Australian National University, Canberra.

Malaspinas, Anna-Sapfo, Michael Westaway, Craig Muller, Vitor Sousa, Oscar Lao, Isabel Alves, Anders Bergstrom, Georgios Athanasiadis, Jade Cheng, Jacob Crawford, et al., 2016. A genomic history of Aboriginal Australia. *Nature* 538: 207–214.

Malinowski, Bronislaw, 1915. The natives of Mailu. *Transactions of the Royal Society of South Australia* 39: 494–706.

Malinowski, Bronislaw, 1922. *Argonauts of the Western Pacific.* Routledge and Kegan Paul, London.

Marsaglia, Kathleen, Kimberly Kramer, Bruno David, and Robert Skelly, 2016. Petrographic analyses of sand temper/inclusions in ceramics of Kikiniu, Kikori River and modern sand samples from the Gulf Province (Papua New Guinea). *Archaeology in Oceania* 51: 131–140.

Mauz, Barbara, Matteo Vacchi, Andrew Green, Goesta Hoffmann, and Andrew Cooper, 2015. Beachrock: A tool for reconstructing relative sea level in the far-field. *Marine Geology* 362: 1–16.

McAlpine, Jimmy, Gael Keig, and R. Falls, 1983. *Climate of Papua New Guinea.* CSIRO, Melbourne.

McNiven, Ian, Bruno David, Ken Aplin, Max Pivoru, Alex Sexton, Jonathon Brown, Chris Clarkson, Kate Connell, John Stanisic, Marshall Weisler, et al., 2010. Historicising the present: Late Holocene emergence of a rainforest hunting camp, Gulf Province, Papua New Guinea. *Australian Archaeology* 71: 41–56.

McNiven, Ian, Bruno David, Thomas Richards, Ken Aplin, Brit Asmussen, Jerome Mialanes, Matthew Leavesley, Patrick Faulkner, and Sean Ulm, 2011. New direction in human colonisation of the Pacific: Lapita settlement of South Coast New Guinea. *Australian Archaeology* 72: 1–6.

McNiven, Ian, Bruno David, Ken Aplin, Jerome Mialanes, Brit Asmussen, Sean Ulm, Patrick Faulkner, Cassandra Rowe, and Thomas Richards, 2012. Terrestrial engagements by terminal Lapita maritime specialists on the southern Papuan coast. In *Peopled Landscapes: Archaeological and Biogeographic Approaches To Landscapes*, edited by S. Haberle and B. David, pp. 121–156. The Australian National University, Canberra.

Mialanes, Jerome, Bruno David, Anne Ford, Thomas Richards, Ian McNiven, Glenn Summerhayes, and Matthew Leavesley, 2016. Imported obsidian at Caution Bay, south coast of Papua New Guinea: Cessation of long distance procurement c. 1900 cal BP. *Australian Archaeology* 82: 248–262.

Moresby, John, 1876. *Discoveries and Surveys in New Guinea and the D'Entrecasteaux Islands.* John Murray, London.

Murray, John, 1912. *Papua or British New Guinea.* T. Fisher Unwin, London.

Nagle, Nano, Mannis van Oven, Stephen Wilcox, Sheila van Holst Pellekaan, Chris Tyler-Smith, Yali Xue, Kaye Ballantyne, Leah Wilcox, Luka Papac, Karen

Cooke, et al., 2017. Aboriginal Australian mitochondrial genome variation – an increased understanding of population antiquity and diversity. *Scientific Reports* 7: article 43041.

Negishi, Yo, and Rintaro Ono, 2009. Kasasinabwana shell midden: The prehistoric ceramic sequence of Wari Island in the Massim, eastern Papua New Guinea. *People and Culture in Oceania* 25: 23–52.

O'Connell, James, Jim Allen, Martin Williams, Alan Williams, Chris Turney, Nigel Spooner, Johan Kamminga, Graham Brown, and Alan Cooper, 2018. When did Homo sapiens first reach Southeast Asia and Sahul? *Proceedings of the National Academy of Sciences of the United States of America* 115: 8482–8490.

O'Connell, James, and Jim Allen, 2012. The restaurant at the end of the universe: Modelling the colonisation of Sahul. *Australian Archaeology* 74: 5–16.

O'Connell, James, and Jim Allen, 2015. The process, biotic impact, and global implications of the human colonization of Sahul about 47,000 years ago. *Journal of Archaeological Science* 56: 73–84.

O'Connor, Sue, Anthony Barham, Ken Aplin, Keith Dobney, Andrew Fairbairn, and Michael Richards, 2011. The power of paradigms: Examining the evidential basis for Early to Mid-Holocene pigs and pottery in Melanesia. *Journal of Pacific Archaeology* 2: 1–25.

O'Connor, Sue, and John Chappell. 2003. Colonisation and coastal subsistence in Australia and Papua New Guinea: Different timing, different modes? In *Pacific archaeology: Assessments and prospects*, edited by C. Sand, pp. 17–32. Service des Musées et du Patrimonie de Nouvelle-Calédonie, Nouméa, New Caledonia.

Ohtsuka, Ryutaro, 1983. *Oriomo Papuans: Ecology of Sago-eaters in Lowland Papua*. University of Tokyo Press, Tokyo.

Osborne, P. L., G. Humphreys, and N. Polunin, 1993. Sediment Deposition and Late Holocene Environmental Change in a Tropical Lowland Basin: Waigani Lake, Papua New Guinea. *Journal of Biogeography* 20: 599–613.

Paijmans, Kees, 1976. *New Guinea Vegetation*. The Australian National University, Canberra.

Pain, Colin, and Pamela Swadling, 1980. Sea level changes, coastal landforms and human occupation near Port Moresby – a pilot study. *Science in New Guinea* 7: 57–68.

Pedro, Nicole, Nicolas Brucato, Veronica Fernandes, Mathilde Andre, Lauri Saag, William Pomat, Celine Besse, Anne Boland, Jean-Francois Deleuze, Chris Clarkson, et al., 2020. Papuan mitochondrial genomes and the settlement of Sahul. *Journal of Human Genetics* 65: 875–887.

Petchey, Fiona, and Roger Green, 2005. Use of three isotopes to calibrate human bone radiocarbon determinations from Kainapirina (SAC), Watom Island, Papua New Guinea. *Radiocarbon* 47: 181–192.

Petchey, Fiona, Matthew Spriggs, Foss Leach, Mike Seed, Christophe Sand, Michael Pietrusewsky, and Katy Anderson, 2011. Testing the human factor: Radiocarbon dating the first peoples of the South Pacific. *Journal of Archaeological Science* 38: 29–44.

Petchey, Fiona, and Sean Ulm, 2012. Marine reservoir variation in the Bismarck Region: An evaluation of spatial and temporal change in ΔR and R over the last 3000 years. *Radiocarbon* 54: 45–58.

Petchey, Fiona, Sean Ulm, Bruno David, Ian McNiven, Brit Asmussen, Helene Tomkins, Nic Dolby, Ken Aplin, Thomas Richards, Cassandra Rowe, et al., 2013. High-resolution radiocarbon dating of marine materials in archaeological contexts: radiocarbon marine reservoir variability between Anadara, Gafrarium, Batissa, Polymesoda spp. and Echinoidea at Caution Bay, Southern Coastal Papua New Guinea. *Archaeological and Anthropological Sciences* 5: 69–80.

Petchey, Fiona, Sean Ulm, Bruno David, Ian McNiven, Brit Asmussen, Helene Tomkins, Thomas Richards, Cassandra Rowe, Matthew Leavesley, Herman Mandui, et al., 2012. 14C marine reservoir variability in herbivores and deposit-feedings gastropods from an open coastline, Papua New Guinea. *Radiocarbon* 54: 967–978.

Pöch Rudolf, 1907. Ausgrabungen alter Topfcherben in Wanigela (Collingwood-Bay). *Mitteilungen der Anthropologischen Gesellschaft* 37: 137–139.

Pretty, Graeme, 1967. Report on an Inspection of Certain Archaeological Sites and Field Monuments In The Territory Of Papua New Guinea, Unpublished report to the Papua and New Guinea Public Museum and Art Gallery.

Ramsey, Christopher, 2009. Bayesian analysis of radiocarbon dates. *Radiocarbon* 51: 337–360.

Rehfeld, Kira, Thomas Munch, Sze Ling Ho, and Thomas Laepple, 2018. Global patterns of declining temperature variability from the Last Glacial Maximum to the Holocene. *Nature* 553: 356–359.

Reimer, Paula, William Austin, Edouard Bard, Alex Bayliss, Paul Blackwell, Christopher Ramsey, Martin Butzin, Hai Cheng, Lawrence Edwards, Michael Friedrich, et al., 2020. The IntCal20 Northern Hemisphere radiocarbon age calibration curve (0–55 cal kBP). *Radiocarbon* 62: 1–33.

Rhoads, James, 1980. Through a glass darkly: Present and past land use of Papuan sagopalm users. Unpublished doctoral thesis, Australian National University, Canberra.

Rhoads, James, 1982. Prehistoric Papuan exchange systems: The Hiri and its antecedents, In *The Hiri in History: Further Aspects of Long Distance Motu Trade in Central Papua*, edited by T. E. Dutton, pp. 131–151. Australian National University, Canberra.

Rhoads, James, 1983. Prehistoric Sites from the Kikori Region, Papua New Guinea. *Australian Archaeology* 16: 96–114.

Rhoads, James, 1994. The Popo site, In *Archaeology of a coastal exchange system: sites and ceramics of the Papuan Gulf*, edited by D. Frankel and J. Rhoads, pp. 53–69. ANU Press, Canberra.

Richards, Thomas, Bruno David, Ken Aplin, and Ian McNiven, 2016. *Archaeological Research at Caution Bay, Papua New Guinea: Cultural, Linguistic and Environmental Setting.* Archaeopress, Oxford.

Roberts, Richard, 1997. Luminescence dating in archaeology: From origins to optical. *Radiation Measurements* 27: 819–892.

Roberts, Richard, and Rhys Jones, 1994. Luminescence dating of sediments: New light on the human colonisation of Australia. *Australian Aboriginal Studies* 2: 2–17.

Ross, Malcolm, 1988. *Proto Oceanic and the Austronesian Languages of Western Melanesia.* Pacific Linguistics, Canberra.

Ross, Malcolm, 2001. Is there an East Papuan phylum? Evidence from pronouns, In *The boy from Bundaberg: Studies in Melanesian linguistics in honour of Tom Dutton*, edited by A. Pawley, M. Ross, and D. Tryon, pp. 301–321. Pacific Linguistics, Canberra.

Rowe, Cassandra, Bruno David, Jerome Mialanes, Sean Ulm, Fiona Petchey, Samantha Aird, Ian McNiven, Matthew Leavesley, and Thomas Richard, 2020. A Holocene record of savanna vegetation dynamics in southern lowland Papua New Guinea. *Vegetation History and Archaeobotany* 29: 1–14.

Rowe, Cassandra, Ian McNiven, Bruno David, Thomas Richards, and Matthew Leavesley, 2013. Holocene pollen records from Caution Bay, southern mainland Papua New Guinea. *The Holocene* 23: 1130–1142.

Ryan, William, Suzanne Carbotte, Justin Coplan, Suzanne O'Hara, Andrew Melkonian, Robert Arko, Rose Weissel, Vicki Ferrini, Andrew Goodwillie, Frank Nitsche, et al., 2009. Global Multi-Resolution Topography (GMRT) synthesis data set. *Geochemistry, Geophysics, Geosystems* 10 1–9.

Saville, William, 1926. *In Unknown New Guinea*. Seely, Service and Company, London.

Seligman, Charles, 1910. *The Melanesians of British New Guinea*. Cambridge University Press, Cambridge.

Shaw, Ben, 2015. The archaeology of Rossel Island, Massim, Papua New Guinea: Towards a prehistory of the Louisiade Archipelago. 2 Vol. Unpublished doctoral thesis, Australian National University, Canberra.

Shaw, Ben, 2019. Archaeology of the Massim Island Region, Papua New Guinea, In *Encyclopedia of Global Archaeology*, edited by C. Smith. Springer Reference, New York. https://doi.org/10.1029/2006JF000677.1007/978-3-030-30018-0

Shaw, Ben, and William Dickinson, 2017. Excavation on Nimowa Island, Louisiade Archipelago, Papua New Guinea: Insights Into Cultural Practices and the Development of Exchange Networks in the Southern Massim Region. *The Journal of Island and Coastal Archaeology* 12: 398–427.

Shaw, Ben, and Simon Coxe, 2021. Cannibalism and developments to socio-political systems from 540 BP in the Massim islands of southeast Papua New Guinea. *Technical Reports of the Australian Museum* 34: 47–60.

Shaw, Ben, Simon Coxe, Jemina Haro, Karen Privat, Simon Haberle, Felicitas Hopf, Emily Hull, Stuart Hawkins, and Geraldine Jacobsen, 2020a. Smallest Late Pleistocene inhabited island in Australasia reveals the impact of post-glacial sea-level rise on human behaviour from 17,000 years ago. *Quaternary Science Reviews* 245: article 106522.

Shaw, Ben, Simon Coxe, Vincent Kewibu, Jemina Haro, Emily Hull, and Stuart Hawkins, 2020b. 2500-year cultural sequence in the Louisiade Archipelago (Massim region) of eastern Papua New Guinea reflects adaptive strategies to re-mote islands since Lapita settlement. *The Holocene* 30: 1075–1090.

Shaw, Ben, Geoff Irwin, Alana Pengilley, and Sarah Kelloway. 2021. Village-specific Kula partnerships revealed by obsidian sourcing on Tubetube Island, Papua New Guinea. *Archaeology in Oceania* 56: 32–44.

Shearman, Philip, 2010. Recent changes in the extent of mangroves in the northern Gulf of Papua New Guinea. *Ambio* 39: 181–189.

Skelly, Robert, Bruno David, Bryce Barker, Alois Kuaso, and Nick Araho, 2010. Migration sites of the Miaro clan (Vailala River region, Papua New Guinea):

Tracking Kouri settlement movements through oral tradition sites on ancient landscapes. *The Artefact* 33: 16–29.

Skelley, Robert, 2014. From Lapita to the Hiri: Archaeology of the Kouri Lowlands, Gulf of Papua, Papua New Guinea. Unpublished doctoral thesis, Monash University, Clayton.

Skelly, Robert, Bruno David, Fiona Petchey, and Matthew Leavesley, 2014. Tracking ancient beach-lines inland: 2600-year-old dentate-stamped ceramics at Hopo, Vailala River region, Papua New Guinea. *Antiquity* 88: 470–487.

Skelly, Robert, and Bruno David, 2017. *Hiri: Archaeology of Long-distance Maritime Trade along the South Coast of Papua New Guinea*. University of Hawaii Press, Honolulu.

Skelly, Robert, Bruno David, Matthew Leavesley, Fiona Petchey, Alu Guise, Roxanne Tsang, Jerome Mialanes, and Thomas Richards, 2018. Changing ceramic traditions at Agila ancestral village, Hood Bay, Papua New Guinea. *Australian Archaeology* 84: 181–195.

Sloss, Craig, Luke Nothdurft, Quan Hua, Shoshannah O'Connor, Patrick Moss, Daniel Rosendahl, Lynda Petherick, Rachel Nanson, Lydia Mackenzie, Alison Sternes, et al., 2018. Holocene sea-level change and coastal landscape evolution in the southern Gulf of Carpentaria, Australia. *Holocene* 28: 1411–1430.

Smith, Anita, and Jim Allen, 1999. Pleistocene shell technologies: Evidence from Island Melanesia, In *Australian Coastal Archaeology*, edited byJ. Hall and I. J. McNiven, pp. 291–297. Australian National University, Canberra.

Smith, Ian, 1973a. The geology of the Calvados Chain, southeastern Papua. *Bureau of Mineral Resources, Geology and Geophysics Australia* 139: 59–67.

Smith, Ian, 1973b. Late Cainozoic volcanism in the southeast Papuan islands. Bureau of Mineral Resources, Geology and Geophysics, Canberra.

Smith, Jan, 1964. *Diastrophic Evolution of Western Papua New Guinea*. University of Tasmania, Tasmania.

Specht, Jim, 1974. An archaeological site at Obu plantation, Central Distict, Papua. *Records of the Papua New Guinea Museum* 4: 44–52.

Specht, Jim, 2012. Caution Bay and Lapita pottery: Cautionary comments. *Australian Archaeology* 75: 3–7.

Spriggs, Matthew, 1991. Nissan, the island in the middle. Summary report on excavations at the North end of the Solomons and the South end of the Bismarcks, In *Report on the Lapita Homeland Project*, edited by J. Allen and C. Gosden, pp. 222–243. Department of Prehistory, Research School of Pacific Prehistory, Australian National University, Canberra.

Spriggs, Matthew, 1997. *The Island Melanesians*. Blackwell Publishers, Oxford.

Sullivan, M., and M. Sassoon, 1987. Prehistoric Occupation of Loloata Island Papua New Guinea. *Australian Archaeology* 24: 1–9.

Summerhayes, Glenn, Judith Field, Ben Shaw, and Dylan Gaffney, 2017. The archaeology of forest exploitation and change in the tropics during the Pleistocene: The case of Northern Sahul (Pleistocene New Guinea). *Quaternary International* 448: 14–30.

Summerhayes, Glenn, Matthew Leavesley, Andrew Fairbairn, Herman Mandui, Judith Field, Anne Ford, and Richard Fullager, 2010. Human Adaptation and Plant Use in Highland New Guinea 49,000 to 44,000 Years Ago. *Science* 330: 78–81.

Summerhayes, Glenn, 2007. Island Melanesian pasts: A view from archaeology, In *Genes, Language and Culture History in the Southwest Pacific*, edited byJ. Friedlaender, pp. 10–35. Oxford University Press, Oxford.

Summerhayes, Glenn, and Jim Allen, 2007. Lapita writ small? Revisiting the Austronesian colonisation of the Papuan South Coast, In *Oceanic explorations: Lapita and Western Pacific Settlement*, edited by S. Bedford, C. Sand, and S. P. Connaughton, pp. 97–122. ANU E-Press, Canberra.

Swadling, Pamela, 1980. Decorative features and sources of selected potsherds from archaeological sites in the Gulf and Central Province. *Oral History* 8: 101–123.

Swadling, Pamela. 1981. The settlement history of the Motu and Koita speaking people of the Central Province, Papua New Guinea, In *Oral Tradition in Melanesia* edited by D. Denoon and R. Lacey. University of PNG and the Institute of Papua New Guinea Studies, Port Moresby.

Swadling, Pamela, 1997. Changing shorelines and cultural orientations in the Sepik-Ramu, Papua New Guinea: Implications for Pacific prehistory. *World Archaeology* 29: 1–14.

Swadling, Pamela, 2016. Mid Holocene social networks in far eastern New Guinea. *Journal of Pacific Archaeology* 7: 7–19.

Swadling, Pamela, and Ombone Kaiku, 1980. Radiocarbon date from a fireplace in the clay surface of an eroded village site in the Papa salt pans, Central Province. *Oral History* 8: 86.

Swadling, Pamela, Polly Weissner, and Akii Tumu, 2008. Prehistoric stone artefacts from Enga and the implication of links between the highlands, lowlands and islands for early agriculture in Papua New Guinea. *Journal de la Société des Océanistes* 126–127: 271–292.

Szabo, Katherine, Bruno David, Ian McNiven, and Matthew Leavesley, 2020. Preceramic shell-working, Caution Bay and the Circum-New Guinea Archipelago, In *Theory in the Pacific, the Pacific in Theory*, edited byT. Thomas, pp. 123–144. Routledge, London.

Thangavelu, Anbarasu, 2015. "Unshelling the past" – An archaeological study of shellfish assemblages from Caution Bay, Papua New Guinea. Unpublished doctoral thesis, University of Southern Queensland, Toowoomba.

Tobler, Ray, Adam Rohrlach, Julien Soubrier, Pere Bover, Bastien Llamas, Jonathon Tuke, Nigel Bean, Ali Abdullah-Highfold, Shane Agius, Amy O'Donoghue, et al., 2017. Aboriginal mitogenomes reveal 50,000 years of regionalism in Australia. *Nature* 544: 180–184.

Torrence, Robin, Sarah Kelloway, and Peter White, 2013. Stemmed tools, social interaction, and voyaging in early-mid Holocene Papua New Guinea. *The Journal of Island and Coastal Archaeology* 8: 278–310.

Urwin, Chris, Quan Hua, and Henry Arifeae, 2021. Combining oral traditions and bayesian chronological modeling to understand village development in the Gulf of Papua (Papua New Guinea). *Radiocarbon* 63: 647–667.

Urwin, Chris, Quan Hua, Robert Skelly, and Henry Arifeae, 2018. The chronology of Popo, an ancestral village site in Orokolo Bay, Gulf Province, Papua New Guinea. *Australian Archaeology* 84: 90–97.

Vanderwal, Ron, 1973. Prehistoric studies in Central Coastal Papua. Unpublished doctoral thesis, Australian National University, Canberra.

Vasey, Daniel, 1982. Subsistence potential of the pre-colonial Port Moresby area, with reference to the hiri trade. *Archaeology in Oceania* 17: 132–142.

Warner, Jeffrey, D Pelowa, Bart Currie, and R Hirst, 2007. Melioidosis in a rural community of Western Province, Papua New Guinea. *Transactions of the Royal Society of Tropical Medicina and Hygiene* 101: 809–813.

White, Peter, 1967. Taim Bilong Bipo: Investigations towards a prehistory of the Papua New Guinea Highlands. Unpublished doctoral thesis, Australian National University, Canberra.

White, Peter, 1972. *Ol Tumbuna: Archaeological Excavations in the Eastern Central Highlands, Papua New Guinea.* Australian National University, Canberra.

White, Peter, and Douglas Hamilton, 1973. Anthropology, In *New Guinea barrier reefs: Preliminary results of the 1969 coral reef expedition to the Trobriand Islands and the Louisiade Archipelago, Papua New Guinea,* edited by W. Manser. University of Papua New Guinea, Port Moresby.

Williams, Francis, 1936a. *Papuans of the Trans-Fly.* Clarendon Press, Oxford.

Williams, Francis, 1936b. Report on stone structures in the Trobriands – Little Stonehenge. *Pacific Islands Monthly* June 1936.

Williams, Martin, Nigel Spooner, Kathryn McDonnell, and James O'Connell, 2021. Identifying disturbance in archaeological sites in tropical northern Australia: Implications for previously proposed 65,000-year continental occupation date. *Geoarchaeology* 36: 92–108.

Wurm, Stephen, 1971. Notes on the Linguistic Situation in the Trans-Fly Area, In *Papers in New Guinea Linguistics 14,* edited by T. Dutton, C. L. Voorhoeve, and S. A. Wurm, pp. 115–172. ANU Press, Canberra.

Yokoyama, Yusuke, Kurt Lambeck, Patrick De Dekker, Paul Johnston, and Keith Fifield, 2000. Timing of the Last Glacial Maximum from observed sea-level minima. *Nature* 406: 713–716.

10 Kisim save long graun

Understanding the nature of landscape change in modelling Lapita in Papua New Guinea

Glenn R. Summerhayes

The recent compilation of Lapita sites across the western Pacific presents a total of over 290 sites, of which 88 came from the Bismarck Archipelago (Bedford et al., 2019: 9, Table 1.1). The list represents a small subsample of sites within the region, as only those with dentate-stamped decorated pottery were included while excluding many plainware sites. Most of these plainware sites were within areas where only Lapita pottery was ever used, and they were no doubt parts of the broader Lapita assemblages. Within the Bismarck Archipelago, our models of settlement, distribution, subsistence, and colonisation have been built on a small number of sites. Only a handful contain remains that indicate subsistence, with most sites found in volcanic locations with acidic soils in which little if any organics survive. Moreover, tectonic activities have impacted the nature of the archaeological assemblages, not to mention the more gradual post-depositional geomorphological processes as well.

It is premature to try to reconstruct Lapita settlement patterns as a social phenomenon without first accounting for the varied landscape changes that have affected our archaeological datasets. Cultural factors must be taken into consideration; however, as noted by Boyd et al. (1999), non-cultural factors also must be taken into account in assessing Holocene archaeological records (Boyd et al. 1999: 283). In their study of the Kandrian area of southwest New Britain (Figure 10.1), they identified eustatic processes and tectonic uplift some 1500 years ago. Ancient Lapita sites with stilt houses would have been affected by the subsequent lowering of sea level and later uplift, raising them above any waterlogged aerobic conditions, thereby not only destroying rich organic remains (Boyd et al. 1999: 287) but also affecting the pottery collections with the leaching of salts making ceramic assemblages friable and liable to disintegrate. Such localised processes, they argue, must be taken into account when explaining major differences between the archaeologically poor record of Kandrian with the rich Arawe archaeological record found only 50 km farther west. This example is used by Boyd et al. (1999) to argue against generalisations between adjacent areas, but more so to alert archaeologists to a need for understanding geomorphological histories of areas under study. Understanding landscape

DOI: 10.4324/9781003139553-10

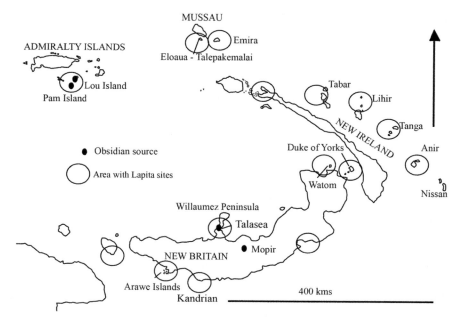

Figure 10.1 Bismarck Archipelago.

change, both natural and artificial, is critical in understanding the nature of the Lapita settlement. This is also seen in the Willaumez peninsula assemblages on the central north coast of New Britain.

This chapter will review geomorphological histories of selected Lapita sites from the Bismarck Archipelago, upon which we model the nature of Lapita settlement. This will be contrasted with three other sites from which little is left after post-depositional events. This contrast between these sites will highlight the importance of untangling geomorphological processes in understanding the nature of Lapita settlements. We will look at how both human and natural processes have set in motion different taphonomic processes, which then differentially affect the sites.

Arawes, Mussau, and Anir: records of exceptional preservation

Boyd et al.'s (1999) concern about comparisons between assemblages from Kandrian and the nearby Arawe Islands are important. Lapita assemblages from both the Arawe Island group of southwest New Britain and Mussau have been used by many to model the nature of Early Lapita settlement and, in particular, stilt house occupation over reef flats and infilling of these settlements due to human activity. Extensive systematic surveying located some of the best-preserved ancient settlements from the western Pacific. An understanding of the geomorphological factors leading to the formation of the deposits was crucial in understanding what made these sites special.

Arawe

Although there is evidence of human presence in the Arawe Islands before Lapita, with Mid-Holocene occupation at Lolmo Cave (Gosden et al. 1994) and purported early occupation at Apalo (Specht et al. 2017), major changes to the landscape are witnessed only later with Lapita occupation. Work by Chris Gosden (an archaeologist) and John Webb (a geomorphologist) demonstrated the existence of stilt village occupation in the Arawe Islands (Gosden and Webb 1994). Lapita sites were located on the lee side of the islands, either on or immediately behind the sandy beaches. A number of sites were excavated, including Apalo (FOJ – site code registered at the National Museum and Art Gallery of Papua New Guinea) on Kumbun Island (Figure 10.2) and Makekur (FOH) on Adwe Island. During the occupation at Apalo, the settlements would have been on the limestone reef platforms some 20 m wide that slope toward the sea, some 30–60 m from the cliff (Gosden and Webb 1994: 33). Gosden and Webb argue that during that period, there was a higher sea stand and that at high tide, houses had between 1.5 and 2 m of water beneath them.

Gosden and Webb (1994: 40) put forward the model that the clearing of forests to create gardens inland led to the subsequent erosion of clays and their deposition behind the reef and stilt settlement locations. Both Kumbun and Adwe islands are raised limestone islands with fragile ecosystems, where any removal of forest cover would create an impact on the environment. The soil/clay subsequently built up behind stilt house occupation, with midden rubbish forming a damming effect against the accumulation of clay at the base of the cliff, thereby infilling the lagoon (Gosden and Webb 1994: 41). In short, it was a change to the landscape caused by human intervention as a result of cultural and economic factors:

> The house piles have a baffling effect; by reducing the strength of the waves and tidal currents underneath the houses, sand that would otherwise have been continuously transported through the area is instead trapped to form sand banks. Any artefact material or rubbish that is dropped or thrown from the houses adds to the sediment accumulating beneath them. (Gosden and Webb 1994: 40)

The archaeological remains were subsequently preserved in what is called a Ghyben-Herzberg lens of brackish water, leading to some spectacular finds as fresh as when they were deposited over 3000 years before. This included the preservation of organics such as seed and wood.

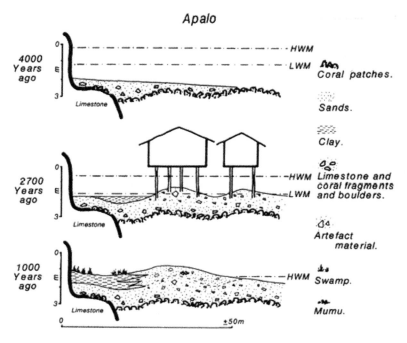

Figure 10.2 Apalo (site FOJ) in the Arawes. Reprinted with permission by Chris Gosden.

Mussau

The second area that has been used to model the nature of Early Lapita settlement and stilt house occupation was found on the small islands just offshore from the island of Mussau. Pat Kirch conducted extensive systematic surveys and excavations to unravel extensive Lapita settlements, which like the Arawe assemblages, were found in pristine condition. Three Lapita villages would have been in sight of each other: Talepakemalai (site code ECA) and Etakosarai (site code ECB) on Eloaua Island, and Etapakengaroasa (site code EHB) on Emanamus Islands, lying just south of the major island of Mussau (Kirch 2001). Kirch called ECA a village (total of 82,000 m²), while ECB and EHB were called hamlets (3000 m² and 1000 m², respectively) (Kirch 2001: 142). On the basis of Pacific sea level studies and an examination of wave-cut notches from Eloaua, Kirch modelled a higher sea stand of 1 to 2 m during the Lapita occupation. As the sea level dropped, the coastal terrace prograded. Kirch argued that the covering of the stilt structures located originally on the "subtidal sandy reef flat" at Talepakemalai had less to do with people than with natural events (Kirch 2001: 132). Kirch provided a convincing argument that a rapid drop in sea level resulted in the erosion of an outer reef and platform, "thus generating an increased quantity of calcareous sediment" (Kirch 2001: 132) which covered the midden deposits. As progradation occurred over time, the beach shifted away, and new stilt structures were built (Figure 10.3). Kirch (2001: 132) argues that by 2500 years BP, the stilt village was abandoned, and as the beach further prograded, the archaeological deposits were "flooded by a Ghyben-Herzberg aquifer," which in turn preserved them.

Both the Arawe and Mussau deposits are remarkably well preserved, allowing a glimpse into Lapita societies that most sites just do not allow. Another site that contained well preserved Lapita assemblages was found in the Anir Island Group.

Anir

The third example is from the Lapita site of Kamgot (ERA), located at the western end of Babase Island, Anir. Anir is composed of two islands, Ambitle and its neighbour Babase island. The Anir Island group lies 60 km off the southeast coast of New Ireland. Babase is mostly volcanic, but its western end is raised limestone. As there were no waterlogged zones at Kamgot preserving structural features of stilt dwellings, Summerhayes et al. (2019: 91) assessed site geomorphological formation processes and the distribution of materials to assess the initial areas of settlement. Kath Szabo's assessment of the molluscan assemblage provided information as to the nature of the underlying substrate. Surface material extends over 400 m in an east-west direction and 60 m in a north-south direction. In total, 77 m² were excavated in 23 test pits over an area of 200 by 100 m. The spread of archaeological materials in the western end of the site is parabolic-shaped in distribution, with Test Pit 17 forming the westernmost limit. No surface or subsurface material was found west of Test Pit 17, where the basal coral ancient reef floor becomes much deeper.

Summerhayes et al. (2019) argue that Kamgot was originally located within an old lagoon, and the site spread over a sandy bar and a reef flat parallel to an outer reef (to the north) some 80 m away, and to the south, a steep limestone rise some 150 m away. Today the site is 100 m in from the high tide mark, with the old raised reef forming a raised beachfront. The initial occupation of Kamgot was directly over the water with upper intertidal habitation following in concert with the building up of the sand spit (Figure 10.4). Like at Apalo, there is an accumulation of clays at the base of the steep rise, and the area is now swampy. Like Apalo, the removal of the forest sitting on top of the limestone resulted in erosion of the clays, resulting in the infilling of the ancient lagoon. Unlike Kumbun Island, however, the island of Babase is actively volcanic, with evidence of tectonic uplift. The base of the major pottery-bearing test pits at Kamgot is some 2.5 m above high tide level (Summerhayes 2000: 171; Summerhayes et al. 2019). Changes to the environment are seen as a combination of human and natural processes.

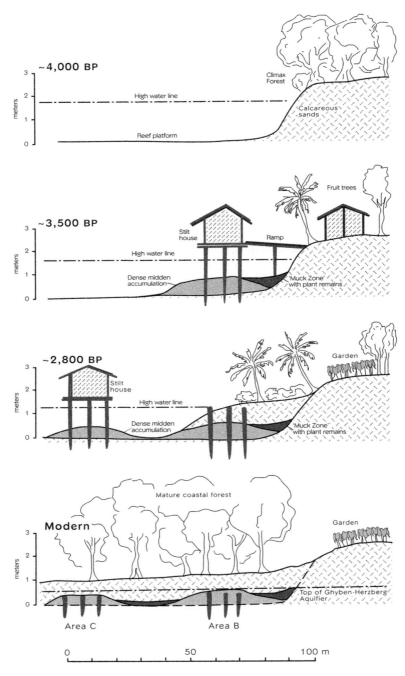

Figure 10.3 Talepakemalai (Site ECA) in Mussau. Reprinted with permission by Patrick V. Kirch.

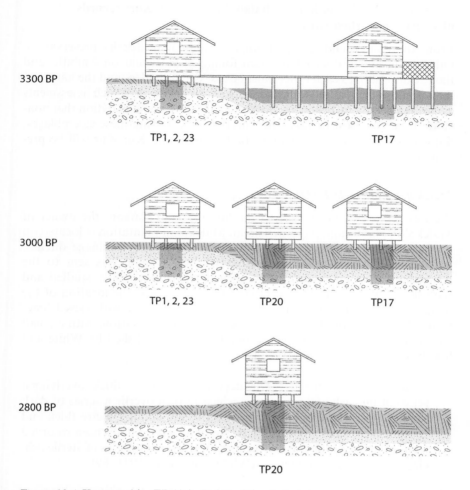

3300 BP

TP1, 2, 23　　　　　　　　　TP17

3000 BP

TP1, 2, 23　　　　TP20　　　　TP17

2800 BP

TP20

Figure 10.4 Kamgot (site ERA) in Babase Island, Anir.

The sites of Talepakemalai (ECA), Adwe (FOH), Apalo (FOJ), and Kamgot (ERA) have yielded a rich archaeological record, providing insights into Early Lapita settlement and subsistence, and providing critical data for modelling colonisation. These newcomers had a great impact on the environment, changing it to sustain a lifestyle. An important point made by Chris Gosden is that "Land use is culturally determined, arising from choices people make about how to provision the social system" (Gosden and Webb 1994: 48). Indeed it would have been strange if no change was measured in the landscape. Yet as will be seen below, other finds indicate that Lapita sites need to be placed into their correct taphonomic and geomorphological context before producing social models regarding settlement practices (see Summerhayes et al. 2009).

Malekolon, Feni Mission, Balbalankin, and Kur Kur: records of heavily disturbed sites

From the Anir island group, Kamgot stands alone for its preservation. Other archaeological sites have been found, in particular on Ambitle, and their relationship to each other must be assessed to understand the nature of the Lapita settlement on Anir and the western Pacific. Yet such assessments can be made only by following Boyd et al. (1999) recommendation that non-cultural factors must be taken into account in assessing these assemblages. Sites of Malekolon, Feni Mission, Balbalankin, and Kur Kur will be presented below.

Malekolen (site code EAQ)

We owe a great deal of debt to the late Graham Carson, the owner of Malekolon plantation, Ambitle Island. Malekolon plantation is located on northeast Ambitle Island (Figure 10.5). While digging drainage ditches, he uncovered sherds of pottery from which a sample was sent to the Australian Museum, Sydney. The pottery was subsequently studied and reported by White and Specht (1971). Of interest was the location of the finds, some 500 m landward from the present coastline with raised limestone cliffs bordering the north and south of the plantation, with a small steam cutting through the plantation. The notes published by White and Specht describe the stratigraphy as:

> a surface set of culturally sterile deposits up to 40 cm thick, overlying a series of fine ash and clay layers which in turn overlie a series of dark sandy levels. These two lower formations are about 1 metre thick and contain sherds and stone artifacts. No other material has been recorded from these levels. Below the occupation levels are a series of sterile ash, clay, and gravel mixed levels. (White and Specht 1971: 89)

The enigma of an inland Lapita site was not solved by Ambrose, who, in 1970 and 1971, undertook excavations at the Malekolon site (EAQ). Ambrose dug 19 m^2 and obtained a radiocarbon date on burnt *galip* nut from basal deposits dating to 2050 ± 210 uncal. years BP (ANU-957; 2700–1540 cal. years BP at 2σ) (Ambrose 1978: 331) and another on charcoal 1340 ± 230 uncal. years BP (ANU-7711; 1775–785 cal. years BP at 2σ) (Ambrose 1976: 104; see also Hogg et al. 2021 for calibrations). Ambrose also undertook subsurface testing using a grid pattern, augering at 20 m intervals, indicating 4000 m^2 of deposit (see also Green 1979: 51).

Although little was published, Dimitri Anson used Ambrose's pottery for his doctoral thesis and his formulation of a "Far Western" Lapita style. Ambrose noted that his radiocarbon dates may not be reliable indicators of occupation due to "flood mixing" (Ambrose 1978: 331), and Anson noted

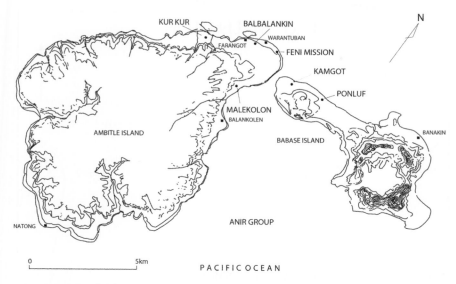

Figure 10.5 Anir Island Group.

that the site was post-depositionally disturbed (1983: 243; 1986: 158). Furthermore, Ambrose's excavations uncovered a variety of pottery decorations, including dentate-impressed, incision, shell impressed ware, and applique, adding to the idea that this jumble of decoration was mixed.

In 1995, Summerhayes returned to Anir to re-excavate Malekolon (EAQ) and to survey for new Lapita sites. The objective at Malekolon was to establish the geomorphological processes behind the site's formation and to understand whether the site was disturbed and why it was located half a kilometre inland. The first priority was a survey of the plantation, which showed that the limestone cliffs on the south and north side of the site formed a "V" shaped valley, with the cliffs joined together on the western perimeter. To the east, the area is bordered by the sea with an outlying reef just offshore.

To assess the impact people had on the landscape and to identify subsequent geomorphological processes that impacted the environment we see today, a series of test pits (five in total) was excavated from the beach to inland into the "V" shaped valley along an east-to-west transect (Figure 10.6). Test pit 1 was located 25 m landward from and 3.3 m above the high water mark. Test pit 2 was 375 m landward from and 3.8 m above the high water mark. Test pit 3 was 444 m landward and 4.6 m above the high water mark, while Test pit 4 was 486 m landward and 5 m above the high water mark. Of particular interest was the location of Test pit 4, which is located 10 m east of where Ambrose found pottery.

Apart from thin topsoil, Test pit 1 contains over 3 m of built-up volcanic ash sitting on the coral bedrock, while Test pits 2 and 3 are mostly clay

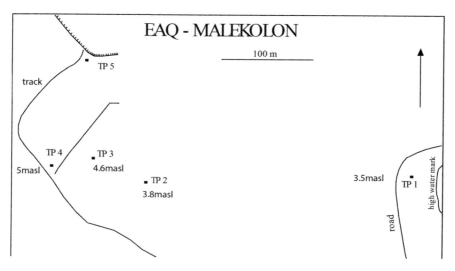

Figure 10.6 Malekolon (site EAQ) in Ambitle Island.

derived from the ash. Test pit 3 has a black sand layer beneath this clay. Test pit 4, which is farther inland, has a stratigraphy showing the light brown ash sitting on top of black beach sand. The black beach sand is 25 cm deep and is found 1 m below the ground surface. It overlays white beach sand. The layer black sand beach layer is slightly sloping to the east. At this level, the site is c. 4–4.4 m above sea level. Even taking into account the high sea stand at the time of occupation, this site demonstrates uplift due to volcanism of at least 2 m.

After analysing the nature of deposits, a tentative history of site formation can be made. Test pit 4 was originally at the edge of an embayment. Occupation was coastal and not half a kilometre inland. At 2300 years ago, the volcano on Ambitle erupted (Licence et al. 1987: 274), and ash was deposited and subsequently eroded into the valley and built up behind the reef. This build-up acted as a dam, with the clays from the top of the island eventually filling in the embayment. Evidence for the damming effect is seen in Test pits 1, 2, and 3.

Artefactual material in Test pit 4 is abundant, with 2559 sherds and 211 obsidian pieces excavated. The bulk of the material is found in the lower 40 cm of brown ash, which overlays the black sand. Also found were a stone adze, a possible stone chisel, and a few pieces of quartz and chert. Decoration on the pottery includes dentate stamping, linear incision, shell impression, striation and nubbins. Test pit 3, which is culturally sterile, also has black sand but below a clay layer. The sand here is 40 cm below that from nearby Test pi 4. This test pit would have been located within the embayment at the time of Lapita occupation (see Summerhayes 2000: 170, table 4). Test pit 5 was located 115 m north of test pit 3 and 14 m south of the cliff line, next to a track leading to the top of an escarpment. It was selected to ascertain if the material was eroding from the top of the escarpment. No artefactual material was found.

Interpretation of the deposits in the above test pits suggests that the Lapita occupation, represented by the cultural materials in Test pit 4, was situated next to an embayment with a fringing reef. Earlier occupation of the site (to be discussed below) was situated on the beach, which due to subsequent progradation and infilling of the valley over time, is represented by deposits situated further inland. Of interest is the nature of post-depositional processes and its timing. Summerhayes obtained two radiocarbon estimates from test pit 4: 1) ANU-11190 (spit 10), charcoal, 2110 ± 240 uncal. years BP, 2727–1570 cal. years BP at 2σ.; and 2) ANU-11193 (spit 11), charcoal: 3220 ± 170 uncal. years BP, 3872–2997 cal. years BP at 2σ. (Summerhayes, 2001a: table 3). When Ambrose's radiocarbon estimates (see above) are considered with these, it confirms Summerhayes' assertion (2004: 147) that ANU-11190 and ANU-957 date the volcanic eruption, while ANU-11193 dates the cultural deposits in Test pit 4. As noted by Hogg et al. (2021: 74), due to the large standard error associated with this date, the upper range limit, when calibrated to 2σ, overlaps with that of the Early period. Hogg et al. (2021: 74) argue that this broad range can be narrowed considerably by reference to obsidian source exploitation within the deposit, which closely aligns with Middle Lapita sites within the Bismarck Archipelago dating to between about 2900 and 2700–2600 BP (Summerhayes, 2004: table 2, 150). Hogg et al. (2021: 74) argue that the most parsimonious explanation is for the cultural material within Test pit 4 to date to the Middle Lapita period. Yet, earlier pottery based on the design was also found at Malekolon, suggesting an earlier occupation, as well. One dentate stamped sherd was dated by Ambrose using thermoluminescence to 3200 years BP, while a second dated of 2500 years BP, all strongly pointing to the early deposits being highly disturbed (Ambrose quoted in Anson, 1983: 12). This led Hogg et al. (2021) to argue that the early deposits were highly disturbed.

In summary, the 1995 fieldwork demonstrated that Test pit 4, which contains Middle Lapita *in situ* deposits, although some 486 m inland from the current coastline, was on the edge of an embayment when the material was deposited. Earlier beach deposits would have been farther inland, and over time the embayment infilled with a prograding shoreline. The earlier occupation had been disturbed 2300 years ago when the volcano on Ambitle erupted (Lindley 2015: 529), and ash was deposited and subsequently eroded into the valley and built up behind the reef.

Feni Mission (site code ERG)

The finding of Lapita pottery at Feni Mission was serendipity. While based at the Feni Catholic Mission in 1995, Summerhayes was asked by the priest

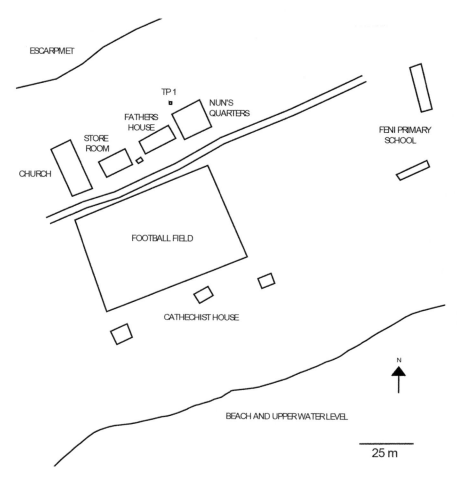

ESCARPMET

TP 1

NUN'S
QUARTERS

FATHERS
HOUSE

STORE
ROOM

CHURCH

FENI PRIMARY
SCHOOL

FOOTBALL FIELD

CATHECHIST HOUSE

N

BEACH AND UPPER WATER LEVEL

25 m

Figure 10.7 Feni Mission (site ERG) in Ambitle Island.

Father Tony Gendusa to dig a garbage pit behind the nuns' quarters. Gendusa was an old-time priest who had been based in the Bismarck Archipelago since the end of World War 2. Subsurface testing was planned for this part of the island to ascertain the rate of uplift, so we obliged with the request and excavated a pit. The mission is situated on the northeastern tip of Ambitle, next to Salat Strait (Figure 10.7 and see Figure 10.5). The mission area is flat and wide, 170 m from the beachfront to the beginning of the slope towards the escarpment. A football field separates the church, store and nuns' quarters from the beachfront, which is 110–120 m distant. Salat Strait separating Ambitle from Babase showed evidence of uplift, with the centre of the strait becoming visible whilst crossing in 1995, whereas it was not visible when visited by Ambrose. Indeed over the time of the field research (1995 through 2002), it was obvious that the strait was becoming more shallow each year.

A single 1 by 1 m test pit was excavated behind the Feni Mission, 4 m behind the nuns' quarters, and 135 m in from the beach and at the height of 5.4 m above sea level (see Figure 10.7; Summerhayes 2000). The stratigraphy is basically clay that becomes stickier with depth. Limestone bedrock is reached at 1.1 m depth. In total, 569 potsherds and 113 pieces of obsidian were found in this test pit. No material was found in the bottom 10 cm. The pottery looks eroded and probably redeposited. Decoration includes dentate stamping, linear incision, applied bands and flat knobs. The dentate decoration is more open and loose, and it fits into the Middle Lapita tradition. A radiocarbon estimate was made on charcoal, with a large standard error of 3090 ± 170 uncal. years BP (ANU-11191), which calibrates to 3690–2850 cal. years BP (Summerhayes 2001b: 56; 2004: 143). The site, however, informs us considerably about the rate of uplift on the northern edge of the island, and taken with the results from Malekolon suggests major tectonic uplift, volcanism, and the lowering of sea level that stranded sites high and dry.

Balbalankin (site code ERC)

The Lapita site of Balbalankin is another example where eustatic processes and tectonic uplift have affected the nature of the archaeological assemblage. Balbalankin is located on the northwestern coast of Ambitle Island to the south of the hamlet of Farangot, on the opposite side of the island from Malekolon. The site is approximately 140–200 m inland from the beach on an area of raised flat garden land, and unlike Malekolon (which is found in a small valley), Balbalankin is backed by an escarpment. The area was littered with surface pottery which was brought to Summerhayes' attention by schoolchildren after a presentation at the Feni school. After a 15-minute surface collection by local children, over 500 sherds (27 decorated) and 26 pieces of obsidian were collected (Summerhayes 2000).

Eight test pits were laid out over a grid (Figure 10.8). The aim was firstly, to find cultural material, and secondly, to understand the depositional history of the site. Test pit 1 was located 180 m inland from the sea's high water mark and 3.2 m above sea level at its surface, while its base was 2.74 m above sea level. Test pits 2 and 3 were set on a north-south axis from Test pit 1. Test pit 2 was located 25 m due south, and Test pit 3 was 26 m due north. Two east-west transects were set out from Test pits 1 and 3, and four more test pits were excavated. Test pits 4 and 6 were set out 25 m east and west from Test pit 1, respectively. Test pits 7 and 5 were set out 28 m and 26 m east and west from Test pit 3 respectively. To ascertain the nature of deposits closer in towards the escarpment, Test pit 8 was set out 50 m due south from test pit 6.

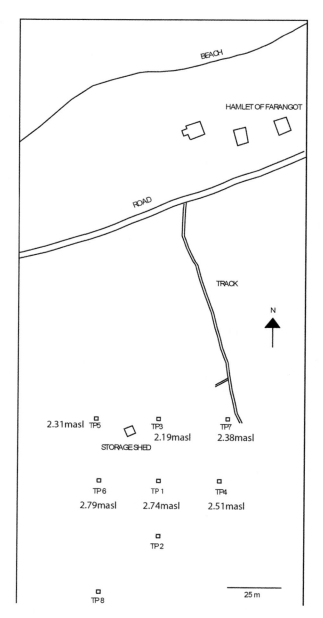

Figure 10.8 Balbalankin (site ERC) in Ambitle Island.

Over 1400 sherds were excavated from Balbalankin (see Summerhayes 2000: table 2). Pottery decoration includes dentate stamping with motifs indicating a Middle Lapita assemblage, linear incision, nubbins, applique and shell impression. Also found was one piece of chert, parts of two *Tridacna* armbands, earth oven stones, and plenty of fishbones. Bone was also recovered from Test pit 7, including a pig cranium cemented in the coral bedrock at the base of the excavation. The Middle Lapita designation is confirmed by the single radiocarbon determination from Test pit 1 of 2620 ± 110 cal. years BP which calibrates to 2950–2365 cal. years BP at 2σ. (ANU-11188 in spit 5, charcoal; Summerhayes, 2001a: table 3).

Of note is that all test pits overlie a basal reef platform at varying depths over the present high water mark. Of importance was the distribution of cultural material, which was primarily found within a white beach sand layer located only in those test pits closer to the sea, in Test pits 1, 3, 4, 5, and 7. Within these test pits, the white beach layer was covered by light brown ash. The depth of the white sand layer varied with the deepest deposits closer to the sea, such as Test pits 3, 5, and 7 (80, 50, and 35 cm respectively), and was completely missing in those test pits closer to the escarpment, namely in Test pits 2, 6, and 8, which had the light brown ash sitting directly over the basal reef bedrock. Artefact densities are directly related to the presence/absence and depth of the artefact-bearing coarse white beach sand layer, which is in turn related to the amount of uplift and subsequent taphonomic processes. The most northerly Test pits, 3, 5, and 7, have the highest artefact concentrations. They also have, as noted above, the thickest deposits of the coarse white beach sand layer and the least tectonic uplift. Measurements from the basal reef demonstrate that Test pits 5, 3, and 7 are 2.2 m, 2.2 m, and 2.4 m above the high water mark. Test pits 1 and 4 were farther inland and had the artefact bearing white beach sand present, where they were 2.7 and 2.5 m above the high water mark. Test pits 2, 6, and 8, on the other hand, have no coarse white beach sand layer and fewer artefacts. Here the base of Test pit 6 was 2.9 m above the high water mark. Measurements from Test pits 2 and 8 were not made; however, it was obvious that the basal reef deposits were higher than those closer to the beach.

It can be seen that uplift is occurring from the centre of this volcanic island. This site is important not only for its assemblage but also for explicating the survival of assemblages by post-depositional processes. Note that within the white sand layer, bivalves were found in the death position in association with finger coral. This indicates that either (a) the finds were originally deposited into the water from above, such as a stilt village, or (b) original occupation was on the shoreline.

From the above, we can surmise that the material was originally deposited in a low energy water environment, similar to that at the Arawe Islands and Talepakemalai (Gosden and Webb 1994; Kirch 2001). On the basis of decoration, the pottery assemblage is "Middle Lapita" (Summerhayes 2007: 148).

Kur Kur (site code ERB)

The hamlet of Kur Kur is located on Ambitle's northwest coast, south of Balbalankin. Pottery sherds and obsidian on the surface were found while undertaking a survey from the beach into the interior. The sherds were found on a track eroding from a clay underlying a yellow tephra, at 160 m landward from the beach. Farther inland, about 500–600 m landward from the beach, the land rises steeply to form a ridge on which an old village (Sibinmou) was situated. The ridge is narrow and slopes on three sides. In 1996, it was decided to excavate the area inland from Kur Kur to locate more material *in situ* and to gain insight into local geomorphological processes on this side of the island.

Prior to excavating, a 700-m-long survey was made along the Fangarn river located 1 km farther north of Kur Kur. The river cuttings are deep, being 10 m high at 700 m inland. The deposits consist of clay, overlain by a thick band of river pebbles, overlain by yellow ash. The thick band of pebbles lies flat just above water level. Ash sits above the pebble layer, and the farther you go inland, the ash deposits increase in thickness to over 10 m in height. Towards the road, the ash is much thinner at around 20 cm thick. Due to the great depth of ash farther inland, it was decided to excavate no farther than halfway to the escarpment. As such, three test pits were set out from the beach to halfway to the ridge, and a farther test pit was located on the ridge where a surface collection of pot and obsidian was made.

The farther inland one proceeds away from the coast, the deposit thickens, and uplift of the coral platform is marked. Test pit 1 is located 60 m in from the high water mark and is 2.8 m above sea level (Figure 10.9). Raised coral bedrock was found at 1.2 m depth and was 1.6 m above the high water mark. Coral bedrock from Test pit 2, located an additional 100 m inland (165 m from the coast), was also found at 1.2 m depth but at 3.95 m above the high water mark. That's an uplift of 2.3 m over a land distance of 100 m. Test pit 3 was dug 282 metres inland from the coast with the surface level some 7.6 m above the high water mark. The site excavation was abandoned at 110 cm depth, with the realisation that the coral base could be 10 m deep based on a nearby stream cutting.

The stratigraphy changes the farther you progress inland. Test pit 1 consisted of a series of ashy soils varying from brown sandy (10–39 cm), white sandy light brown (39–49 cm) to a yellow-brown ashy soil with stones throughout (49–62 cm), overlying a yellow ashy soil sitting directly over coral (63–120 cm). Test pit 2, on the other hand, does have yellow-orange brown ash just below the surface; however, the base is sloping from east to west, 34 cm to 44 cm below the surface. This site is directly over a hard brown clay sitting on the coral. Potsherds and obsidian are found in this brown clay underlying the ash (potsherds = 371, obsidian = 20). The potsherds are highly weathered and friable, with little decoration. Test pit 3, like Test pit 2, is made of a series of yellow ashes overlying dark brown clay at

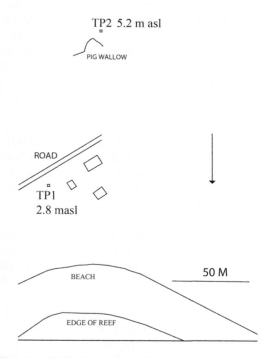

TP3 7.6 masl

TP2 5.2 m asl

PIG WALLOW

ROAD

TP1
2.8 masl

BEACH

50 M

EDGE OF REEF

Figure 10.9 Kur Kur (site ERB) in Ambitle Island.

70–83 cm below the surface with the layer sloping from east to west. Like Test pit 2, weathered potsherds were found (only 4 pieces) in the brown clay with a dense lens of waterworn pebbles.

The stratigraphy and finds tells us that this whole area has been disturbed by massive erosion of clay from the interior slopes as a result of the volcanic eruption noted with Malekolon, with secondary rewashing of ash on top. Test pits 2 and 3 have potsherds in a hard clay deposit directly beneath volcanic ash. The surface of the clay is sloping steeply, suggesting that the ceramics were in the process of re-deposition at the time when the ash fell. If any Lapita settlements existed in the past, then they would be highly disturbed, as seen here at Kur Kur or underneath metres of clay farther inland.

Lapita: a coastal settlement pattern?

It has been argued that Early Lapita sites in the Bismarck Archipelago, with evidence of waterlogged deposits, indicate the presence of stilt villages built over the intertidal zone at Talepakemalai in Mussau and at Makekur and Apalo in the Arawe Islands (Gosden and Webb 1994, Kirch 2001, Summerhayes et al. 2019). The question is: With so few sites available, how does this extend to other sites? From the evidence presented above, although most of the sites from Ambitle were heavily impacted by environmental change, all were located on the coastal beach or reef flats, confirming that they also represent communities over intertidal zones. All were sitting either on basal beach sands and reef platforms or in the case of Malekolon on the black sand beach of an embayment. Regarding other non-waterlogged deposits where stilt settlements were based, palaeo-shoreline reconstructions include Kamgot, as discussed above, but also the Early Lapita site of Tamuarawai on Emirau (Summerhayes et al. 2010), and later sites from the Solomons (see Felgate 2007; Sheppard and Walter 2009). From Remote Oceanic islands, examples include the later Lapita site of Nukuleka in Tonga (Burley 2016) and perhaps Bourewa in Fiji, although more evidence is needed (Nunn 2007: 167). These sites and their distributions have suggested (Summerhayes et al. 2019: 89) that Lapita sites, and Lapita lifeways, are closely linked with the sea.

Only in one other area was it argued that the Early Lapita settlement was inland, namely in the Willaumez Peninsula. The finding of Lapita pottery found inland from the present coastline led Specht and Torrence (2007) to make the point that Early Lapita was not coastal oriented. They note that "model(s) of settlements concentrated on beaches and small offshore islands are likely to be a product of taphonomic factors and biased research practices. Consequently the model is seriously flawed" (Specht and Torrence 2007: 73). As noted in Summerhayes (2010: 25–26), major gaps in our knowledge in this area remain, and many of the inland sites would have been on the coast 3000 years ago. The volcanic history of the Willaumez Peninsula demonstrates a number of violent eruptions producing major landscape change and infilling of embayments and passages between what were once islands forming the peninsula. Torrence et al. (2009: 519) point out that a large embayment (they named Kulu Bay) existed on the western side of the Willaumez Peninsula, which extended the coastline 17 km inland. They argue that the bay would have been transformed into an estuary after a volcanic eruption (Wk1) between 6150 and 5770 cal. years BP. The area was subsequently infilled either soon after the WK2 eruption or thereafter by a "massive discharge". Further work is needed to establish exactly where the coastline would have been 3300 years ago in this region. As noted by

Summerhayes (2010: 26), the Willaumez Peninsula could have been either a series of islands separated from the mainland or a series of estuaries with sites located on higher ground to the north and south. With a higher sea stand of 1.5 to 2 m at 3300 years ago and the strong probability of uplift in this highly volcanic region, reconstructions of past coastlines are a major priority. The few Early Lapita sites in this region are indeed coastal, and evidence for upland locations (i.e. on hills overlooking the coast) are later in time. Also, this is not to say that pottery was not traded inland as recent work from plainware assemblages in the highlands of Kaironk Valley (See Gaffney et al. 2015) or plainware sherds found inland of southern New Britain (Summerhayes 2001b) demonstrate, although examples of this are rare.

Conclusion

Understanding the nature of how cultural practices combine with "natural processes" to shape taphonomy and site preservation is crucial in understanding the nature of the archaeological assemblage and must be taken into account in assessing Holocene archaeological records (Boyd et al. 1999: 283). The sites of Malekolon, Mission, Balbalankin, and Kur Kur remind us that most Lapita sites are of a completely different nature to Kamgot, Talepakemalai, Adwe, and Apalo. The paucity of material does not equate with a small hamlet or insignificant settlement. What they all have in common is their coastal location, even when today they may be situated half a kilometre inland. Their presence informs us of past settlement, and examination of obsidian and pottery inform us of past connections between communities. The environmental reconstructions allow us to look at a landscape with settlements on coastal lagoons much closer to the steep hills on the northern site of Ambitle Island. Lapita occupation inland from this western coastline would be covered by metres of deposit caused by erosion from the volcanic events some two millennia ago. These reconstructions also raise the point that other sites were totally covered along the rest of Ambitle Island. Although we mark just five sites from Anir, this could well be 50 or more. How would this fit into a colonisation model where all coasts were inhabited? It also makes one reflect on southwest New Britain and islands to their west which are archaeologically unknown or areas to the east like Gasmata where any site that had existed may have been destroyed with volcanic uplift and erosion. These blanks on the map do not equate with the absence of Lapita settlements. When all the dots are eventually joined, we may see an intensity of sites not expected.

Acknowledgements

Thanks to Mike Carson for the invitation to submit a chapter and for his patience. I would like to acknowledge the National Museum and Art Gallery of Papua New Guinea, and the National Research Institute of Papua New

Guinea, for facilitating research. Also, a big thanks to the communities of Anir and the Provincial Government of New Ireland for working in the Feni Islands. I also appreciate very much Professors Chris Gosden and Pat Kirch for permission to reproduce Figures 2 and 3, respectively.

References

Ambrose, Wal R., 1973. 3,000 years of trade in New Guinea Obsidian. *Australian Natural History* 17: 370–373.

Ambrose, Wal R., 1976. Intrinsic hydration rate dating of Obsidian. In *Advances in Obsidian Glass Studies Archaeological and Geochemisrty*, edited by R. E. Taylor, pp. 81–105. Noynes Press, Park Ridge, New Jersey.

Ambrose, Wal. R., 1978. The loneliness of the long distance trader in Melanesia. *Mankind* 11(3): 326–333.

Anson, Dimitri, 1983. Lapita pottery of the Bismarck Archipelago and its affinities. Unpublished doctoral thesis, University of Sydney, Sydney.

Anson, Dimitri, 1986. Lapita pottery of the Bismarck Archipelago and its affinities. *Archaeology in Oceania* 21: 157–165.

Bedford, Stuart, Matthew Spriggs, Dave Burley, Christophe Sand, Peter Sheppard, and Glenn R., Summerhayes, 2019. Debating Lapita: Distribution, chronology, society and subsistence. In *Debating Lapita: Chronology, Society and Subsistence*, edited by Stuart Bedford and Matthew Spriggs, pp. 5–33. Terra Australis 52. ANU E Press, Canberra.

Boyd, William E., Jim Specht, and John Webb, 1999. Holocene shoreline change and archaeology on the Kandrian coast of West New Britain. In *Australian Coastal Archaeology*, edited by Jay Hall and Ian McNiven, pp. 281–289. Archaeology and Natural History publications, Department of Archaeology and Natural History, Research School of Pacific and Asian Studies, Australian National University, Canberra.

Burley, David V., 2016. Reconsideration of sea level and landscape for first Lapita settlement at Nukuleka, Kingdom of Tonga. *Archaeology in Oceania* 51:84–90.

Felgate, Matthew, 2007. Leap-frogging or limping? Recent evidence from the Lapita littoral fringe, New Georgia, Solomon Islands. In *Oceanic explorations: Lapita and Western Pacific settlement*, edited by Stuart Bedford, Christophe Sand, and Sean P. Connaughton, pp. 123–140. Terra Australis 26. ANU E Press, Canberra. http://doi.org/10.22459/TA26.2007

Gaffney, Dylan, Glenn R. Summerhayes, Anne Ford, Judith Field, Tim Denham, William R. Dickinson, and James Scott, 2015. Early Austronesian influences in the New Guinea Highlands at 3,000 years ago. *PLOS ONE.* http://doi.org/.1371/journal.pone.0134498

Gosden, Chris, and John Webb, 1994. The creation of a Papua New Guinean landscape: Archaeological and geomorphological evidence. *Journal of Field Archaeology* 21: 29–51.

Gosden, Chris, John Webb, Brendan Marshall, and Glenn Summerhayes, 1994. Lolmo Cave: A mid to late Holocene site, the Arawe Islands, West New Britain Province, Papua New Guinea. *Asian Perspectives* 33: 97–119.

Green, Roger C., 1979. Lapita. In *The Prehistory of Polynesia*, edited by Jessie Jennings, pp. 27–60. The Australian National University Press, Canberra.

Hogg, Nicholas, Glenn R. Summerhayes, and Elaine Chen, 2021. Moving on or settling down? Studying the nature of mobility through Lapita pottery from the Anir Islands, Papua New Guinea, *Records of the Australian Museum* 34: 71–86. http://doi.org/.3853/j.1835-4211.34.2021.174471-86

Kirch, Patrick V., 2001. *Lapita and its transformations in Near Oceania: Archaeological investigations in the Mussau Islands, Papua New Guinea, 1985–1988.* Contribution 59. Archaeological Research Facility, University of California at Berkeley, Berkeley.

Licence, P., Terrill, J. and Fergusson, L., 1987. Epithermal gold mineralisation, Ambitle Island, Papua New Guinea. In *Proceedings of the Pacific Rim Congress 87*, Gold Coast, 26–29 August 1987, pp. 273–278. The Australian Institute of Mining and Metallurgy, Melbourne.

Lindley, David, 2015. Late Quaternary geology of Ambitle Volcano, Feni Island Group, Papua New Guinea. *Australian Journal of Earth Sciences* 62: 529–545.

Nunn, Patrick D., 2007. Echoes from a distance: Research into the Lapita occupation of the Rove Peninsula, southwest Viti Levu, Fiji. In *Oceanic explorations: Lapita and Western Pacific settlement*, edited by Stuart Bedford, Christophe Sand and Sean P. Connaughton pp. 163–176. Terra Australis 26. ANU E Press, Canberra.

Sheppard, Peter, and Richard Walter, 2009. Inter-tidal Late Lapita sites and geotectonics in the Western Solomon Islands. In *Lapita: Ancestors and descendants*, edited by Peter Sheppard, Tim Thomas, and Glenn R. Summerhayes, pp. 73–100. New Zealand Archaeological Association Monograph 28. New Zealand Archaeological Association, Auckland.

Specht, Jim, and Robin Torrence, 2007. Pottery of the Talasea Area, West New Britain. *Technical Reports of the Australian Museum* 20: 131–196.

Specht, Jim, Chris Gosden, Carol Lentfer, Geraldine Jacobsen, Peter J. Matthews, and Sue Lindsay, 2017. A Pre-Lapita Structure at Apalo, Arawe Islands, Papua New Guinea. *Journal of Island and Coastal Archaeology* 12: 151–172.

Summerhayes, Glenn R., 2000. Recent archaeological investigations in the Bismarck Archipelago, Anir – New Ireland province, Papua New Guinea. *Bulletin of the Indo-Pacific Prehistory Association* 19: 167–174

Summerhayes, Glenn R., 2001b. Lapita in the far west: Recent developments. *Archaeology in Oceania* 36(2): 53–63.

Summerhayes, Glenn R., 2001a. Defining the chronology of Lapita in the Bismarck Archipelago. In *The Archaeology of Lapita Dispersal in Oceania*, edited by Geoff Clark, Atholl J. Anderson, and Tarisi Vunidilo, pp. 25–38. Terra Australis 17. Pandanus Books, Research School of Pacific and Asian Studies, Australian National University, Canberra.

Summerhayes, Glenn R., 2004. The nature of prehistoric obsidian importation to Anir and the development of a 3,000 year old regional picture of obsidian exchange within the Bismarck Archipelago, Papua New Guinea. In *Archaeologist and Anthropologist in the Western Pacific: Essays in Honour of Jim Specht*, edited by Val J. Attenbrow and Richard Fullagar, pp. 145–156. Records of the Australian Museum Supplement, Sydney.

Summerhayes, Glenn R., 2007. The rise and transformation of Lapita in the Bismarck Archipelago. In *From Southeast Asia to the Pacific. Archaeological Perspectives on the Austronesian Expansion and the Lapita Cultural Complex*, edited by Scarlett Chui and Christophe Sand, pp. 129–172. Academia Sinica, Taipei.

Summerhayes, Glenn R., 2010. Lapita Interaction – an update. In *2009 International Symposium on Austronesian Studies*, edited by Masegseg Z. Gadu and Hsiu-man Lin, pp. 11–40. Taiwan National Museum of Prehistory, Taidong.

Summerhayes, Glenn R., Matthew Leavesley, and Andrew Fairbairn, 2009. Impact of human colonisation on the landscape – a view from the Western Pacific. *Pacific Science* 63 (4):725–745.

Summerhayes, Glenn. R., Kath Szabo, Matthew Leavesley, and Dylan Gaffney, 2019. Kamgot at the Lagoon's Edge: Site Position and Resource Use. In *Debating Lapita: Chronology, Society and Subsistence*, edited by Stuart Bedford and Matthew Spriggs, pp. 89–103. Terra Australis 57. ANU E Press, Canberra

Summerhayes, Glenn R., Lisa Matisoo-Smith, Herman Mandui, Jim Allen, Jim Specht, Nick Hogg, and Sheryl McPherson, 2010. Tamuarawai (EQS): An Early Lapita Site on Emirau, New Ireland, PNG. *Journal of Pacific Archaeology* 1: 62–75.

Torrence, Robin, Vince Neal, and William E. Boyd, 2009. Volcanism and historical ecology on the Willaumez Peninsula, Papua New Guinea. *Pacific Science* 63: 507–535.

White, J. Peter, and Jim Specht, 1971. Prehistoric pottery from Ambitle Island, Bismarck Archipelago. *Asian Perspectives* 14: 88–94.

11 How island peoples adapt to climate change

Insights from studies of Fiji's hillforts

Patrick D. Nunn, Elia Nakoro,
Roselyn Kumar, Meli Nanuku, and
Mereoni Camailakeba

Palaeogeography in island worlds

Oceanic islands are those found in the ocean basins, commonly distant from continental shores. Almost all have become settled by people only in the past three millennia, often much more recently. Owing to an absence of large rivers, most such islands have only comparatively small areas of coastal lowland – their most changeable parts. This fact means that palaeogeographic changes have not generally been as extensive as along continental fringes. Yet changes have occurred, and their effects have been amplified (compared to those along continental fringes) by the comparative smallness and remoteness of many island contexts, two geographical factors that significantly reduce the adaptation options available to island societies under climate-driven resource stress, for instance (Nunn 2007b; Nunn and Kumar 2018; Robinson 2019; Weir 2020).

Most people living on oceanic islands for most of their pre-globalisation histories occupied island coasts and, as today, subsisted largely from foods obtained from nearby terrestrial and marine ecosystems. Given the concentration of inhabited oceanic islands in the tropical oceans, much of that subsistence involved marine organisms, especially those obtainable from shallower nearshore areas (Szabó and Amesbury 2011; Thomas 2014). The sensitivity of nearshore marine ecosystems to externally forced changes, such as sea-level fall and tectonic uplift, forced livelihood changes on island peoples in the past (Albert et al. 2007; Nunn and Carson 2015b) just as it is expected to in the future (Connell 2015; Leal Filho 2020).

This chapter argues that the plausible similarity in establishment ages of hillforts in Fiji (and most other Asia-Pacific island groups) points to the existence of an external region-wide driver of the conflict implicit in their establishment. We regard the most likely such driver to be the "AD 1300 Event", a ~70-cm cooling-driven sea level fall across the Indo-Pacific region, approximately AD 1250–1350, which caused a massive drop in the amount of food customarily available to coastal dwellers, leading them to compete

DOI: 10.4324/9781003139553-11

for the remaining food and eventually to abandon exposed coastal settlements for the comparative safety of ones on hilltops or inland caves. A test of this hypothesis, comprehensively explained in Nunn (2007b), is whether or not hillfort-establishment ages are consistent with the proposed chronology, specifically whether they cluster in the immediate aftermath of the AD 1300 Event.

This chapter looks first at island hillforts across the tropical Indo-Pacific region and reviews the evidence that their establishment may have been driven by climate change. There is then a section describing Fiji's hillforts, including the four case-study areas where the authors have mapped and excavated hillforts; much-unpublished data is presented, especially from Kadavu. We then look at the climate-conflict nexus in Fiji, evaluating on the basis of the presented evidence whether or not there is likely to be a causative link. Finally, we discuss what can be learned from the study of Fiji hillforts about both the past and the future, especially the degree to which oceanic islands might be considered uncommonly vulnerable to climate change, a recurrent theme in recent global assessments (Klöck and Nunn 2019; Nurse et al. 2014).

Did climate change drive the establishment of pre-contact Island hillforts in the Indo-Pacific?

The remains of hillforts once occupied by people exist on high islands across the Indo-Pacific, from Timor in the west to Rapa (French Polynesia) in the east, from Micronesia in the north to New Zealand in the south (Kennett and McClure, 2012; O'Connor et al. 2012; Rainbird 1996; Schmidt 1996). Reviews of hillforts for particular island groups have been made, but only a few consider the entire region and evaluate the case for region-wide synchronicity of hillfort establishment and thus a common region-wide driver (Field and Lape 2010; Groube 1970; Ladefoged and Pearson 2000; Lape and Chao 2008; Nunn 2007b; Nunn 2000; Nunn and Carson 2015a; Nunn et al. 2007).

Well-studied examples include the island of Timor, where a post-AD 1000 shift in settlement "towards fortified and defensively-oriented" sites (Lape and Chao 2008: 11) was initially found. Later research incorporated radiocarbon ages of diagnostic fortified sites (*pa'amakolo*) and suggested these began to be established within the period AD 1334–1373 (O'Connor et al. 2012: 208). At the opposite end of the Pacific, extensive work on fortifications (*pare*) on the island of Rapa in French Polynesia suggest these were first constructed along ridgelines within the period AD 1300–1400, expanding particularly in two subsequent phases (Anderson and Kennett 2012). On islands within most of the Pacific, there is also clear evidence that hillforts were constructed at a similar time. In addition to four datasets from Fiji, reviewed below, those from upland inland fortified sites (*'olo*) in the islands of Samoa suggest that these started to be built within the period AD 1270–1310 (Pearl 2004), a conclusion confirmed by later work (Addison and Asaua 2006).

The synchronicity of hillfort establishment ages on islands across the Asia-Pacific strongly implies that the reason behind hillfort construction was not local but region-wide. Several authors have recognised this, arguing that the most plausible region-wide driver is a climatic one, perhaps ENSO-linked drought (Field 2008; Field and Lape 2010) or cooling-driven sea-level fall (Nunn 2007b; Nunn and Carson 2015a). The former may have been a significant subregional driver of change, as may tsunami impact (Goff et al. 2012), but only the latter appears to satisfy the requirement for an approximately synchronous region-wide cause of change.

The current model to explain hillfort establishment on Pacific islands as a result of sea-level fall involves (1) cooling-driven sea-level fall during the AD 1300 Event (AD 1250–1350), (2) the subsequent exposure of reef surfaces, lagoonal turbidity increases, coastal-lowland water-table falls, all of which combined to reduce the amount of food available to coastal dwellers by an estimated 80% (Nunn 2007b). The ensuing food crisis, at least decades-long, led to conflict between coastal peoples, groups of which abandoned their (exposed) coastal settlements to move to (less-exposed) sites inland, sites which included caves, rock shelters, and hilltop fortifications (Figure 11.1).

The AD 1300 Event was an idea introduced about 20 years ago (Nunn 2000) that linked the climatic evidence for cooling and sea-level fall across the Pacific (AD 1270–1475) to environmental changes that led to food crises for people, especially on Pacific islands, and forced transformational change exemplified by the abandonment of coastal settlements and the establishment of inland/upslope settlements like hillforts (see also – Nunn 2007a, 2007b; Nunn and Campbell 2020). Prior to the AD 1300 Event, from about AD 750, the Pacific region was affected by the Little Climatic Optimum (Medieval Warm Period or Medieval Climate Anomaly) during which many Pacific island societies flourished and the weather was conducive to long-distance voyaging. After the AD 1300 Event, until the first half of the 19th century, the Pacific region was affected by the Little Ice Age that saw societal fragmentation and conflict – the likely legacy of the AD 1300 Event – and the cessation of long-distance cross-ocean contact.

Fiji's hillforts: case studies

Many of Fiji's hillforts (*koronivalu*) were reported to be occupied during the first seven decades of the 19th century – when Europeans spread throughout the islands – often becoming centres of resistance against the British colonisers and their supporters (Parry 1987). Hillforts were ubiquitous; one colonial official found that "almost every important hilltop in Western Viti Levu [Island] is crowned with an entrenchment of some kind" (Thomson 1908: 89). The function of hillforts to protect their occupants and deter aggressors appears clear; an Australian journalist in AD 1870 described what he found in Viti Levu Bay in northeast Viti Levu Island:

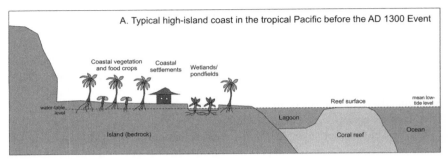

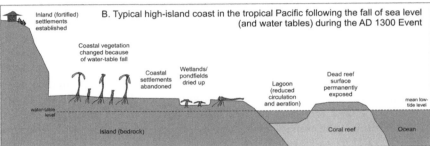

Figure 11.1 Model illustrating the proposed effects of climate (cooling) driven sea-level fall during the AD 1300 Event (~AD 1250–1350) on coastal environments and settlements of tropical Pacific Islands (after Nunn and Carson 2015a). A: Prior to the AD 1300 Event, nearshore marine ecosystems had adapted to the higher sea level, lagoons and reefs were optimally productive; onshore coastal settlements flourished and utilised food crops growing on coastal plains and in wetland and pondfields. B: Following the AD 1300 Event, as a result of sea-level fall of ~60 cm, reef surfaces were exposed, lagoons became turbid and unflushed, and inland water tables had fallen, depriving plants of access to fresh groundwater; the resulting food crisis led to conflict between coastal dwellers, which resulted in the abandonment of coastal settlements in favour of fortified upland settlements, mostly hillforts in Fiji, from around AD 1400.

in the distance the houses of the mountaineers, perched curiously on the apex of rocky pinnacles, a position singularly secure from invasion.... Walking around the base of the bay in a northeasterly direction we came to a thickly populated town on the crest of a hill, east of which could be seen another native village, built on an immense rock. The natives who were all clustered together on a little plateau in front of the town received us very nervously.... (Britton 1870: 55)

For many years, Fiji's hillforts have been ignored by researchers interested in these islands' ancient history but also by most local residents. The reasons for this situation are many, but it seems clear that the comparative visibility of many coastal-edge archaeological sites explains the focus on these. And for many local communities, hillforts remain little-known because people have no reason to either visit them or try to find out about them. Yet ranged against this is the fact that the "hillfort period", plausibly several hundred years in duration, represents a sizeable and significant piece of Fiji history. The Fiji Museum is mandated to illuminate and conserve all parts of Fiji history, which explains its involvement in the projects reported on here for almost two decades.

Some early researchers in Fiji described some of its hillforts in detail. One account from the western end of Kadavu Island (see below and Figure 11.3) described the ridge-line hillfort of Naborua as follows.

> Three long rocky headlands jut out from Nabukelevu, the extinct volcano, like the buttress roots of rain-forest trees, narrow with blunt ends and a precipice on each side. The fortress was perched on the central ridge. Its only path led along the mountain side, very narrow so that approach had to be in single file. The fort had a death drop on all sides. (Tippett 1958: 147)

Ridge-line fortifications, especially triple-junction ones, are common in the dissected volcanic terrain on many Fiji islands (and similar elsewhere) (Clunie 1977; Palmer 1969). A stylised example is shown in Figure 11.2. It is based on that described by Palmer (1969) from Nananu-i-Ra Island, that at Nabouwalu (in Bua; Nunn, Nakoro, et al. 2019), and that at Bogikoro on the Vatia Peninsula (Robb and Nunn 2012). Note the common use of defensive ditches (when active, with sharpened stakes in their muddy floors) and stone walls along ridgelines, sometimes with large house foundations nearby suggesting a guard post. The large sub-circular centres of such triple junctions often supported a central mound (perhaps for lookout, perhaps to accommodate an elite group) surrounded by stone walls and wide ditches with narrow raised causeways. House foundations (*yavu*) are scattered around these sites, suggesting they housed sizeable numbers of people.

Our research has focused on four areas of Fiji, shown in Figure 11.3. The earliest investigations built on the work of Julie Field in the Sigatoka Valley (Field 2003, 2004) and utilised environmental proxies from downslope areas to confirm the likely establishment ages and occupation ranges of various hillforts. The next major project was in the adjoining Ba Valley and nearby Vatia Peninsula, once described as having "the largest concentration of hill fortifications in northern Viti Levu [Island]" (Parry 1997: 119). The third study focused on the hillforts of western Bua on Vanua Levu Island, selected because the earliest written (European) accounts of hillforts in Fiji come from here, produced by people attracted by the area's reputation for sandalwood (*yasi*). The fourth study comes from the island of Ono, just off the eastern end of elongate Kadavu Island in the south of the Fiji group, where no such research had been undertaken before 2019.

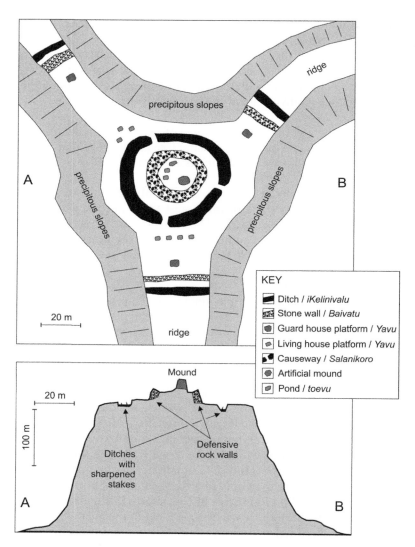

Figure 11.2 Typical ridge-line fortification in Fiji to illustrate the common elements, based on that described by Palmer (1969) from Nananu-i-Ra Island, that at Nabouwalu (in Bua; Nunn, Nakoro, et al. 2019), and that at Bogikoro on the Vatia Peninsula (Robb and Nunn 2012). A: Plan view shows how the highest part of a triple ridge junction was selected and natural defensive attributes enhanced by walls and ditches and down-slope "first lines of defence". B: Section view emphasises the common topographical form of Fiji hillforts in such locations.

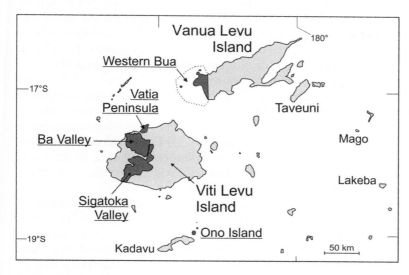

Figure 11.3 Map of the Fiji Islands, showing the five main areas in which hillfort research has been carried out; note that the Ba Valley and Vatia Peninsula hillforts are treated as one unit in this chapter.

Research into hillforts elsewhere in Fiji has been largely piecemeal and opportunist while yet revealing interesting data. Studies include ones on some Lau Islands in eastern Fiji (Best 1992, 1993), on Taveuni Island (Frost 1974), the Yasawa and Mamanuca islands of western Fiji (Smith and Cochrane 2011), and the low hillfort of Koivuanabuli on Mali Island, off the north-central Vanua Levu mainland (Burley et al. 2016). The seminal works of John Parry (1977, 1987, 1997) used aerial photography to identify fortified sites, high and low, in several parts of Viti Levu Island.

Sigatoka Valley hillforts and their environmental proxies

The Sigatoka Valley drains the leeward side of Viti Levu Island. Its low-lying axial part is generally narrower than the Ba Valley (see next section), and there is a steep rise to the higher parts of the catchment, which are carved into isolated peaks and sharp ridgelines. The other major difference from the Ba Valley is the existence of isolated limestone outcrops in the Sigatoka Valley that proved attractive locations for fortifiable inland settlement. Those at Bukusia and Korovatuma were described as being:

> natural fortresses, on the summit of precipitous crags covered with forest and honey-combed with caves… [with]… earthworks, and fences both of stone and bamboo, wherever natural defence was absent. (Gordon 1879: 294–295)

The main survey of the inland Sigatoka Valley settlement was undertaken by Field (2003) and involved determining establishment ages for ten inland upland (hillfort) sites. Subsequent interpretations of these dates and their links to contemporaneous environmental change were reported by Kumar et al. (2006).

Establishment ages are plotted against age and major subdivisions of climate in Figure 11.4. Of the earliest dates shown, that for Tatuba (see below and Figure 11.6A) probably represents a continuation of the sporadic occupation that began in perhaps the first millennium AD, while that for Qoroqorovakatini (see below and Figure 11.6B) may represent the same. Mass inland colonisation is inferred to have commenced during the AD 1300 Event with the occupation of Korokune and Nokonoko, both of which are relatively close to the coast. There followed general colonisation of these inland sites thereafter, something that was sustained throughout the Little Ice Age.

There are two lines of independent proxy evidence for environmental change during the AD 1300 Event from the Sigatoka Valley. The first is that dates for charcoal-rich bands in alluvial sediments in the middle and lower parts of the valley are concentrated during the AD 1300 Event and early Little Ice Age (Nunn and Kumar 2004), implying that this marked a time of increased upland vegetation burning, plausibly for cropping by newly-settled human arrivals. That hillfort occupants did plant on steep slopes in the immediate vicinity is shown by this account from the early period of European settlement in Fiji:

> Jealousy that made every village distrustful of its neighbours compelled the inhabitants to fortify themselves on the most inaccessible heights, and prevented them from cultivating any land beyond the few feet around each man's dwelling; if more were required, the cultivator, afraid to descend into the plain discovered some spot in the recesses of the mountains where he might plant his yams secure from molestation. (Anonymous 1864)

The second line of proxy evidence are the dates from palaeosols within sand dunes along the seaward edge of the Sigatoka Delta, which suggest that the main period of dune accumulation began during the AD 1300 Event, both as a result of increased terrigenous sediment inputs to the delta front but also because of a recently-lowered sea level (de Biran, 2001; Kumar et al. 2006).

Ba Valley and Vatia Peninsula hillforts

The Ba Valley drains the northwestern (dry) side of Viti Levu Island. While the axial part of the valley is broad and low-relief in nature, its sides rise steeply, and many isolated peaks and sharp ridgelines exist. A project funded by the Vetlesen Foundation (2008–2010) sought to identify former upland settlement sites in such locations. Four such sites were dated: Naqara, Nayavutu, Tubabaka and Vatusososo. The dates are shown in Figure 11.5A.

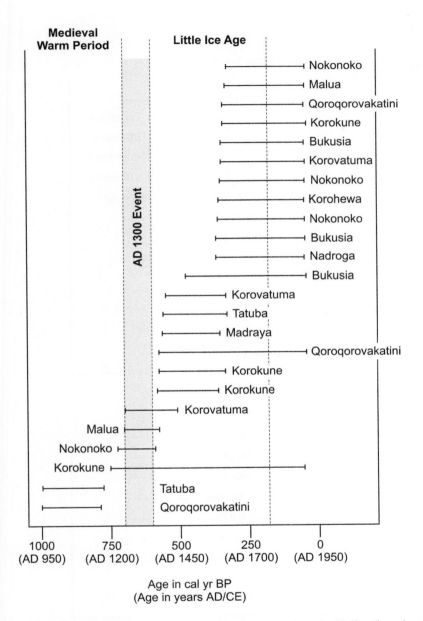

Figure 11.4 Establishment ages for hillforts in the Sigatoka Valley (based on data from Field 2003).

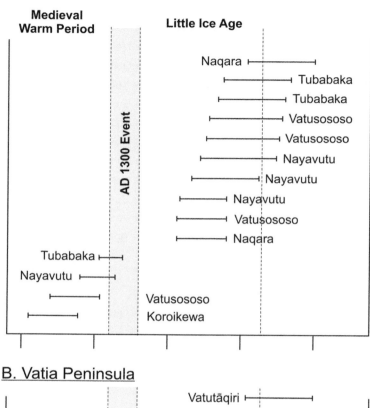

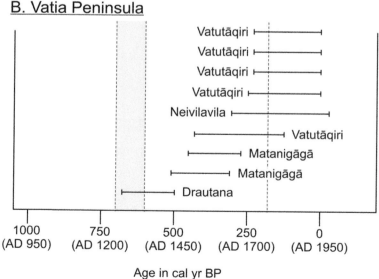

Figure 11.5 Establishment ages for hillforts in the Ba Valley and Vatia Peninsula (based on data from Nunn 2012).

Dates suggest a two-phase occupation of inland upland sites in the Ba Valley (Nunn 2012). The earliest phase probably occurred during the AD 1300 Event, and involved occupation of the two Ba Valley sites closest to the coast. There then appears to have been a second, more sustained movement inland that saw all dated sites in the Ba Valley being occupied during the Little Ice Age, dated to approximately AD 1350–1800 in the tropical Pacific (Nunn 2007b). While it is legitimate to question whether the apparent hiatus is real or a data artefact, the general picture that emerges is one that shows that inland settlement beginning only during and in the aftermath of the AD 1300 Event.

A comparable picture has emerged from studies of hillforts on the Vatia Peninsula, a high peaky andesite promontory with "the largest concentration of hillforts" in the Fiji Islands (Parry 1997: 119). Several hillforts were identified and mapped here (Robb and Nunn 2012). Particular attention was given to the Vatutāqiri (Vatuvatuvā) hillfort, a ridge-top fortification with several concentric stone walls surrounding an artificial mound, pictured in Figure 11.6C (Robb and Nunn 2014b). Radiocarbon ages (see Figure 11.5B) suggest a gradual establishment of hillforts here slightly earlier or coeval with the later phase in the nearby Ba Valley.

There is independent evidence for environmental change during the AD 1300 Event in this area provided by a single radiocarbon age of AD 1290–1440 from basal sediments on Tavuca Island, off the mouth of the Ba River (Nunn, McKeown, et al. 2019). Tavuca is a sand cay that is interpreted on the basis of this date to have begun accumulating on a coral reef platform newly emerged following the sea-level fall during the AD 1300 Event. Slightly farther offshore is the reef islet named Sucutolu (Rodda 1990), for which an emergence date of AD 1110–1400 is known (originally reported by Nunn 2000). It has also been conjectured that mangroves spread across the growing Ba Delta during much of the Little Ice Age following the sea-level fall of the AD 1300 Event (Nunn 2000).

Western Bua hillforts

Owing to foreign interest in sandalwood and *bêche-de-mer*, some of the earliest written descriptions of Fiji are from western Bua. They include accounts of hillforts as early as 1808 and the following from the 1840s:

> There is a high and insulated peak north of Dimba-dimba [Naicobocobo] Point, which has a town perched on its very top [Seseleka Hillfort]. The bay of Ruke-ruke [Rukuruku] has a reef across its mouth, leaving only a narrow ship channel into it. They [the ships] anchored under Ivaca Peak, a high and bold bluff, whose height ... is one thousand five hundred and sixty-three feet [476 m]. On its top is also a town [Uluinasiva Hillfort].
> (Wilkes 1845: 217)

Figure 11.6 Examples of Fiji hillforts. A: Remains of the stone wall at the mouth of
Tatuba Cave, leading to the hillfort of the same name, upper Sigatoka
Valley. B: The summit of the limestone outcrop at Qalimare is the site of
the Qoroqorovakatini hillfort, middle Sigatoka Valley. C: Boulder-
facing on the (artificial) mound named Vatuvatuvā at the highest point
of Vatutāqiri hillfort, Vatia Peninsula. D: Summit of the Seseleka hillfort
with artificial pond (*toevu*) in foreground, western Bua. E: The Korolevu
hillfort, northwest Bua, was the refuge for people living around its base.
F: The dolmen on the summit of Madre hillfort, Ono Island, Kadavu.
All photographs by the authors.

Over a three-year period, 16 hillforts were mapped and excavated within western Bua, including five on offshore Yadua Island (Martin et al. 2019; Nunn, Nakoro, et al. 2019). Four clusters deemed to represent separate polities were identified, the best-studied being the Seseleka one which involved a single large high hillfort (Seseleka – see Figure 11.6D) surrounded by lower (tributary) hillforts intended to act as lines of first defence, lookouts (especially below the cloud line), and to provide food (crops like taro, seafood especially shellfish) to the elite group living higher up. Water may also have been carried upslope, although the presence of rainwater ponds (*toevu*) on and around the summit of Seseleka suggests its occupants sought to be self-sufficient in this. The presence of shellfish remains across the 2.5 ha Seseleka summit site testify to the importance of marine foods in its occupants' diets, a similar situation to that for Ba Valley hillforts (Robb and Nunn 2014a). On the summit of Seseleka are house mounds (*yavu*) along its slightly lower western side and lookout mounds along its slightly higher eastern side. Shell dates and those from *toevu* sediments suggest the site was occupied as early as AD 1670–1680 (Nunn, Nakoro, et al. 2019). The approach to Seseleka is along narrow steep-sided ridges (*tua*) cross-cut by defensive ditches and cross-axis stone walls that would have considerably enhanced the natural defensive attributes of Seseleka.

Radiocarbon ages for western Bua hillforts (Figure 11.7) show all those dated were probably occupied during the Little Ice Age. Owing to the state of preservation of sites, it is likely that samples do not represent establishment

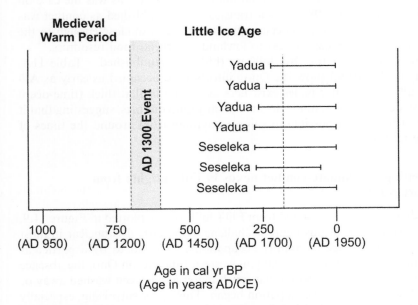

Figure 11.7 Radiocarbon ages for western Bua hillforts (Nunn, Nakoro, et al. 2019). Establishment ages may be earlier.

ages, which were earlier. A good example is the dates from the Seseleka *toevu* (AD 1670–1680 to 1950) that may mark the time when this hillfort was abandoned, the pond no longer regularly cleaned.

Ono (Kadavu) hillforts

Unpublished 2019 research on the hillforts of the Kadavu island group in southern Fiji represents the only research ever carried out on this topic here. While hillforts exist elsewhere in these high volcanic islands, this preliminary research was carried out on two occasions on the high dissected offshore volcanic island of Ono (not to be mistaken for the Ono Island in the southern Lau group, southeast Fiji). The five hillforts identified and mapped on Ono are shown in Figure 11.8A. The centre of the island is a volcanic crater, so the four highest and largest hillforts (not Naivarorobelō) are spread along its high steep-sided rim rather than within the crater's low interior.

Both their relationship with the topography as well as oral traditions suggest that the four major hillforts (Korovou, Madre, Qilai, Uluisolo) when occupied, all functioned independently of each other; there is evidence that Madre incorporated a ceremonial site (see Figure 11.6F). Many people inhabiting today's (exclusively coastal) villages on Ono trace their ancestry back to people living on nearby hillforts; that is Vabea to Korovou, Nabouwalu to Madre, Naqara to Qilai, and Narikoso to Uluisolo. The hillfort of Naivarorobelō is lower, closer to the sea, and appears to have been established later than the main hillforts, possibly – as was the case on Lakeba Island (Best 1984) – as a transitory site established as conflict was waning, fortified to give its occupants security yet also close enough to the coast to allow them easy access to lowland and reefal food resources.

Radiocarbon ages for Ono hillforts (hitherto unpublished – Table 11.1, see Figure 11.8B) all show the Ono hillforts were occupied as early as AD 1654, although similar issues exist here as in Bua in that thick (time-deep) cultural deposits could not be located on any Ono hillforts, suggesting that it may not have been possible to sample anything from around the times of their establishment.

Evaluating the climate-conflict nexus in Fiji: insights from Hillfort Ages

All radiocarbon ages obtained from Fiji's hillforts are plotted in Figure 11.9. The interpretation of this figure is challenging because it is unclear in most instances which stage of hillfort occupation is referred to by a particular date. In many instances, especially in western Bua and on Ono, the absence of cultural sequences suggests that much material has been washed away or slipped downslope since occupation began. This is not surprising, especially given the often featureless lightly-vegetated nature of many hillfort summits

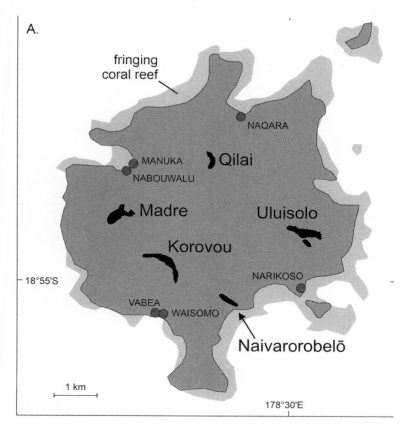

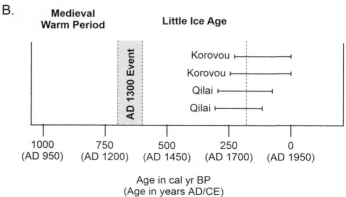

Figure 11.8 A: Map of Ono Island (Kadavu, Fiji) showing the six modern coastal villages and the five hillforts identified and mapped in 2019. B: Radiocarbon ages (from Table 11.1) for Korovou and Qilai hillforts.

Table 11.1 Calibrated radiocarbon dates from fortified hilltop sites in Ono Island, Kadavu, Fiji. All dates from the University of Waikato Radiocarbon Dating Laboratory were corrected, using local marine correction factor (ΔR) of 11 ± 26 ^{14}C years (Petchey et al. 2008) and calibrated using the Marine13 marine curve (Reimer et al. 2013) in OxCal v4.3.2 (Bronk Ramsey 2017)

Hillfort	Sample	Laboratory number	Sample material	Depth below surface (cm)	$d^{13}C$	Conventional radiocarbon age (BP)	Calibrated radiocarbon age (cal BP) at 95.4% probability	Calendar age at 95.4% probability
Qilai	Qilai-B	Wk-50145	Anadara antiquata	surface	1.9 ± 0.3	574 ± 29	285–122	AD 1665–1828
Qilai	Qilai-E	Wk-50146	Anadara antiquata	10	2.3 ± 0.3	597 ± 28	296–139	AD 1654–1811
Korovou	Korovou-A	Wk-50147	Anadara antiquata	surface	1.1 ± 0.3	489 ± 34	229–0	AD 1721–1950
Korovou	Korovou-C	Wk-50148	Anadara antiquata	surface	1.8 ± 0.4	509 ± 37	247–0	AD 1703–1950

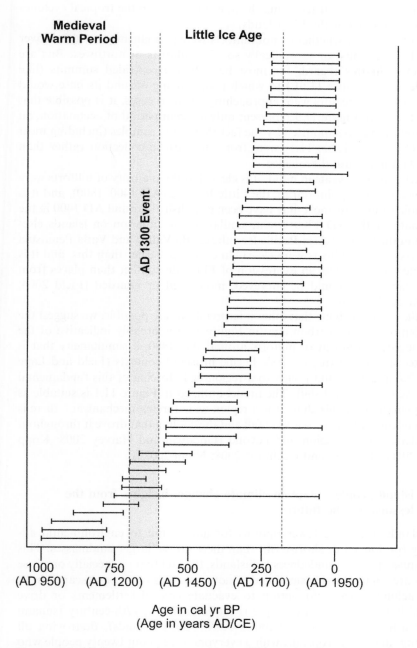

Figure 11.9 All 58 radiocarbon ages for the hillforts described in this study.

and the steep slopes surrounding them, not to mention the tropical cyclones that routinely batter the Fiji Islands.

Another issue is whether or not hillforts were permanently occupied (over several generations) or sporadically so. The hillforts of northwest Bua are instructive in this regard, for most have bare steep-sided summits (like Korolevu, see Figure 11.6E) to which people living around its base would retreat when aggressors were approaching. In such cases, it is possible that hillforts ages in Figure 11.9 represent only a recent period of occupation, an interpretation also supported by the fact that many samples (including most from Ono – see Table 11.1) were from the surface collection rather than from within a cultural sequence.

Bearing these caveats in mind, it is clear that the majority of hillforts were occupied at some point during the Little Ice Age (AD 1400–1800), and it is plausible to suppose most may have been established around AD 1400 in the aftermath of the AD 1300 Event, similar to the situation on islands elsewhere in the tropical Asia-Pacific (see above, Ba Valley and Vatia Peninsula Hillforts). Some hillforts were clearly in existence earlier than this, and it is possible they were less an expression of Fiji-wide conflict than places from which landscapes could be surveyed, monitored or guarded (Field 2004; Kumar et al. 2006).

While acknowledging that other interpretations are possible, we suggest the distribution of radiocarbon ages in Figure 11.9 is broadly indicative of the Fiji-wide establishment of hillforts around AD 1400, a simultaneity that is repeated across the tropical Asia-Pacific in island contexts (Field and Lape 2010; Nunn 2007b) and points to a region-wide forcing of this fundamental and sustained cultural shift. The model outlined in Figure 11.1 is suitable as an explanation for this shift. It is a region-wide forcing mechanism – there is independent evidence of sea-level fall and the cooling that drove it throughout the Pacific Islands region and beyond (Goodwin and Harvey 2008; Kopp et al. 2016; Morrison and Cochrane 2008; Nunn 2007b).

How island peoples adapt to climate change: insights from the past, lessons for the future

On islands, there are fewer options for adaptation to environmental adversity (prolonged or short-term) than there are on larger landmasses. This is because of the "boundedness" of islands, the fact that – especially on those which are comparatively small and remote – there is often "no escape" from approaching threats, no option to evacuate coastal settlements or drive several hours inland. An example is the impact of a 17th-century tsunami that washed across low Pukapuka Atoll (Cook Islands), destroying all dwellings and food crops, drowning everyone save about twenty people who re-established Pukapukan society with great difficulty (Beaglehole and Beaglehole 1938). Or more recently, when in April 2020, Tropical Cyclone Harold drove across Ono Island (discussed above), flattening almost every house, destroying most food gardens, leaving the people – with the help of government and aid donors – to pick up the pieces.

While on some islands like these, people can shelter from extreme events in caves or the most solid buildings, the point is that their livelihoods are disproportionately exposed by being islanders. There are no alternative food or water resources on islands, often none on similarly-impacted nearby islands, which is why many Pacific societies evolved adaptive measures like the routine production of surpluses and forms of food preservation so they might endure in the aftermath of such impacts (Aalbersberg et al. 1988; McMillen et al. 2017; Medina Hidalgo et al. 2020; Nunn and Kumar 2019).

But what happened during the AD 1300 Event was neither short-lived nor readily recovered from. It was a prolonged multidecadal cooling and sea-level fall that shifted the environmental baseline for island peoples across the tropical Pacific. It would have become clear after a few generations that there was no "going back" to the "times of plenty" that characterised the Medieval Warm Period, which preceded the AD 1300 Event (Nunn et al. 2007). What succeeded it was the Little Ice Age, a global event that saw temperatures fall with devastating effects on societies across the world (Fagan 2000; Tuchman 1978), and which in tropical Pacific Island groups like Fiji saw not just cooler, less predictable climate but also a lowered sea level that plausibly explains a sustained situation of food insecurity that was survivable only through conflict (see Figure 11.1). This may be why around AD 1400–1500 throughout the Fiji Islands and all other high tropical Pacific Island groups, conflict commenced and eventually forced the abandonment of the coastal settlements where most people had lived until this point, driving people on high islands inland and upslope (sometimes offshore) to establish themselves in fortifiable locations (Nunn 2007b).

Today the environmental baseline is again shifting; current impacts of climate changes are effectively irreversible, there is no going back to the times of plenty, and climate scientists and others are now calling for transformational adaptation as the best long-term option for humanity (Holler et al. 2020; McNamara et al. 2020; Wilson et al. 2020). In contrast to shorter-term incremental adaptation, transformational adaptation accepts the baseline change and requires humans – in every part of the world – to change their expectations, aspirations, their interactions with the natural world by acknowledging its limits.

In the past on oceanic islands like those in the tropical Pacific, the few alternative options for human subsistence meant that, following severe and prolonged (climate-driven) impacts on the natural environment, transformational adaptation was forced upon people – as unwilling and as unaccustomed to this as we are today. People living on high Pacific islands 600–700 years ago abandoned coastal settlements in favour of those high up, inland, in fortifiable locations: a transformation unparalleled in the previous, more than two millennia of settlement in the western Pacific islands. People started fighting one another, perhaps even started eating one another, as a result: transformations probably unheard of a hundred years or so earlier (Nunn 2007b).

People's diets in the Pacific islands also changed 600–700 years ago as a result of processes summarised in Figure 1. Prior to the AD 1300 Event, most Pacific islanders lived along island coasts, their diets often largely comprised of (nearshore) marine foods (Kinaston, et al. 2014; Szabó and Amesbury 2011). Within a few hundred years, inland foods likely comprised the largest component of human diets (Robb and Nunn 2014a; Stantis et al. 2016; Valentin et al. 2011), a transformation of diet that has parallels with the situation in many modern Pacific Island contexts over the past century (Baumhofer et al. 2020; Medina Hidalgo et al. 2020; Savage et al. 2020).

And then, we consider the impacts that climate-driven changes had on human societies 600–700 years ago, transforming these from largely peaceable ones to largely "warlike" ones, at least as commonly characterised in the minds of their earliest literate chroniclers. Today climate-driven stress on people's livelihoods takes different yet comparable forms, including a rise in domestic violence and in suicide and despair that sometimes follows particular events (Bourque and Willox 2014; Miles-Novelo and Anderson 2019; Scott-Parker and Kumar 2018).

All such issues resonate globally. All societies, all contexts, are "bounded" to some degree. In the past, geographical constraints to their growth were overcome by occupying others' spaces. Yet for several decades now, it has been manifest that "spaceship Earth" is the ultimate limit to growth so that sustainable futures must be re-configured, just as they were on Pacific islands during the Little Ice Age. As this chapter was being finalised, the world was in the grip of COVID-19, borders have been closed, and most nations are contemplating a less-globalised world in the future than that to which we had all been accustomed, even a year ago. There are parallels in the Pacific. The AD 1300 Event shut down long-distance voyaging; it fragmented large interconnected coastal settlements and replaced them with small less-connected fortified hillforts; it replaced "times of plenty" with "times of less", also the likely short-term outcome of COVID-19 on our world.

Conclusion

The study of Fiji's hillforts shows that most were likely established about the same time, around AD 1400 at the start of the Little Ice Age, plausibly in response to a climate-driven sea-level fall that considerably depleted the foods obtainable from coastal areas, both onshore and offshore. This led to a prolonged food crisis that in turn led to conflict and the abandonment of (undefendable) coastal settlements in favour of ones in (fortifiable) locations, including hillforts. This scenario appears broadly applicable to tropical island contexts across the entire Asia-Pacific region, a fact that in turn strengthens the case for a regional (rather than island-specific) cause of hillfort establishment and the conflict likely to have unpinned it.

This model is consistent with what else we know about (pre-modern) island people's responses to environmental adversity, whether this comes in the form of short-lived extreme events (natural disasters like tropical cyclones or tsunamis) or prolonged escalating change (like sea-level change, cooling, droughts). Since options for human subsistence on (smaller, more remote) islands are generally limited, responses are often transformative, acknowledging the shift in the environmental baseline. This applied to the AD 1300 Event (~AD 1250–1350) when the transition between the Medieval Warm Period and the ensuing Little Ice Age was marked by cooling and sea-level fall, the latter significantly impacting food sources for coastal dwellers and forcing their re-location inland. It also applies to today, not just as rising sea level forces inland re-location but also as climate change more broadly forces changes to livelihoods, including diets and behaviours that would not otherwise have occurred.

This study shows that the study of the past has much to offer the understanding of the future. Fiji's hillforts may be largely forgotten by most modern Fijians, but they exist here, as elsewhere in islands in the Asia-Pacific, as a tangible reminder of the fragility of island food systems and the attendant challenges posed by climate change (and its manifestations) to island people's livelihoods.

Acknowledgements

We acknowledge the hospitality and collaboration of rural Fiji people in all the areas where we have studied Fiji hillforts over the past 20 years. We thank the staff of the Fiji Museum, especially the current Director Sipiriano Nemani, for their help and the Ministry of iTaukei Affairs for support. Funding for different components of this project has come mostly from the G. Unger Vetlesen Foundation and the New Colombo Plan of the Australian Government.

References

Aalbersberg, William G. L., C. E. A. Lovelace, K. Madhoji, and Susan V. Parkinson, 1988. Davuke, the traditional Fijian method of pit preservation of staple carbohydrate foods. *Ecology of Food and Nutrition* 21 (3): 173–180.

Addison, David J., and Tautala S. Asaua, 2006. One hundred new dates from Tutuila and Manu'a: Additional data addressing chronological issues in Samoan prehistory. *Journal of Samoan Studies* 2: 95–117.

Albert, Simon, James Udy, Graham Baines, and Debra McDougall, 2007. Dramatic tectonic uplift of fringing reefs on Ranongga Is., Solomon Islands. *Coral Reefs* 26 (4): 983–983.

Anderson, Atholl J., and Douglas J. Kennett, editors, 2012. *Taking the High Ground: The Archaeology of Rapa, a Fortified Island in Remote East Polynesia.* Terra Australis 37. Australian National University E Press, Canberra.

Anonymous, 1864. COC (Consular Outward Correspondence). Document on file, National Archives of Fiji, Suva.

Baumhofer, N. Kau'i, Sela V. Panapasa, E. Francis Cook, Christina A. Roberto, and David R. Williams, 2020. Sociodemographic factors influencing island foods consumption in the Pacific Islander Health Study. *Ethnicity & Health* 25 (2): 305–321. https://doi.org/10.1080/13557858.2017.1418300

Beaglehole, Ernest, and Pearl Beaglehole, 1938. *Ethnology of Pukapuka*. Bernice Pauahi Bishop Museum Bulletin 150. Bishop Museum Press, Honolulu.

Best, Simon B., 1984. Lakeba: The prehistory of a Fijian island. Unpublished doctoral dissertation, University of Auckland, Auckland.

Best, Simon B., 1992. Fortifications in Fiji and Samoa: Comparisons and predictions. *Archaeology in New Zealand* 35 (1): 40–44.

Best, Simon B., 1993. At the halls of the mountain kings: Fijian and Samoa fortifications: comparisons and analysis. *Journal of the Polynesian Society* 102: 385–447.

de Biran, Antoine, 2001. The Holocene geomorphic evolution of the Sigatoka Delta, Viti Levu, Fiji Islands. Unpublished doctoral dissertation, University of the South Pacific, Suva.

Bourque, Francois, and Ashlee Cunsolo Willox, 2014. Climate change: The next challenge for public mental health? *International Review of Psychiatry* 26 (4): 415–422. https://doi.org/10.3109/09540261.2014.925851

Britton, Henry, 1870. Fiji in 1870, being the letters of "The Argus" special correspondent. Samuel Mullen, Melbourne.

Bronk Ramsey, Christopher, 2017. Methods for summarizing radiocarbon datasets. *Radiocarbon* 59 (6): 1809–1833.

Burley, David V., Travis Freeland, and Jone Balenaivalu, 2016. Nineteenth-century conflict and the Koivuanabuli fortification complex on Mali Island, Northern Fiji. *Journal of Island and Coastal Archaeology* 11 (1): 107–121. https://doi.org/10.1080/15564894.2015.1050132

Clunie, Fergus G. A., 1977. *Fijian Weapons and Warfare*. Bulletin of the Fiji Museum. Fiji Museum, Suva.

Connell, John, 2015. Vulnerable Islands: Climate change, tectonic change, and changing livelihoods in the Western Pacific. *Contemporary Pacific* 27 (1): 1–36.

Fagan, Brian M., 2000. *The Little Ice Age: How Climate Made History, 1300-1850*. Basic Books, New York.

Field, Julie S., 2003. The evolution of competition and cooperation in Fijian prehistory: Archaeological research in the Sigatoka Valley, Fiji. Unpublished doctoral dissertation, University of Hawai'i, Honolulu.

Field, Julie S., 2004. Environmental and climatic considerations: A hypothesis for conflict and the emergence of social complexity in Fijian prehistory. *Journal of Anthropological Archaeology* 23 (1): 79–99. https://doi.org/10.1016/j.jaa.2003.12.004

Field, Julie S., 2008. Explaining fortifications in Indo-Pacific prehistory. *Archaeology in Oceania* 43 (1): 1–10.

Field, Julie S., and Peter V. Lape, 2010. Paleoclimates and the emergence of fortifications in the tropical Pacific islands. *Journal of Anthropological Archaeology* 29 (1): 113–124. https://doi.org/10.1016/j.jaa.2009.11.001

Frost, Everett Lloyd, 1974. *Archaeological Excavations of Fortified Sites on Taveuni, Fiji*. Social Science Research Institute, University of Hawai'i, Honolulu.

Goff, James, Bruce G McFadgen, Catherine Chagué-Goff, and Scott L. Nichol, 2012. Palaeotsunamis and their influence on Polynesian settlement. *The Holocene* 22 (9): 1067–1069. https://doi.org/10.1177/0959683612437873

Goodwin, Ian D., and Nick Harvey, 2008. Subtropical sea-level history from coral microatolls in the Southern Cook Islands, since 300 AD. *Marine Geology* 253 (1–2): 14–25. https://doi.org/10.1016/j.margeo.2008.04.012

Gordon, Arthur, 1879. *Letters And Notes Written during the Disturbances in the Highlands (known as the "devil Country") of Viti Levu, Fiji, 1876*. Clark, Edinburgh.

Groube, Les M., 1970. The origin and development of earthwork fortifications in the Pacific. In S*tudies in Oceanic Culture History, Volume 1*, edited by Roger C. Green and Marion Kelly, pp. 133–164. Pacific Anthropological Records 11. Department of Anthropology, Bernice Pauahi Bishop Museum, Honolulu.

Holler, Joseph, Quinn Bernier, J. Timmons Roberts, and Stacy-ann Robinson, 2020. Transformational adaptation in least developed countries: Does expanded stakeholder participation make a difference? *Sustainability* 12 (4): 1657. https://doi.org/10.3390/su12041657

Kennett, Douglas J., and Sarah B. McClure, 2012. The archaeology of Rapan fortifications. In *Taking the High Ground: The Archaeology of Rapa, a Fortified Island in Remote East Polynesia*, edited by Atholl J. Anderson and Douglas J. Kennett, pp. 203–234. Terra Australis 37. Australian National University E-Press, Canberra.

Kinaston, Rebecca, Hallie Buckley, Frédérique Valentin, Simon Bedford, Matthew Spriggs, Stuart Hawkins, and Estelle Herrscher, 2014. Lapita diet in Remote Oceania: New stable isotope evidence from the 3000-year-old Teouma site, Efate Island, Vanuatu. *PloS One* 9 (3): e90376. https://doi.org/10.1371/journal.pone.0090376

Klöck, Carola, and Patrick D. Nunn, 2019. Adaptation to climate change in Small Island Developing States: A systematic literature review of academic research. *Journal of Environment and Development* 28 (2): 196–218. https://doi.org/10.1177/1070496519835895

Kopp, Robert E., Andrew C. Kemp, Klaus Bittermann, Benjamin P. Horton, Jeffrey P. Donnelly, W. Roland Gehrels, Carling C. Hay, Jerry X. Mitrovica, Eric D. Morrow, and Stefan Rahmstorf, 2016. Temperature-driven global sea-level variability in the Common Era. *Proceedings of the National Academy of Sciences of the United States of America* 113: E1434–E1441. https://doi.org/10.1073/pnas.1517056113

Kumar, Roselyn, Patrick D. Nunn, Julie S. Field, and Antoine de Biran, 2006. Human responses to climate change around AD 1300: A case study of the Sigatoka Valley, Viti Levu Island, Fiji. *Quaternary International* 151: 133–143. https://doi.org/10.1016/j.quaint.2006.01.018

Ladefoged, Thegn N., and Richard Pearson, 2000. Fortified castles on Okinawa Island during the Gusuku Period, AD 1200–1600. *Antiquity* 74 (284): 404–412.

Lape, Peter V., and Chin-Yung Chao, 2008. Fortification as a human response to late Holocene climate change in East Timor. *Archaeology in Oceania* 43 (1): 11–21.

Leal Filho, W., editor, 2020. *Managing Climate Change Adaptation in the Pacific Region*. Springer Nature, Cham, Switzerland.

Martin, Pierick C.M., Patrick D. Nunn, Niko Tokainavatu, Frank R. Thomas, Javier Leon, and Neil Tindale, 2019. Last-millennium settlement on Yadua Island, Fiji: Insights into conflict and climate change. *Asian Perspectives* 58 (2): 316–330.

McMillen, Heather L., Tamara Ticktin, and Hannah Kihalani Springer, 2017. The future is behind us: Traditional ecological knowledge and resilience over time on Hawai'i Island. *Regional Environmental Change* 17 (2): 579–592. https://doi.org/10.1007/s10113-016-1032-1

McNamara, Karen E., Rachel Clissold, Ross Westoby, Annah Piggott-McKellar, Roselyn Kumar, Tahlia Clarke, Frances Namoumou, Francis Areki, Eugene Joseph, Olivia Warrick, and Patrick D. Nunn, 2020. An assessment of community-based adaptation initiatives in the Pacific Islands. *Nature Climate Change* 10: 628–639.

Medina Hidalgo, Daniela, Isaac Witten, Patrick D. Nunn, Sarah Burkhart, Jacinta R. Bogard, Harriot Beazley, and Mario Herrero, 2020. Sustaining healthy diets in times of change: Linking climate hazards, food systems and nutrition security in rural communities of the Fiji Islands. *Regional Environmental Change* 20 (73). https://doi.org/10.1007/s10113-202-01653-2

Miles-Novelo, Andreas, and Craig A. Anderson, 2019. Climate change and psychology: Effects of rapid global warming on violence and aggression. *Current Climate Change Reports* 5 (1): 36–46. https://doi.org/10.1007/s40641-019-00121-2

Morrison, Alex E., and Ethan E. Cochrane, 2008. Investigating shellfish deposition and landscape history at the Natia Beach site, Fiji. *Journal of Archaeological Science* 35 (8): 2387–2399. https://doi.org/10.1016/j.jas.2008.03.013

Nunn, Patrick D., 2000. Environmental catastrophe in the Pacific Islands around A.D. 1300. *Geoarchaeology* 15 (7): 715–740.

Nunn, Patrick D., 2007a. The A.D. 1300 event in the Pacific Basin. *Geographical Review* 97 (1): 1–23.

Nunn, Patrick D., 2007b. *Climate, Environment and Society in the Pacific during the Last Millennium*. Elsevier, Amsterdam.

Nunn, Patrick D., 2012. Na koronivalu ni Bā: Upland settlement during the last millennium in the Bā River Valley and Vatia Peninsula, northern Viti Levu Island, Fiji. *Asian Perspectives* 51 (1): 1–21.

Nunn, Patrick D., and Roselyn Kumar, 2004. Alluvial charcoal in the Sigatoka Valley, Viti Levu Island, Fiji. *Palaeogeography, Palaeoclimatology, Palaeoecology* 213 (1–2): 153–162. https://doi.org/10.1016/S0031-0182(04)00381-5

Nunn, Patrick D., and Mike T. Carson, 2015a. Collapses of island societies from environmental forcing: Does history hold lessons for the future? *Global Environment* 8: 109–131.

Nunn, Patrick D., and Mike T. Carson, 2015b. Sea-level fall implicated in profound societal change about 2570 cal yr BP (620 BC) in western Pacific island groups. *Geo: Geography and Environment* 2 (1): 17–32. https://doi.org/10.1002/geo2.3

Nunn, Patrick D., and Roselyn Kumar, 2018. Understanding climate-human interactions in Small Island Developing States (SIDS): Implications for future livelihood sustainability. *International Journal of Climate Change Strategies and Management* 10 (2): 245–271. https://doi.org/10.1108/IJCCSM-01-2017-0012

Nunn, Patrick D., and Roselyn Kumar, 2019. Measuring peripherality as a proxy for autonomous community coping capacity: A case study from Bua Province, Fiji Islands, for improving climate change adaptation. *Social Sciences* 8 (225). https://doi.org/10.3390/socsci8080225

Nunn, Patrick D., and John R. Campbell, 2020. Rediscovering the past to negotiate the future: How knowledge about settlement history on high tropical Pacific islands might facilitate future relocations. *Environmental Development* 35 (100546). https://doi.org/10.1016/j.envdev.2020.100546

Nunn, Patrick D., Rosalind Hunter-Anderson, Mike T. Carson, Frank Thomas, Sean Ulm, and Michael J. Rowland, 2007. Times of plenty, times of less: Last-millennium societal disruption in the Pacific Basin. *Human Ecology* 35 (4): 385–401. https://doi.org/10.1007/s10745-006-9090-5

Nunn, Patrick D., Michelle McKeown, Adrian McCallum, Peter Davies, Eleanor H. John, Reemal Chandra, Frank R. Thomas, and Sharon N. Raj, 2019. Origin, development and prospects of sand islands off the north coast of Viti Levu Island, Fiji, Southwest Pacific. *Journal of Coastal Conservation* 23 (6): 1005–1018. https://doi.org/10.1007/s11852-019-00707-w

Nunn, Patrick D., Elia Nakoro, Niko Tokainavatu, Michelle McKeown, Paul Geraghty, Frank R. Thomas, Pierick Martin, Brandon Hourigan, and Roselyn Kumar, 2019. A Koronivalu kei Bua: Hillforts in Bua Province (Fiji), their chronology, associations, and potential significance. *Journal of Island and Coastal Archaeology*: accepted and published online in 2019. https://doi.org/10.1080/15564 894.2019.1582119

Nurse, Leonard, Roger McLean, John Agard, Lino P. Briguglio, Virginie Duvat, Netatua Pelesikoti, Emma Tompkins, and Arthur Webb, 2014. Small islands. In *Climate Change 2014: Impacts, Adaptation, and Vulnerability. Part B: Regional Aspects. Contribution of Working Group II to the Fifth Assessment Report of the Intergovernmental Panel on Climate Change*, edited by Vicente R. Barros, C. B. Field, David J. Dokken, Michael D. Mastrandrea, Katarine J. Mach, T. Eren Bilir, Monalisa Chatterjee, Kristie L. Ebi, Yuka Otsuki Estrada, Robert C. Genova, Betelhem Girma, Eric S. Kissel, Andrew N. Levy, Sandy MacCracken, and Patricia R. Mastrandrea. Cambridge University Press, Cambridge.

O'Connor, Susan, Andrew McWilliam, Jack N. Fenner, and Sally Brockwell, 2012. Examining the origin of fortifications in East Timor: Social and environmental factors. *Journal of Island and Coastal Archaeology* 7 (2): 200–218. https://doi.org/10.1080/15564 894.2011.619245

Palmer, Bruce, 1969. Fortified sites on ridge-junctions, Fiji. *New Zealand Archaeological Association Newsletter* 12: 15–19.

Parry, John T. 1977. *Ring-ditch fortifications: Ring-ditch fortifications in the Rewa Delta, Fiji: Air photo interpretation and analysis*. Fiji Museum, Suva.

Parry, John T., 1987. *The Sigatoka Valley: Pathway into Prehistory*. Bulletin of the Fiji Museum. Fiji Museum, Suva.

Parry, John T., 1997. *The North Coast of Viti Levu, Ba to Ra. Air Photo Archaeology and Ethnohistory*. Bulletin of the Fiji Museum, Volume 10. Fiji Museum, Suva.

Pearl, Frederick B., 2004. The chronology of mountain settlements on Tutuila, American Samoa. *Journal of the Polynesian Society* 113 (4): 331–348.

Petchey, Fiona, Atholl Anderson, Alan Hogg, and Albert Zondervan, 2008. The marine reservoir effect in the Southern Ocean: An evaluation of extant and new ΔR values and their application to archaeological chronologies. *Journal of the Royal Society of New Zealand* 38 (4): 243–262. https://doi.org/10.1080/0301422 0809510559

Rainbird, Paul, 1996. A place to look up to: A review of Chuukese hilltop enclosures. *Journal of the Polynesian Society* 105 (4): 461–478.

Reimer, Paula J., Edouard Bard, Alex Bayliss, J. Warren Beck, Paul G. Blackwell, Christopher Bronk Ramsey, Caitlin E. Buck, Hai Cheng R. Lawrence Edwards, Michael Friedrich, Pieter M. Grootes, Thomas P. Guilderson, Haflidi Haflidason,

Irka Hajdas, Christine Hatté, Timothy J. Heaton, Dirk L. Hoffmann, Alan G. Hogg, Konrad A. Hughen, K. Felix Kaiser, Bernd Kromer, Sturt W. Manning, Mu Niu, Ron W Reimer, David A. Richards, E. Marian Scott, John R. Southon, Richard A. Staff, Christian S. M. Turney, and Johannes van der Plicht, 2013. IntCal13 and Marine 13 radiocarbon age calibration curves 0–50,000 years cal BP. *Radiocarbon* 55 (4): 1869–1887.

Robb, Kasey F., and Patrick D. Nunn, 2012. Nature and chronology of prehistoric settlement on the Vatia Peninsula, northern Viti Levu Island, Fiji. *Journal of Island and Coastal Archaeology* 7 (2): 272–281. https://doi.org/10.1080/15564894.2011.614320

Robb, Kasey F., and Patrick D. Nunn, 2014a. Changing role of nearshore-marine foods in the subsistence economy of inland upland communities during the last millennium in the tropical Pacific Islands: Insights from the Ba River Valley, Northern Viti Levu Island, Fiji. *Environmental Archaeology* 19 (1): 1–11. https://doi.org/10.1179/1749631413Y.0000000012

Robb, Kasey F., and Patrick D. Nunn, 2014b. Vatutāqiri: An immense stone-walled fortification on the Vatia Peninsula, northern Viti Levu Island, Fiji. *People and Culture in Oceania* 30: 1–19.

Robinson, Stacy-ann, 2019. Mainstreaming climate change adaptation in small island developing states. *Climate and Development* 11 (1): 47–59. https://doi.org/10.1080/1 7565529.2017.1410086

Savage, Amy, Hilary Bambrick, and Danielle Gallegos, 2020. From garden to store: local perspectives of changing food and nutrition security in a Pacific Island country. *Food Security* 12: 1331–1348. https://doi.org/10.1007/s12571-020-01053-8

Schmidt, Matthew, 1996. The commencement of pa construction in New Zealand prehistory. *Journal of the Polynesian Society* 105 (4): 441–460.

Scott-Parker, Bridie, and Roselyn Kumar, 2018. Fijian adolescents' understanding and evaluation of climate change: Implications for enabling effective future adaptation. *Asia Pacific Viewpoint* 59 (1): 47–59. https://doi.org/10.1111/apv.12184

Smith, Cecilia, and Ethan Cochrane, 2011. How is visibility important for defence? A GIS analysis of sites in the western Fijian Islands. *Archaeology in Oceania* 46 (2): 76–84.

Stantis, Christina, Hallie R. Buckley, Rebecca L. Kinaston, Patrick D. Nunn, Klervia Jaouen, and Michael P. Richards, 2016. Isotopic evidence of human mobility and diet in a prehistoric/protohistoric Fijian coastal environment (c. 750–150 BP). *American Journal of Physical Anthropology* 159 (3): 478–495. https://doi.org/10.1 002/ajpa.22884

Szabó, Katherine, and Judith R. Amesbury, 2011. Molluscs in a world of islands: The use of shellfish as a food resource in the tropical island Asia-Pacific region. *Quaternary International* 239: 8–18. https://doi.org/10.1016/j.quaint.2011.02.033 |10.1016/j.quaint.2011.02.033

Thomas, Frank R., 2014. Shellfish gathering and conservation on low Coral Islands: Kiribati Perspectives. *Journal of Island and Coastal Archaeology* 9 (2): 203–218. https://doi.org/10.1080/15564894.2014.921959

Thomson, Basil, 1908. *The Fijians: A Study in the Decay of Custom.* Heinemann, London.

Tippett, Alan R., 1958. The nature and social function of Fijian war. *Transactions and Proceedings of the Fiji Society* 5: 137–155.

Tuchman, Barbara, 1978. *A Distant Mirror: The Calamitous 14th Century.* Knopf, New York.

Valentin, Frédérique, Estelle Herrscher, Fiona Petchey, and David Addison, 2011. An analysis of the last 1000 Years human diet on Tutuila (American Samoa) using carbon and nitrogen stable isotope data. *American Antiquity* 76 (3): 473–486. https://doi.org/10.7183/0002-7316.76.3.473

Weir, Tony, 2020. Adaptation in small islands: research themes and gaps. In *Managing Climate Change Adaptation in the Pacific Region,* edited by Walter Leal Filho, pp. 45–68. Springer, Hamburg, Germany.

Wilkes, Charles, 1845. *Narrative of the United States Exploring Expedition during the Years 1838, 1839, 1840, 1841, 1842. Volume 3.* Lea and Blanchard, Philadelphia.

Wilson, Robyn S., Atar Herziger, Matthew Hamilton, and Jeremy S. Brooks, 2020. From incremental to transformative adaptation in individual responses to climate-exacerbated hazards. *Nature Climate Change* 10 (3): 200–208. https://doi.org/10.1038/s41558-020-0691-6

12 3500 years in a changing landscape

The House of Taga in the Mariana Islands, Western Micronesia

Mike T. Carson and Hsiao-chun Hung

In the Mariana Islands of the northwest tropical Pacific, the House of Taga shares much in common with other impressive monumental sites of the world, where people can see and experience gargantuan stonework ruins that dominate a landscape (Figures 12.1 and 12.2). The site on the surface exhibits the exquisite values of ancient architectural design and engineering, and the place strongly embodies local indigenous heritage, history, and identity. Moreover, deeper layers beneath the surface have revealed an older history of a long-term evolving social and ecological landscape.

The House of Taga ranks among the most iconic monumental sites in the Asia-Pacific region, specifically representing the largest ever standing structure of the traditional *latte* stonework architecture of the Mariana Islands. In addition to the construction qualities of the site, the indigenous unwritten histories refer to the chief Taga, who built his house here probably during the AD 1600s. Several other legends surround the life of Taga, the purpose of the site's uniquely large construction, and the relation with a larger surrounding the village of Sanhalom. The House of Taga and Sanhalom village were abandoned as residential places since about AD 1700, but the site has persisted in its identity as a symbol of indigenous Chamorro heritage.

Excavations at the site have revealed a series of archaeological layers, deep beneath the ground and dating long prior to the time of the *latte* stonework ruins of the House of Taga. The deepest and oldest cultural layer has been dated about 1500–1300 BC.

Successive generations of people lived at this place continually through more than three millennia, in total living through multiple environmental events and cultural periods. Throughout this time span, the landscape underwent significant change in its physical shape due to lowering of sea level, growth of new coral reefs and ecosystems, shifting positions of accessible freshwater sources, and rearrangement of inhabitable lands and resource zones. Concurrent with those transformations, successive generations of people adjusted the size and distribution of their residential settlement, land-use patterns, housing construction, burial practice, and other aspects of the natural-cultural landscape system.

With the large scale site excavation in 2011 and 2013, considerable information has been learned about how people lived here. This chapter

DOI: 10.4324/9781003139553-12

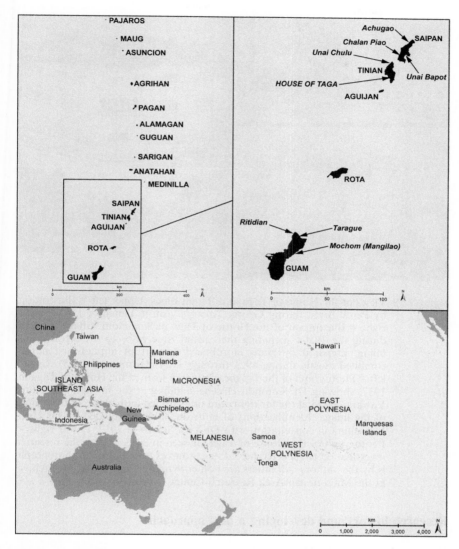

Figure 12.1 Position of the House of Taga Site in Tinian, Mariana Islands, situated within the Asia-Pacific region.

presents the research history and then the full chronological sequence of the site layers and landscape at the House of Taga. The research history clarifies how the site's deep chronological potential was recognised and eventually studied in detail. The chronological sequence narrates how the natural and cultural settings changed through time in variable aspects.

Figure 12.2 Views of the House of Taga at different times. Upper left = Illustration by Percy Brett, during George Anson's visit at Tinian in 1742. Upper right = Illustration of the House of Taga in Sanhalom Village, as seen during the 1920s, including the coastal zone prior to the artificial in-filling, prepared by Hans Hornbostel as part of unpublished notes, compiled mostly during 1921 through 1924 (Hornbostel 1925). Lower left = Photograph of the Japanese research team at the House of Taga in 1937, showing Dr Kotondo Hasebe standing at the far left, Dr Ichiro Yawata seated at the far right, and unidentified individuals in the centre of the image. Identifications of individuals were provided by Dr Hiro Kurashina in consultation with Dr Yosi Sinoto in 2012. Lower right = Photograph of the House of Taga, as seen in 2013, during the research investigation as reported here. Image sources = Lower right photograph is by the authors. All others are non-copyrighted material at the archive at the Micronesian Area Research Center, University of Guam.

Research history and developing a new approach

Stonework ruins of the House of Taga

As mentioned, the House of Taga was the largest standing structure of the *latte* style of stone pillar housing in the Marianas. The site was named after a traditional chief named Taga, who has been credited with the impressive construction and with remarkable feats of strength during his lifetime. The indigenous Chamorro name of "As Taga" literally translates as "Place of Taga", and an alternative "Guma Taga" translates as "House of Taga". The traditions about Taga point to the AD 1600s as the approximate period of

Taga, corresponding with the last time when people actively inhabited the traditional *latte* houses and villages.

Latte sites represent a unique form of construction, not seen anywhere else in the world. The uniqueness involves sets of upright stone pillars (known as *haligi*) that supported hemispherical capital stones (known as *tasa*), in turn supporting the wood and thatch components of a house (or *guma* in the indigenous Chamorro language). These houses were used for dwelling, storage, and other functions of communities or villages (*songsong*), constructed throughout diverse habitats of every island of the Marianas (Laguana et al. 2012; Thompson 1940).

Most of the knowledge about the site has come from observing the ruins, assessing the ethnohistorical evidence, and gathering general knowledge about *latte* in other sites. By comparison with the 4.5-m-tall stone ruins of the House of Taga, most *latte* in the Marianas were made in much smaller proportions, generally about 1 m in height for the total of the pillars (*haligi*) and capitals (*tasa*). Only one other *latte* potentially could have been larger than the House of Taga, but it was not completed. The stone pieces of this one potentially larger *latte* never were removed from their limestone quarry site of As Nieves in the nearby island of Rota (April 2004).

Clearly, the House of Taga was made intentionally to impress everyone who visited the place. Given the site's extraordinary size and monumentality, the most obvious research questions have involved the architectural design and cultural context.

Archaeological investigations at the House of Taga have involved documentation and measurements of the stone components, as well as studies of the surface-visible layer of dense pottery and other artefacts. In the late 1880s, Antoine-Alfred Marche recorded the first scientific description of the megalithic ruins, including the oldest known photographs that showed most of the *haligi* in upright positions and with their over-capping *tasa* still in place at that time (Marche 1889, 1982). Some decades later, during the 1920s through 1930s, Hans Hornbostel (1925) and Kotondo Hasebe (1938) reported dense concentrations of broken pottery on the surface, and their excavations exposed a single cultural layer that extended from the surface down to 2 m depth, containing numerous burial features among the general habitation debris.

After World War II, only sparse traces of pottery remained on the surface, and most of the *haligi* and *tasa* had fallen to the ground. In this setting, Alexander Spoehr (1957: 85–98) excavated more of the cultural deposit within the footprint of the House of Taga. Those excavations contributed to Spoehr's definition of the *latte* period of the region, now known as referring to the time range of approximately AD 1000 through 1700.

At that time of Spoehr's field research, the excavations did not reveal a deeper layer beneath the House of Taga. Rather, the House of Taga itself disclosed only its surface-visible stonework ruins of the AD 1600s, and a single cultural layer was associated with the surface context.

Identifying the subsurface potential

The notion of finding older subsurface layers in any place of the Marianas had not been known until the publication of Alexander Spoehr's (1957) excavations and first radiocarbon dates that extended at least as early as 1000 BC in the island of Saipan. Following Spoehr's (1957) work, an exploratory excavation was performed about 30–40 m inland from the House of Taga (Pellett and Spoehr 1961). This excavation revealed for the first time that several other cultural layers existed beneath the surface-associated *latte* layer. Moreover, at the time of the work, this single trench had yielded the earliest and richest artefact assemblage yet known in the Marianas. Previously, this early period had been known only through extremely sparse findings, such as in a subsurface layer at Chalan Piao in Saipan (Spoehr 1957). In this case, the findings inland from the House of Taga had constituted the first detailed glimpse into the Marinas region's oldest artefact assemblage, especially noting the beautifully decorated earthenware pottery and shell ornaments.

The early subsurface layers at the site were not dated by radiocarbon in the work by Pellett and Spoehr (1961) due to a lack of charcoal that was preferred for dating at the time, but the decorated pottery was recognised as belonging to the earliest cultural context of the Marianas, dated at Chalan Piao at least as early as 1000 BC. The dating at Chalan Piao had been based on a single shell because of a lack of charcoal in that site as well. Meanwhile, other excavations in different parts of Pacific Oceania had begun to document somewhat similar decorated pottery, associated with radiocarbon dates around 1000 BC, for example, at the "Lapita Type Site" in New Caledonia and at other sites of Island Melanesia and West Polynesia (Gifford and Shutler 1956).

The identified ancient subsurface layers near the House of Taga, as well at Chalan Piao, were disassociated from surface-visible stonework ruins of *latte*, and therefore they must have related with an ancient context of a substantially different physical landscape than was the case for the surface-related ruins. In those two instances, the subsurface layers were more than 1 m deep, and they were about 80–100 m inland from the present-day shorelines. In order to find more of these layers, new excavations would need to consider where to search, in the absence of obvious signs on the ground surface.

During the 1970s, novel research began to clarify where to find the early cultural layers of the Mariana Islands. Dean Thompson (1977) conducted test pit excavations in selected areas where permission and access were possible around the island of Saipan, searching for indications of ancient buried layers of the ancient period around 1000 BC or older. Those test excavations verified that the early varieties of red-slipped pottery mostly were found about 1–2 m deep and about 80–120 m inland from modern-day shorelines. Additionally, during the 1970s, excavations near the *latte* site of Unai Bapot in Saipan uncovered another exceptionally deep and early cultural layer, yielding red-slipped pottery in a context without datable charcoal but pre-dating the next superimposed charcoal-rich layer around 1000 BC (Marck 1978).

By the 1980s, the oldest cultural layers in the Marianas could be understood as positioned in deeply buried contexts, disassociated with surface-visible site ruins, and pre-dating the next overlaying charcoal-rich contexts around 1000 BC. Over the next two decades, excavations uncovered more of these deepest and oldest cultural layers at several sites, including Tarague (Kurashina and Clayshulte 1983; Kurashina and Clayshulte, 1981; Ray 1981) and the Mangilao Golf Course in Guam (Dilli et al. 1998), Unai Chulu in Tinian (Craib 1993; Haun et al. 1999), and Achuago in Saipan (Butler 1994, 1995). Additional work at Chalan Piao in Saipan confirmed the early dated context with the red-slipped pottery (Moore et al. 1992). Reliable charcoal was sparse or absent for radiocarbon dating in those site layers, but the available charcoal samples plus associated marine shell samples pointed to dates at least as old as 1500 BC.

While the early Marianas site layers were verified in multiple sites of at least three separate islands (Guam, Tinian, and Saipan), the ancient landscape contexts were unclear until advancements were made in understanding the sea level history and change in coastal morphology. At the Tarague site in Guam, Kurashina and Clayshulte (1983) noticed that the coastal morphology had transformed substantially since the time of initial occupation. At the Achugao site in Saipan, Butler (1994, 1995) proposed that the ancient sea level must have been higher than the present conditions.

Measurements of ancient sea level became possible with exposed coral reefs and tidal notches, now positioned about 1.5–2 m higher than their present-day counterparts around Guam and the Mariana Islands (Dickinson 2000, 2003; Kayanne et al. 1993). Radiocarbon dating of the exposed reef surfaces had produced dates mostly prior to 1000 BC, and this dating corresponded with a mid-Holocene highstand approximately 3000 through 1000 BC in this region. The precise measurements and dates varied slightly around the Marianas and more broadly across Pacific Oceania (Dickinson 2003).

The sea-level history now has been well established in the Marianas region, based on observations and radiocarbon dating directly at the ancient archaeological sites and at other non-site natural coastlines of the islands (Figure 12.3). Since 2005, a research project has documented several of these instances, in contexts pre-dating 1000 BC, as part of an effort to depict what the ancient landscapes looked like and where archaeological sites may have been situated (Carson 2011, 2014a). This project led to the discovery of the deeply buried site layers in three localities of Ritidian in Guam, dated around 1500 BC (Carson 2012, 2017). The palaeolandscape context furthermore clarified the findings at Unai Bapot in Saipan, where new excavations confirmed the earliest palaeo-beach habitation at least as early as 1500 BC (Carson 2008; Carson and Hung 2017).

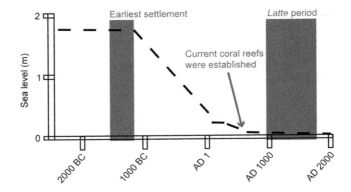

Figure 12.3 Sea-level history, compared with major portions of archaeological sequence in the Mariana Islands. Measurements and dating follow prior reporting (Carson 2011, 2014b, 2016; Dickinson 2000, 2003; Kayanne et al. 1993).

The regional palaeolandscape research has been reported in detail elsewhere (Carson 2011, 2014a, 2014b: 2016, 2020), wherein the major findings clarified that:

a. The ancient sea level at 3000 through 1100 BC was about 1.8 m above the present level in most places, and certainly, these conditions existed at the time when people first lived in the Mariana Islands around 1500 BC;
b. The contexts of the ancient shorelines pre-dated about 1100 BC, and a sea-level drawdown period started thereafter;
c. Only a few of those ancient shorelines supported ancient habitations, dated generally in the range of 1500–1100 BC;
d. The ancient shoreline habitation layers now have been buried at least 1 m deep, and they have been stranded about 100 m inland from present-day shorelines;
e. In the places where people lived at those ancient shoreline sites, they built houses on posts raised above the inter-tidal or shallow subtidal zones;
f. In those ancient shoreline contexts, charcoal samples were poorly preserved and unreliable for radiocarbon dating;
g. Outside the contexts with unreliable early charcoal, direct pairing of charcoal with *Anadara* sp. shells in different layer contexts of multiple sites consistently confirmed a minimal marine reservoir correction (ΔR) of -49 ± 61 radiocarbon years, including one instance as old as 1437–1288 BC at Unai Bapot in Saipan and another instance as old as 1413–1266 BC at the House of Taga in Tinian; and
h. According to the oldest reliable site dating, people first lived at these ancient shorelines by 1500 BC and perhaps slightly earlier.

The above-outlined findings supported the 2011–2013 investigation at the House of Taga in terms of understanding where to excavate, how to obtain reliable radiocarbon dates for the ancient layers, and how to contextualise the ancient landscape settings of each layer. The new site excavations helped to refine the regional picture of palaeolandscape research, as well.

2011–2013 investigation

The 2011–2013 research was possible only after first knowing about the initial findings as reported by Pellett and Spoehr (1961), in combination with new knowledge about the regional and global sea-level history, change in coastal morphology, and related issues about the palaeolandscape sequence, as mentioned above. This larger and longer research history in the Marianas in many ways influenced the design of the new investigation.

The choice of location for the new excavation deliberately avoided the surface-visible stonework remains, instead aiming to investigate the area of a series of deep cultural layers. The excavation successfully found the area of the dense cultural deposit, approximately in the position as had been reported previously (Figure 12.4). This location is about 100 m landward from

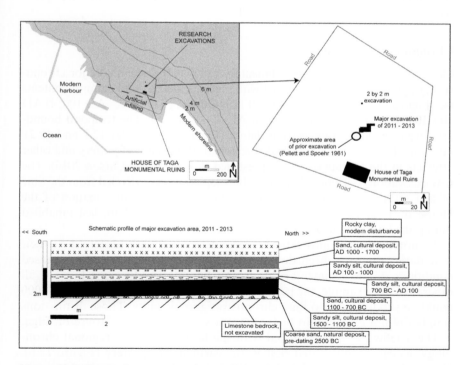

Figure 12.4 Excavation areas and overall stratigraphic sequence at the House of Taga Site, 2011 and 2013.

where the modern shoreline would have been situated, discounting the modern infilling along the coastal zone (see Figure 12.4).

Within the area of the verified dense and early habitation, the housing structures and residential debris appeared to have been aligned with the ancient shoreline, such that the major axis was parallel with the shoreline. Corroborating this picture, about 15 m farther landward from the major excavation area in 2011–2013, a test pit of 2 by 2 m found that the lowest cultural layer there was sparse and positioned directly over the basal limestone, and this position could reflect the landward limit of the early habitation zone.

Within the primary habitation area, the excavations in total exposed more than 90 sq m of a contiguous area. This exposure provided by far the largest single view of the earliest habitation layer in the Mariana Islands. This amount of excavation may seem small by international standards, but so far, it ranks as the largest single contiguous hand-excavated area in the region.

The excavations revealed a surface layer of massive disturbance and bulldozing all across the surface, related with the bulldozer clearing of the land after World War II and again later. The base of the land-clearing disturbance, however, entailed an abrupt transition over a partly truncated underlying cultural deposit of the traditional *latte* period (approximately AD 1000 through 1700). Beneath this partly truncated later, the deeper and older layers of the site were undisturbed.

Chronological overview

Given the parameters of the sedimentary units at the site, six major chronostratigraphic units were recognised with ancient artefacts and cultural contents (see Figure 12.4): 1) 1500–1100 BC; 2) 1100–700 BC; 3) 700 BC–AD 100; 4) AD 100–1000; 5) AD 1000–1700; and 6) AD 1700–Present. The temporal boundaries were based on radiocarbon dating (Figure 12.5; Tables 12.1 and 12.2), augmented by the well established chronological sequence of pottery and other artefacts in the Marianas region (presented thoroughly by Carson 2016). The new excavation confirmed that the archaeological layers extended from the time of first settlement around 1500 BC all the way through the creation of the House of Taga and the associated *latte* village of Sanhalom, last inhabited during the late AD 1600s.

As noted, the chronostratigraphic units were slightly different in their boundaries than has been documented at other sites due to the localised conditions of the sedimentary units and boundaries. In some but not all other sites, for example, the periods of 700 BC–AD 100 and of AD 100–1000 could be subdivided into finer chronostratigraphic units. Those distinctions simply did not exist within the sedimentary deposits at the House of Taga.

The artefact sequence aligned with the larger Marianas regional chronology, as can be illustrated best in the earthenware pottery (Figure 12.6). The pottery chronology has been the most robust among the Marianas artefacts due to the abundance of this material in every cultural context

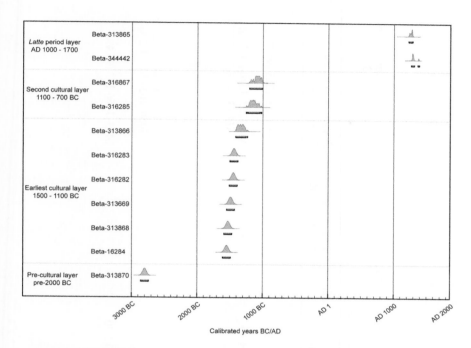

Figure 12.5 Probability distributions of radiocarbon dating output from 2011–2013
excavations at the House of Taga site. Raw data are shown in
Tables 12.1 and 12.2. Calibrations were by the OxCal program (Bronk
Ramsey and Lee 2013), using the INTCAL13 and MARINE13 calibration curves (Reimer et al. 2013).

prior to about AD 1700. While some aspects of the chronological distinctions were achieved more precisely at other sites, the records at the House of
Taga were remarkably productive and informative, especially for the earliest
cultural layers. The temporal boundaries of those two oldest cultural layers
(1500–1100 BC and at 1100–700 BC) were just as refined at the House of
Taga as at other sites.

The physical landscape chronology has been illustrated in Figure 12.7,
using the same time periods as had been identified in the localised chrono-
stratigraphy. This framework, together with the pottery chronology (see
Figure 12.6), can facilitate the next sections of presenting the site's ancient
contexts and findings in their chronological order. The output was based on
the measured positions and depths of the chronostratigraphic layers (see
Figure 12.4), combined with knowledge of the local and regional sea-level
history (see Figure 12.3) and observations of the sedimentary matrix in each
layer.

Table 12.1 Radiocarbon dating results from 2011–2013 excavations at the House of Taga in Tinian. Marine reservoir corrections followed the calculations as shown in Table 12.2. Calibrations were by the OxCal program (Bronk Ramsey and Lee 2013), using the INTCAL13 and MARINE13 calibration curves (Reimer et al. 2013).

Context	Beta-#	Material	δ13C (‰)	δ18O (‰)	Conventional age (years BP)	Marine reservoir correction (ΔR)	2-Sigma calibration (calendar years)
Latte period cultural layer in enlarged coastal plain							
Heated-rock oven (hearth)	313865	Charcoal, chunk	−23.5	–	760 ± 30	–	AD 1222–1285 (95.4%)
Subadult burial	344442	Incisor tooth	−19.6	–	710 ± 30	–	AD 1262–1310 (80.1%); AD 1362–1387 (15.3%)
Cultural layer in stable backbeach setting							
Rubbish pit	316867	Charcoal, narrow twigs	−23.7	–	2900 ± 30	–	1264–1045 BC (95.4%)
Heated-rock oven (hearth)	316285	Charcoal, narrow twigs	−23.7	–	2940 ± 30	–	1258–1245 BC (1.9%); 1230–1046 BC (92.2%); 1030–1020 BC (1.2%)
Earliest cultural layer in palaeo-seashore context							
Heated-rock oven (hearth)	313866	Charcoal, narrow twigs	−25.8	–	3070 ± 30	–	1413–1266 BC (95.4%)

						Calibrated date
Heated-rock oven (hearth)	316283	*Anadara* sp. shell	0	—	3390 ± 30	−28 ± 48	1481–1190 B.C. (95.4%)
Heated-rock oven (hearth)	316282	*Anadara* sp. shell	−1.3	—	3400 ± 30	−49 ± 61	1529–1190 BC (95.4%)
Rubbish pit	313669	*Anadara* sp. shell	−0.3	—	3440 ± 30	−49 ± 61	1596–1249 BC (95.4%)
Heated-rock oven (hearth)	313868	*Anadara* sp. shell	−0.7	—	3480 ± 30	−49 ± 61	1633–1291 BC (95.4%)
Heated-rock oven (hearth)	316284	*Anadara* sp. shell	+0.3	—	3500 ± 30	−49 ± 61	1660–1317 BC (95.4%)
Pre-cultural palaeo-beach deposit						
Layer matrix	313870	*Acropora* sp. branch coral	−2.9	—	4750 ± 30	−49 ± 61	3321–2926 BC (95.4%)

Table 12.2 Marine reservoir correction for *Anadara* sp. shells, in paired contexts with charcoal samples, in Marianas archaeological sites of Ritidian in Guam, House of Taga in Tinian, and Unai Bapot in Saipan. Marine reservoir corrections were calculated with the Deltar online software package (Reimer and Reimer, 2017). Calibrations were by the OxCal program (Bronk Ramsey and Lee 2013), using the INTCAL13 and MARINE13 calibration curves (Reimer et al. 2013).

Context	Beta-#	Material	δ13C (‰)	δ18O (‰)	Conventional age (years BP)	Marine reservoir correction (ΔR)	2-Sigma calibration (calendar years)
Ritidian, Guam, Stable backbeach layer, 90–100 cm (Carson 2010)	263449	*Anadara* sp. shell	+2.1	Not measured	2810 ± 40	−70 ± 80	867–412 B.C. (95.4%)
Ritidian, Guam, Stable backbeach layer, 92 cm (Carson 2010)	263448	Carbonised *Cocos nucifera* (coconut) endocarp	−24.5	–	2510 ± 40	Not applicable	796–509 B.C. (95.4%)
Ritidian, Guam, Stable backbeach layer, 105–110 cm (Carson 2010)	239578	*Anadara* sp. shell	+1.5	Not measured	3140 ± 40	1 ± 56	1165–806 B.C. (95.4%)
Ritidian, Guam, Stable backbeach layer, 98–105 cm (Carson 2010)	239577	Carbonised *Cocos nucifera* (coconut) endocarp	−25.4	–	2810 ± 40	Not measured	1073–1066 B.C. (0.7%); 1057–843 B.C. (94.7%)

Sample	Lab no.	Material					Calibrated date
Ritidian, Guam, Stable backbeach layer, 80 cm (Carson 2017a, 2017b)	424685	*Anadara* sp. shell	−1.2	−2.1	2870 ± 30	−56 ± 66	788–420 B.C. (95.4%)
Ritidian, Guam, Stable backbeach layer, pit feature from origin at 80 cm (Carson 2017a, 2017b)	433372	Charcoal	−26.3	–	2470 ± 30	Not applicable	768–476 B.C. (92.4%); 464–453 B.C. (1.2%); 445–431 B.C. (1.8%)
House of Taga, Tinian, Hearth Feature A, 170 cm (Carson 2020)	316283	*Anadara* sp. shell	0	Not measured	3390 ± 30	−28 ± 48	1481–1190 B.C. (95.4%)
House of Taga, Tinian, Hearth Feature A, 170 cm (Carson 2020)	313866	Charcoal, narrow twigs	−30.1	–	3070 ± 30	Not applicable	1413–1266 B.C. (95.4%)
Feature C hearth, originating from Layer VI-A (Carson and Hung 2017)	461342	*Anadara* sp. shell	+0.2	−1.1	3370 ± 30	−90 ± 48	1522–1230 B.C. (95.4%)
Feature C hearth, originating from Layer VI-A (Carson and Hung 2017)	448705	Charcoal	−25.9	Not measured	3110 ± 30	Not applicable	1437–1288 B.C. (95.4%)
TOTAL POOLED						**−49 ± 61**	

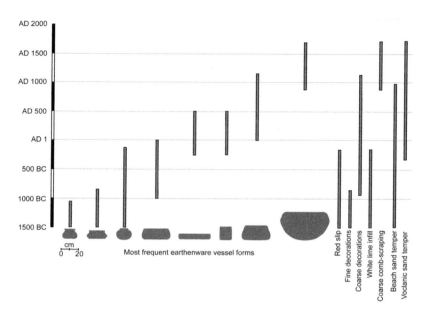

Figure 12.6 Chronological summary of the Marianas regional pottery sequence.

1500–1100 BC

Around 1500 BC, the shorelines of the Mariana Islands supported the first-ever permanent settlements in the remote-distance Islands of Pacific Oceania (see Figure 12.1). The first settlers in the Mariana Islands lived in post-raised houses in at least a few of the ancient shorelines, and one of those oldest habitations was at the House of Taga site (Figure 12.8). At this site, in particular, the four oldest dating results from secure archaeological contexts had overlapped at 1529–1317 BC (see Figure 12.5). The single oldest cultural sample was from a heated-rock oven (or "hearth feature"), calibrated at 1660–1317 BC for a piece of *Anadara* sp. shell (Beta-316284 in Table 12.1). The next three oldest dates for *Anadara* sp. shells were from heated-rock ovens and a rubbish pit, calibrated at 1633–1291 BC (Beta-313868), 1596–1249 BC (Beta-313869), and 1529–1190 BC (Beta-316282).

The shell dating is considered reliable for this particular case of *Anadara* sp. clamshells in the Mariana Islands sites due to several paired dating results of *Anadara* sp. shells with charcoal in different archaeological contexts that consistently pointed to a narrow range of marine reservoir correction (ΔR) at -49 ± 61 years (see Table 12.2). Specifically, at the House of Taga excavation, one dated pairing was in a rock-heated oven feature, wherein the charcoal date was calibrated at 1413–1266 BC (Beta-313866, paired with Beta-316283 in Table 12.2). A similarly ancient paired was calibrated at 1437–1288 BC at the Unai Bapot site in Saipan (Beta-448705, paired with

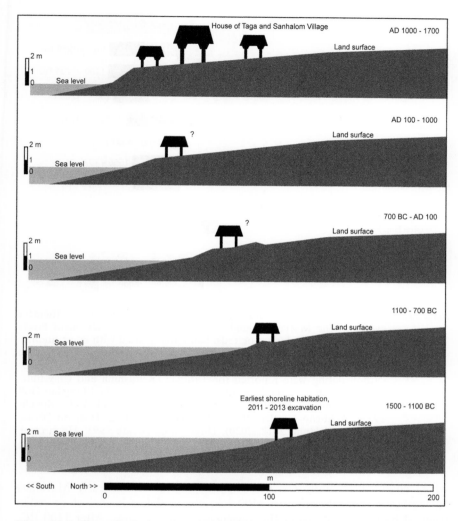

Figure 12.7 Chronological sequence of landscape change at the House of Taga Site. The vertical scale is exaggerated in comparison with the horizontal scale. Locations of ancient houses are approximated within each time period.

Beta-461342 in Table 12.2). These two pairings were extremely rare findings in the Marianas, wherein charcoal and *Anadara* sp. shells were retained inside broken pottery fragments, furthermore protected inside heated-rock oven features. Otherwise, the pairing contexts were possible only in the slightly later layers of emergent stable backbeach settings after 1100 BC.

The marine reservoir correction (ΔR) of -49 ± 61 years has been helpful for allowing direct dating of *Anadara* sp. shells in the older site layer contexts

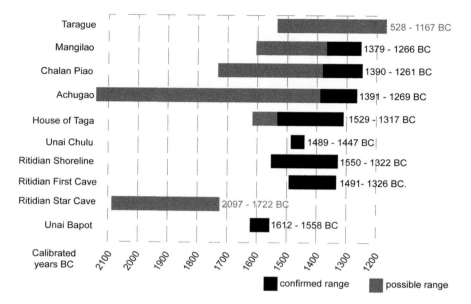

Figure 12.8 Summary of earliest confirmed radiocarbon dating of initial settlement events in the Mariana Islands. Actual durations of settlement layers were extended for longer periods. Data are updated with latest findings (Carson 2020) and based on critical reviews of the source data (Carson 2014a; Carson and Kurashina 2012). Original source data of radiocarbon dating were reported for Tarague (Kurashina and Clayshulte 1983; Kurashina and Clayshulte, 1983), Mochom or the Mangilao Golf Course (Dilli et al. 1998), Chalan Piao (Moore et al. 1992; Spoehr 1957), Achugao (Butler 1994, 1995), House of Taga (Carson 2014a; Carson and Hung 2012, 2020), Unai Chulu (Craib 1993; Haun et al. 1999), Ritidian (Carson 2010, 2017), and Unai Bapot (Carson 2008; Carson and Hung 2017).

where charcoal generally has been absent. Overall in the early Marianas palaeo-beach contexts prior to 1100 BC, charcoal has been lost or redistributed by the conditions of the ancient shoreline settings. After 1100 BC, charcoal became abundant and more reliable for radiocarbon dating in the emergent stable backbeach settings. Only in rarely protected instances of the pre-1100 BC contexts charcoal has been dated successfully, including one instance as noted at the House of Taga site and another instance at Unai Bapot.

The 2011–2013 investigation uncovered sets of post moulds, heated-rock ovens, rubbish pits, and stonework remnants that depicted the interconnected activity areas of the oldest living surface at the site (Figure 12.9). The ancient habitation area was emplaced directly at the ancient shoreline, partly extending into the inter-tidal or shallow subtidal zone. Only the

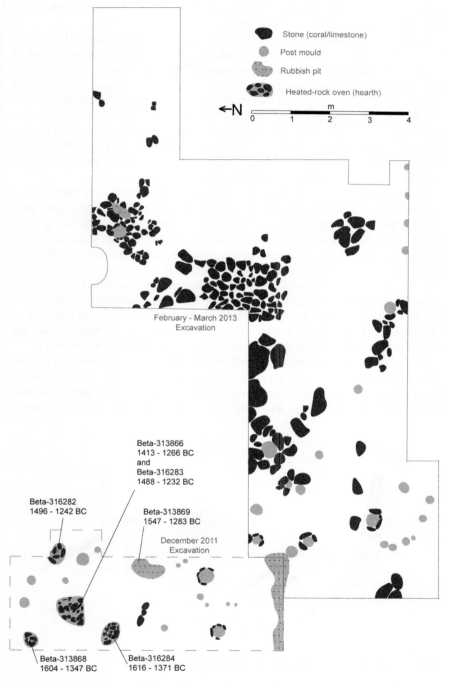

Figure 12.9 Plan map of the earliest settlement layer at the House of Taga Site, 1500–1100 BC.

landward portion had retained the traces of heated-rock ovens in positions that were outside the usual tidal range.

In practical terms, the stonework components of the site had stabilised the loose sandy sedimentary surface, similar to the modern engineering use of "riprap" to stabilise shoreline areas. Some of these stonework constructions were built around the housing post moulds, reflecting an intention to create stable space directly around the ancient houses. People likely used these stable surfaces for supporting various activities at or around their houses. Furthermore, many of the post moulds had been lined with bracing stones, thereby increasing the stability in the sandy sedimentary matrix. Within the interstices of the stonework, small objects often were recovered during the excavation, mostly including tiny shell beads.

The food midden in this layer was typical of the earliest settlement period in the Marianas (Figure 12.10), dominated by shellfish remains and preserving very few animal bones. The sparse animal bones referred to fish, birds, and turtles. The shellfish included mostly *Anadara* sp. clamshells, and indeed these clams have comprised the dominant food remains in all of the known earliest habitation sites of the region. A few centuries later, however, *Anadara* sp. clams were unable to survive in more than minimal populations after the drawdown in sea level following 1100 BC (Amesbury 1999, 2007). Other shellfish indicators of early settlement contexts were the remains of chiton, limpets, and sea urchins that tended to be over-harvested after the first few centuries of people living in any place (Carson 2016: 96–99).

Along with the protein-rich seafood at the coastal and nearby marine zones, the terrain of Tinian and the other islands already had supported luxurious forest growth before people had arrived to live here. The first island residents certainly modified the natural forests, and they likely

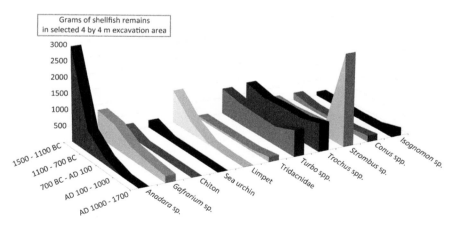

Figure 12.10 Summary of shellfish remains at House of Taga.

selected the most useful plants such as coconut, native breadfruit, cycads, and others. They may have imported limited plants that survived the long journey of more than 2000 km, but the direct evidence for those imports has been unclear except for the likelihood of importing betelnut palms (Athens and Ward 2006), probably from Island Southeast Asia.

The initial forest modification has been documented in the lake-bottom and swamp-bottom sedimentary catchments of island-wide palaeoenvironmental records, specifically referring to an abrupt horizon of charcoal influx, increase in disturbance-related grasses, and decline in native palms all around 1500 BC, with particular evidence in Tinian at 1661–1278 BC (Athens and Ward 1998) and in islands of Guam and Saipan at approximately the same time range (Athens and Ward 1993, 1999, 2006). While some interpretations have suggested older dating closer to 2200–2000 BC for the studies in Guam and Saipan (Athens and Ward 2004; Athens and Athens et al. 2004), a closer examination of the sedimentary units and dating samples has indicated that the earlier studies by the same researchers were indeed correct about the dating around 1500 BC (Carson 2016: 100–102).

Direct evidence of plant foods in archaeological sites has been imprecise in the most ancient contexts. During exploratory trial analysis, stone tools from the earliest layers revealed possible starches and other microfossils when these objects were examined through 30x magnification under cross-polarised light. Currently, specific starchy foods are difficult to ascertain due to the uncertainties of matching starch grains with known taxa after probable damage during pounding, cooking, and age within a buried archaeological layer. Furthermore, more numbers of samples would be preferred for a systematic study and including comparisons with the matrix of surrounding sediments. Nonetheless, studies of retained starches and other microfossils in later contexts of Guam and Saipan have shown the most probable use of banana, breadfruit, and possibly taro or yam (Horrocks et al. 2015).

The Marianas archaeological record never included formal agricultural fields or domesticated animals until the time of foreign influences of the AD 1500s and later. In these contexts, the food supplies must have been obtained through low-impact modifications of the natural forest habitat, along with accessing the shellfish, fish, and other resources from coastal and marine zones. Outside the Marianas, most other Pacific Islands showed widespread deforestation, creation of formal agricultural fields, and massive depopulation of birds and other animals during the centuries following initial habitation (Kirch 1997; Steadman 1995), but no such records existed in the Marianas until much later periods of population expansions after AD 1000. Instead, the Marianas records showed minimal anthropogenic impacts, notably while people did not install permanent residential sites in the island interiors until after AD 100 or in the farther northern islands until after AD 1000.

The earliest settlements must have targeted places with access to freshwater sources, although no permanent streams or lakes ever existed around the House of Taga. In this case, water access here may have relied to some

Figure 12.11 Historical wells at and around the House of Taga site in Tinian. Left
image: Hsiao-chun Hung at the Taga Well, around present-day 3 m
elevation nearby the site. Right image = historical well of the early
through middle 1900s, in process of excavation during 2013, showing
reference scale bars in 20-cm increments.

extent on rainwater, plus access into the seepage flows of the island's aquifer
outflowing just above sea level at the coast. Adjacent with the site today, a
stone-lined well is remembered as a relic from the time of Taga's residence
during the 1600s (left image in Figure 12.11), and other wells were dug here
during the early through middle 1900s (right image in Figure 12.11). These
wells had tapped into the freshwater lens floating over the modern sea level.
The same lens at 1500–1100 BC, however, would have been positioned 1.8 m
higher and associated with the coastal setting of that time (see Figure 12.7).

Overall, the earliest inhabitants collected shellfish as their primary protein, along with variable amounts of turtles, fish, birds, and probably bats
from different habitats. Plant foods involved at least some starchy foods,
likely from the gardening areas and native forests, augmented by cultural
manipulations of the forest ecology.

While those first generations of islanders adapted to their preferred
shoreline niches, they produced abundant artefact assemblages of pottery,
personal ornaments, and assorted stone and shell tools. The earliest pottery
consisted of earthenware, mostly with a red slip and rarely with a black
burnished outer surface. The vessels were overall small in size, up to a
maximum of 20 cm diameter in most cases, and they were made with thin-
sided walls that could be as thin as 1 mm in certain parts. A few pieces
retained paddle-impressed or carved paddle impressions on their exteriors,
possibly as part of a deliberate artistic texturing rather than purely as a
product of the manufacturing process.

Among the earliest pottery, the decorated pieces showed dentate-stamped,
circle-stamped, hand-drawn circles, and fine line-incised designs, often
highlighted by white lime infill (Figure 12.12). The designs mostly were
placed in rows or zones of filled spaces, juxtaposed against non-filled rows or

Figure 12.12 Examples of decorated pottery from the earliest occupation layer at the House of Taga Site, 1500–1100 BC.

zones. The white lime was composed of slaked lime from burned shells or coral, also used as an ingredient in chewing betelnut.

The earliest decorative techniques in the Marianas pottery matched with the observations in slightly older pottery of the northern through central Philippines of Island Southeast Asia dated perhaps as early as 1700 BC and certainly established by 1500 BC (Hung 2008; Hung et al. 2011). The diagnostic characteristics include the use of dentate-stamped and circle-stamped patterns, highlighted by white lime infilling over a red-slipped surface of earthenware vessels.

The vessel shapes were more varied in the Philippines, and the assemblages in the Marianas, therefore, represented a subset of this diversity. At least a few aspects of the Marianas pottery, such as the particular decorative patterns, have involved localised expressions that were either rare or unknown in the Philippines. The combinations of technical and artistic choices were remarkably similar in each of the known earliest settlement sites across the Marianas.

The personal portable or wearable ornaments mostly were made of shells, in forms of beads, discs, bands, bangles, circlets, and pendants (Figure 12.13).

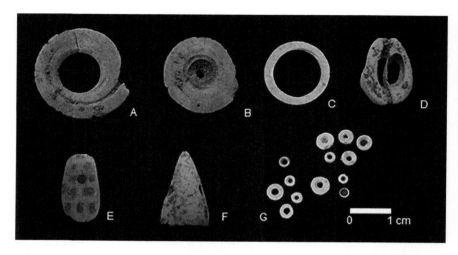

Figure 12.13 Examples of shell ornaments from the earliest occupation layer at the House of Taga Site, 1500–1100 BC. A, B, C = circlets or discs. D = *Cypraea* sp. shell bead. E, F = shell pendants. G = small round beads, made of *Conus* spp. or other shells.

They were made through actions of slicing the shells, drilling perforations, and grinding or polishing the surfaces into the final forms. Perhaps the most distinctive pieces were the small cowrie shells (*Cypraea* sp.) that were sliced lengthwise and then ground or polished, so far without definite parallels in other documented collections outside the Marianas. The *Cypraea* sp. beads disappeared after 1100 BC, along with the disappearance of shell discs, bands, circlets, and small pendants. At the same timing after 1100 BC, the simple small round beads were made in significantly lesser quantities.

In addition to the pottery and personal ornaments, the ancient artefacts included fishing gear and general-use tools. Fishing gear (Figure 12.14) included mostly small rotating hooks made of nacreous shells, along with small net weights made of crystalline limestone. Rare drilled cowrie shell caps were used with the region's oldest known octopus lure contraptions (Carson and Hung 2021). Stone and shell tools were large pounders, adze blades, rare axe or chisel blades, and simple flaked tools for general cutting and slicing (Figure 12.15).

The settlements in the Marianas by 1500 BC had required the longest distance of sea-crossing voyage of their time from the nearest and most likely source in Island Southeast Asia, in excess of 2000 km (Craib 1999). Historical linguistic studies have pointed to the Malayo-Polynesian language grouping in the northern through central Philippines as the most likely

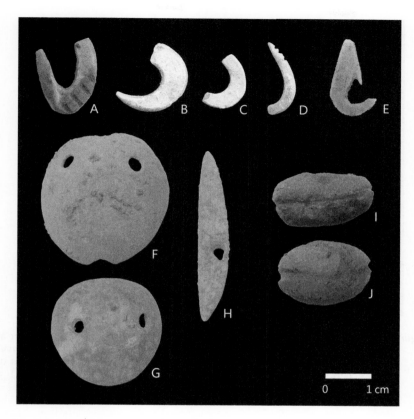

Figure 12.14 Examples of fishing gear from the earliest occupation layer at the House of Taga Site, 1500–1100 BC. A, B, C, D, E = fragments of shell fishing hooks. F, G = *Cypraea* sp. shell component of an ancient octopus lure rigging. H = Bone, modified with pointed ends and drilled hole, possibly used as a part of fishing-related gear or tool. I, J = Probable net sinkers or other sinker devices, made of crystalline limestone.

source of the Chamorro language of the Marianas (Blust 2000, 2019). Similarly, ancient DNA studies of skeletal remains from Guam in the Marianas have confirmed that the foundational population included the indigenous "Haplogroup E" lineages that most likely descended from ancestral populations in the Philippines (Pugach et al. 2020). These findings accord with the cross-regional comparisons of the earliest pottery assemblages in the Marianas and elsewhere at 1500 BC, as was mentioned earlier in this chapter.

Figure 12.15 Examples of shell and stone tools from the earliest occupation layer at the House of Taga Site, 1500–1100 BC. A = Shell adze, made of *Cassis cornuta* shell. B = Flakes of varied types of limestone. C = Flakes of chert. D =Fragment of stone adze. E = Fragment of large stone axe. F = Fragment of stone pounder, made of porous volcanic stone.

1100–700 BC

Around 1100 BC, the inhabited seashore niche began to change at the House of Taga, as occurred at the other known early habitations of the Marianas at the same time (see Figure 12.7). The changing conditions were instigated by a drawdown in sea level that started at 1100 BC and continued for the next several centuries (see Figure 12.3), directly affecting the shapes of the coastal landforms and the associated coastal ecologies at the shoreline niches where people had been living. At the House of Taga and other sites, the initial drawdown had created stranding of the older habitation surfaces, often marked by the accumulation of broken branch corals and other tidal surge debris that no longer could be removed through the regular tidal cycle.

The lower sea level after 1100 BC created broader beaches, as well as spaces where stable backbeaches and ridges could form. At the House of Taga site, this context after 1100 BC supported a new backbeach, dated by charcoal in a rubbish pit at 1264–1045 BC (Beta-316867 in Table 12.1) and in a heated-rock oven at 1260–1050 BC (Beta-316285 in Table 12.1). As was seen in the other Marianas coastal sites, the new stable beach settings after 1100 BC retained abundant charcoal for radiocarbon dating.

The abruptness of the stratigraphic layer after 1100 BC indicated a significant shift in the local environment, and the associated dating allowed a constraint of the underlying initial settlement layer. The dating samples in this stratigraphic layer were obtained from secure contexts of cultural features within a broader arrangement in this layer (Figure 12.16). Moreover, these features overlaid and necessarily post-dated the older features in the lower cultural layer, thereby distinguishing two chronologically different contexts of the site inhabitation.

Concurrent with the progradation of the beach, the access to the freshwater lens adjusted accordingly with the slightly lowered sea level (see Figure 12.7). The shellfish records showed a decline in the previously preferred *Anadara* sp. clams, while people adjusted with shifting attention to other shellfish taxa (see Figure 12.10). Meanwhile, the targeted harvesting around the habitation areas likely contributed to a reduction in the amounts of chiton, limpets, and sea urchins in the food debris.

The change in the physical landscape context coincided with the disappearance of the earliest forms of decorated pottery and personal ornaments (see Figure 12.6). The later expressions involved coarser and thicker pottery, with greatly simplified broad incisions and circle-stamped designs. People made fewer of the small round shell beads during the next centuries, while they no longer produced the earlier forms of *Cypraea* sp. shell beads or the other assorted pendants, bands, rings, and circlets that had characterised the period of 1500–1100 BC.

700 BC–AD 100 and AD 100–1000

For the long time span of 700 BC through AD 1000, the 2011–2013 excavations revealed two major chronostratigraphic units, wherein the pottery traditions corresponded with ranges of approximately 700 BC–AD 100 and then AD 100–1000. These two contexts could be defined only in broad terms, in contrast to the more refined chronological bracketing in other early sites of the Marianas.

During these two approximate time periods, people had not created housing footprints or other formal structural features emplaced within the area of the 2011–2013 excavation. Rather, the exact housing locations must have been in slightly different positions, resulting in only generalised cultural debris of broken artefacts and midden within the same area of the older housing of the previous two cultural layers. Currently, the distance is difficult to calculate between here and the postulated nearby housing areas, except to note that the distance was within the range of the habitual discard of broken artefacts and food midden.

Given the inconsistencies in the stratigraphic units and the absence of secure contexts of features such as hearths and pits, tentative dating was

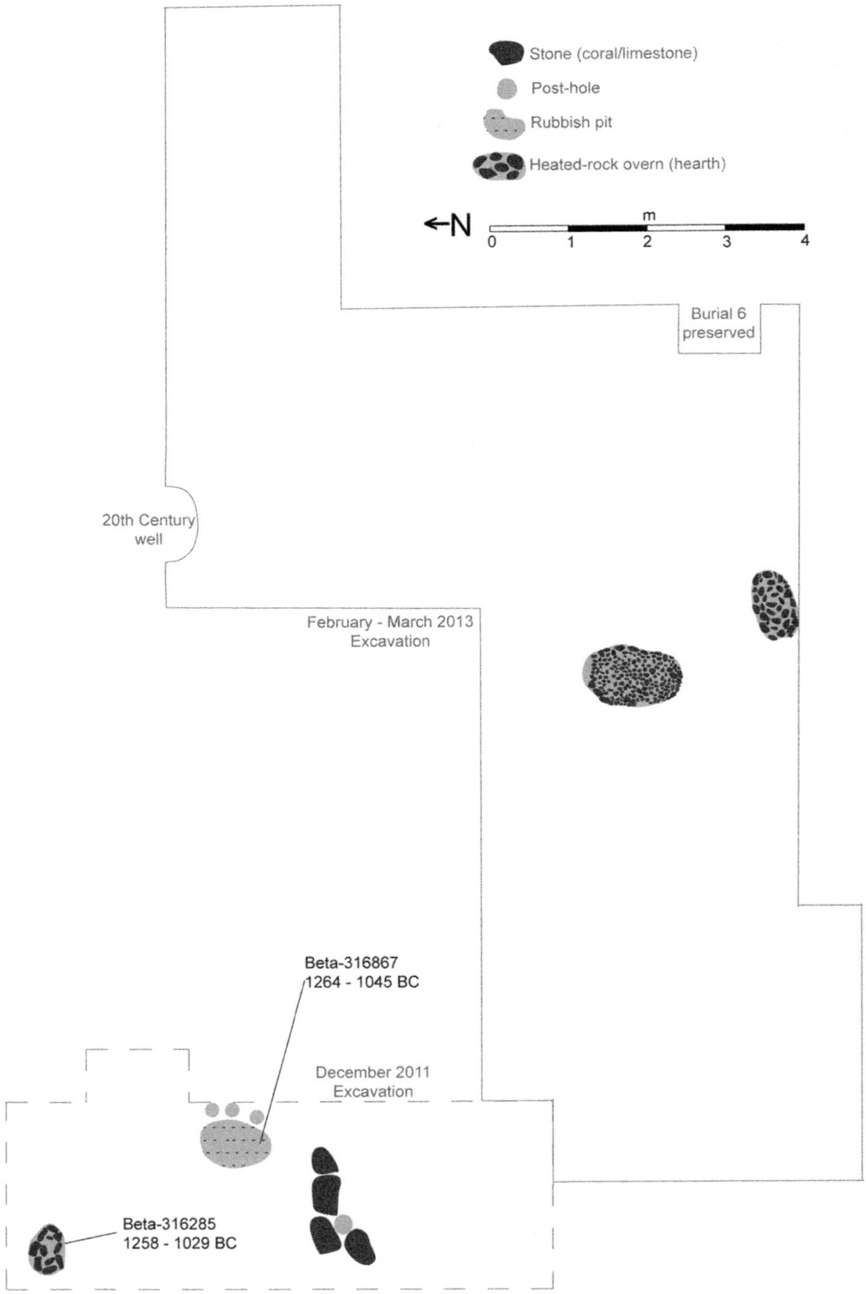

Figure 12.16 Plan map of the second occupation layer at the House of Taga Site, 1100–700 BC.

performed on the basis of comparing the pottery forms with the larger Marianas regional sequence (see Figure 12.6).

During these several centuries, the drawdown in sea level had slowed temporarily around AD 100–200, but otherwise, it continued a gradual drawdown to approach the modern level by some point within the range of AD 500–1000 (see Figure 12.3). Overall, people had adjusted with using the broadening beaches, although the coral reefs and coastal habitats continued to be somewhat unstable throughout most of these centuries. A new coral reef ecosystem started to form, though, during the later centuries around AD 500–1000, after the sea level had stabilised.

This larger contextual information is based on observations and measurements in several sites of Guam and Saipan (Carson 2016: 183–219), and the parameters in principle would apply in Tinian and other islands of the Marianas. In the other known sites of these chronological contexts, consistently people had shifted to live in slightly different locations, distinguished in at least two major periods before and after approximately AD 100. In all of these cases, the uncertainties of the coastal ecology likely prompted people to shift their habitation areas at least slightly, while simultaneously, people began to create more numbers of residential sites in the inland or upland settings of the larger southernmost islands of the Marianas.

Within the range of the two periods from 700 BC through AD 1000, the site-specific stratigraphic units at the House of Taga were consistent with the expectations of the elevation and distance of the coastal terrain from the shoreline (see Figure 12.7). These reconstructions may yet be refined if new excavations can target the precise locations of the housing features and living surfaces of these relevant centuries. Meanwhile, the general parameters are congruent with the overall regional information, including the slight shift in residential focus and coastal activities.

AD 1000–1700

By AD 1000, the sea level had stabilised near the present level (see Figure 12.7), and the new coral reef ecosystem had been established. The coastal terrain had developed mostly the same zones of stable backbeach as can be seen today, along with the position of the freshwater lens floating above the current sea level. A stone-lined well reportedly was constructed during this period (see left image in Figure 12.11), allowing access into the now deeply buried water lens at the site's position, now stranded landward from the shoreline and removed from the direct access to the water seepage flows just above sea level.

If the reported association of the AD 1600s is accurate for the stone-lined well, then this feature would represent the only known case of its kind in any indigenous pre-Spanish context of the region. The well was constructed by first excavating a crater-like depression, then later lining the depression with stones toward creating a more vertical shaft-like interior of the well. This technology was possible with traditional stone, shell, and wood tools and raw materials. An uppermost course of metal-cut or machine-cut stone, with binding mortar, was added at a later time, and a protective metal fence now surrounds the well.

The uniqueness of the stone-lined well may be viewed as consistent with the uniqueness of the exceptional size of the *latte* of the House of Taga. These qualities could suggest an intensive investment in this place beyond the usual practices as seen in other residential sites of the same time period. Similarly, the broken pottery and other artefacts previously were reported as extraordinarily dense at the surface of the site (Hasebe 1938; Hornbostel 1925; Spoehr 1957), indicative of intensive activity here during the later decades of occupation, just prior to the site abandonment.

The megalithic components of the House of Taga were built toward the end of this period, probably during the 1600s. Similar but smaller structures of the same *latte* architectural traditions likely were built earlier, closer to AD 1000. By the 1600s, the *latte* village of Sanhalom, including the monumental House of Taga itself, extended over a large area (see upper right image in Figure 12.2).

Within the area of the 2011–2013 excavation, the cultural layer of approximately AD 1000–1700 contained abundant broken pottery. The pottery included exceptionally large sizes of bowls (see Figure 12.6), made with coarse paste, large temper inclusions, and thick walls, in the forms of large incurving bowls with thickening rims. Decoration was minimal, and the pottery overall fit the expectations of communal storage and serving of food and water.

As with the pottery findings, the shellfish remains of this layer were typical of the associated time period. The older emphasis on *Anadara* sp. clams had disappeared several centuries earlier, and then people targeted other shellfish taxa. By this time, *Strombus* sp. shellfish were the dominant taxa in the food middens, and more bones of fish, turtles, and birds were preserved in this later-aged context.

This particular layer had been truncated in its uppermost portion due to later land-clearing activities associated with World War II and other later activities. The land-altering events had removed most of the *latte* ruins and habitation debris of the village of Sanhalom overall. As a result of those land-clearing events, the only surviving *latte* ruins were at the impressive House of Taga itself, plus a few scattered remnants in displaced positions.

Although formal *latte* ruins were not present here, the surviving lower and, therefore, older portion of this layer had preserved several remnants of structural features and burials (Figure 12.17). An incisor tooth from a subadult burial was dated by radiocarbon at AD 1210–1380 (Beta-344442 in Table 12.1). A charcoal sample from a heated-rock oven was dated at AD 1220–1280 (Beta-313865 in Table 12.1). These dating results have been among the oldest so far reported for general *latte* contexts in the Marianas region, in this case referring to the general association of the pottery and artefact assemblage rather than to the actual remains of *latte* structures.

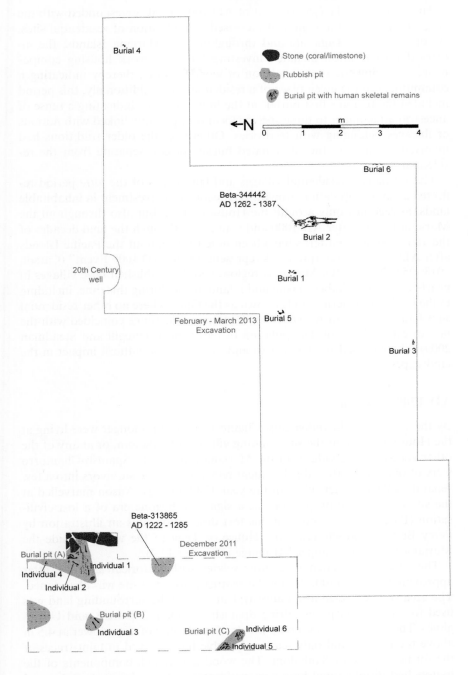

Figure 12.17 Plan map of the ethnohistorical *latte*-associated occupation layer at the
House of Taga Site, AD 1000–1700.

The findings in the *latte* context of AD 1000–1700 corresponded with the general pattern in the region of increased formalisation of residential sites. At the House of Taga site and throughout the Mariana Islands, the increased formalisation included investment in stonework housing components, replacing the older tradition of wooden posts, thereby indicating a commitment to the given place of a residential site. Additionally, this period included the region's first burials at the housing areas, indicating a sense of ancestral attachment to these specific locations, possibly linked with notions of lineages belonging with territories. Otherwise, the older traditions had involved burials in other designated burials places, separate from the residential housing.

The formalised residential villages and landscapes of the *latte* period reflected a major population growth, expansion, and investment in inhabitable lands, as seen not only around the House of Taga but also throughout the Mariana Islands after AD 1000 and continuing through the final decades of the 1600s. Similar patterns have been noted throughout the Pacific Islands after AD 1000 and interpreted as representing an "AD 1000 Event" (Carson 2018: 255–306). In the Marianas region, people established *latte* villages in nearly every inhabitable coastal and inland locale during this time, including in the farther northern islands known as the Gani, where no other residential sites had existed prior to AD 1000. This same time range coincided with the first direct evidence of depopulation of native birds (Pregill and Steadman 2009), likely caused by the growth and expansion of cultural impact in the landscapes.

AD 1700–Modern

By the year 1700, the indigenous Chamorro people no longer were living at the House of Taga, at the surrounding village of Sanhalom, or at any of the other ancestral *latte* villages of the Marianas, due to the Spanish-Chamorro wars of the late 1600s and subsequent relocation of the survivors into a few separate villages. Thereafter, in the year 1742, George Anson marvelled at the site's impressive proportions as a sign of a bygone era of a lost civilisation (Barratt 1988), and soon his text description plus an illustration by Percy Brett would showcase the House of Taga to the world outside the Mariana Islands (see upper left image in Figure 12.2).

The surface-associated layer now relates with historical land use after approximately AD 1700, during the centuries after the site was abandoned. The area was respected as an ancestral site, while the surrounding land was used for various purposes during Spanish, German, Japanese, and US regimes. The surviving stone components of the House of Taga tower at 4.5 m above the ground and only scattered rubble has survived in faint traces of the former village of Sanhalom. The wood and thatch components of the houses had disintegrated long ago, and notably, they already had decayed by the time of Anson's visit in 1742.

On the surface of the site today, the surviving stonework ruins of the House of Taga continue to impress the site's visitors. Currently, people can appreciate the site as an extraordinary example of the *latte* architectural tradition. Within this scope of public appreciation of the surface-visible site ruins, the knowledge about the older subsurface cultural layers and landscape chronology would strengthen the significance of this site. This knowledge provides an important understanding of the long-term adaptations of people with the changing landscape over the course of more than 3000 years.

Conclusions

At the place now known as the House of Taga in Tinian of the Mariana Islands, people lived here continually for more than three millennia, yet at least five major events or periods brought major change to the landscape setting.

1. When people initially lived in this place close to 1500 BC, they were the first individuals ever to interact with this landscape as an inhabited environment. During the next few centuries, those original settlers and their descendants targeted a specialised shoreline niche and habitat ecology. They established the foundation of the region's natural-cultural landscape heritage through the inherited skills and knowledge from their homeland in Island Southeast Asia, combined with their creative minds in the newly inhabited lands.
2. Starting about 1100 BC, a drawdown in sea level began to create a loss of precisely the shoreline ecological niches that had been targeted by the original island settlers. While people adjusted with this changing setting in practical terms of shifting their housing and modifying their seafood-harvesting activities, they simultaneously produced notably different forms and styles of pottery, personal ornaments, and other cultural expressions. Additionally, the lowering of sea level resulted in changing the positions of many of the accessible freshwater seeps, flows, or pools in low-lying coastal areas.
3. The sea level continued to lower overall until approaching the current level around AD 500–1000, and several aspects of the landscape changed during this long time span. Different parts of the landscape system changed at separate yet concurrent paces and scales, including the growth of new coral reefs and related ecosystems, progradation and stabilisation of coastal plains, the establishment of culturally influenced forests, and new arrangements of residential settlements.
4. Around AD 1000, the coastal conditions mostly resembled the present-day setting, and much of the historically known and modern landscape conditions could be applied as far back as this point. This context persisted with variable degrees of internal change through the late AD

1600s, including the creation of the monumental House of Taga and the Sanhalom village complex in Tinian. This timing coincided with a major change in a formalisation of residential settlement structures, not only in the local expression of *latte* houses and villages in the Mariana Islands but also cross-regionally throughout the Pacific, involving the creation of stonework villages, monuments, at-house burial traditions, and indications of lineage-based territoriality.

5. By AD 1700, Spanish interests in the Marianas culminated in massive loss of life among the indigenous Chamorro people, relocation of survivors into a few easily controlled villages, and development of foreign traditions of land-use patterns, religious beliefs, political structure, and generally new ways of interacting with the local landscapes. Sites such as the House of Taga and the surrounding Sanhalom village were abandoned as residential places, but they persisted as culturally important and perhaps sacred components of the landscape. During the next centuries until now, foreign actions and influences have continued to shape and re-shape the landscape context. Indigenous perspectives have gained more prominence recently, including stronger recognition of the House of Taga as a heritage symbol.

In conclusion, the heritage values of the site on the surface can be understood more clearly and appreciated more deeply in relation to the longer chronology of how people lived in this place. This study can contribute to modern and future issues of global concerns in changing landscapes.

Acknowledgements

The site investigations were performed in cooperation with the Commonwealth of the Northern Mariana Islands Historic Preservation Office. Funding is acknowledged with the Chiang Ching-kuo Foundation for International Scholarly Exchange (RG021-P-10) and with the Australian Research Council (DP150104458 and DP190101839).

References

Amesbury, Judith R., 1999. Changes in species composition of archaeological marine shell assemblages in Guam. *Micronesica* 31: 347–366.

Amesbury, Judith R., 2007. Mollusk collecting and environmental change during the prehistoric period in the Mariana Islands. *Coral Reefs* 26: 947–958.

April, Victoriano N., 2004. *Latte Quarries of the Mariana Islands*. Latte: Occasional Papers in Anthropology and Historic Preservation, Number 2. Agana Heights, Guam Historic Resources Division, Department of Parks and Recreation, Guam

Athens J. Stephen, Michael F. Dega, and Jerome V. Ward, 2004. Austronesian colonisation o the Mariana Islands: The palaeoenvironmental evidence. *Bulletin of the Indo-Pacific Prehistory Association* 24: 21–30.

Athens, J. Stephen, and Jerome V. Ward, 1993. Paleoenvironment of the Orote Peninsula. In *The archaeology of Orote Peninsula: Phase I and II, archaeological survey of areas proposed for projects to accommodate relocation of Navy activities from the Philippines to Guam, Mariana Islands*, edited by James Carucci, pp. 153–197. Report prepared for U.S. Department of the Navy, Pacific Division, Naval Facilities Engineering Command. International Archaeological Research Institute, Inc., Honolulu.

Athens, J. Stephen, and Jerome V. Ward, 1998. Paleoenvironment and prehistoric landscape change: A sediment core record from Lake Hagoi, Tinian, CNMI. Report prepared for U.S. Department of the Navy, Pacific Division, Naval Facilities Engineering Command. International Archaeological Research Institute, Inc., Honolulu.

Athens, J. Stephen, and Jerome V. Ward, 1999. Paleoclimate, vegetation, and landscape change on Guam: The Laguas core. In *Archaeological inventory survey of the Sasa Tenjo Vista Fuel Tanks Farm, Piti District, Territory of Guam*, edited by Boyd M. Dixon, J. Stephen Athens, Jerome V. Ward, Tina Mangieri, and Timothy Rieth, pp. 121–151. Report prepared for U.S. Department of the Navy, Pacific Division, Naval Facilities Engineering Command. International Archaeological Research Institute, Inc., Honolulu.

Athens, J. Stephen, and Jerome V. Ward, 2004. Holocene vegetation, savannah origins and human settlement on Guam. In *A Pacific Odyssey: Archaeology and Anthropology in the Western Pacific: Papers in Honour of Jim Specht*, edited by Val Attenbrow and Richard Fullagar, pp. 15–30. Records of the Australian Museum, Supplement 29. Australian Museum, Sydney.

Athens, Stephen J., and Jerome V. Ward, 2006. *Holocene Paleoenvironment of Saipan: Analysis of a Core from Lake Susupe*. Micronesian Archaeological Survey Report Number 35. Commonwealth of the Northern Mariana Islands Division of Historic Preservation, Saipan.

Barratt, Glynn, editor. 1988. *H. M. S. Centurion at Tinian, 1742: The Ethnographic and Historic Records*. Micronesian Archaeological Survey Report 26. Division of Historic Preservation, Commonwealth of the Northern Mariana Islands, Saipan.

Blust, Robert, 2000. Chamorro historical phonology (Mariana Islands, proto-Austronesian). *Oceanic Linguistics* 39: 83–122.

Blust, Robert, 2019. The Austronesian homeland and dispersal. *Annual Review in Linguistics* 5: 417–434.

Bronk Ramsey, Christopher, and Sharen Lee, 2013. Recent and planned developments of the program OxCal. *Radiocarbon* 55: 3–4.

Butler, Brian M., 1994. Early prehistoric settlement in the Mariana Islands: New evidence from Saipan. *Man and Culture in Oceania* 10: 15–38.

Butler, Brian M., editor, 1995. *Archaeological Investigations in the Achugao and Matansa Areas of Saipan, Mariana Islands*. Micronesian Archaeological Survey Report 30. Commonwealth of the Northern Mariana Islands, Division of Historic Preservation, Department of Community and Cultural Affairs, Saipan.

Carson, Mike T., 2008. Refining earliest settlement in Remote Oceania: renewed archaeological investigations at Unai Bapot, Saipan. *Journal of Island and Coastal Archaeology* 3: 115–139.

Carson, Mike T., 2010. Radiocarbon chronology with marine reservoir correction for the Ritidian archaeological site, northern Guam. *Radiocarbon* 52: 1627–1638.

Carson, Mike T., 2011. Palaeohabitat of first settlement sites 1500–1000 B.C. in Guam, Mariana Islands, western Pacific. *Journal of Archaeological Science* 38: 2207–2221.

Carson, Mike T., 2012. An overview of the *latte* period. *Micronesica* 42: 1–79.

Carson, Mike T., 2014a. *First Settlement of Remote Oceania: Earliest Sites in the Mariana Islands.* Springer Briefs in Archaeology. Springer, New York.

Carson, Mike T., 2014b. Palaeo-terrain research: Finding the first settlement sites of remote Oceania. *Geoarchaeology* 29: 268–275.

Carson, Mike T., 2016. *Archaeological Landscape Evolution: The Mariana Islands in the Asia-Pacific Region.* Springer International, Cham, Switzerland.

Carson, Mike T., 2017. *Rediscovering Heritage through Artefacts, Sites, and Landscapes: Translating a 3500-year Record at Ritidian, Guam.* Access Archaeology. Archaeopress, Oxford.

Carson, Mike T., 2018. *Archaeology of Pacific Oceania: Inhabiting a Sea of Islands.* Routledge World Archaeology Series. Routledge, New York and London.

Carson, Mike T., 2020. Peopling of Oceania: Clarifying an Initial settlement horizon in the Mariana Islands. *Radiocarbon* 62: 1733–1754.

Carson, Mike T., and Hiro Kurashina, 2012. Re-envisioning long-distance Oceanic migration: Early dates in the Mariana Islands. *World Archaeology* 44: 409–435.

Carson, Mike T. and Hsiao-chun Hung, 2012. Archaeological Research Excavations at the Landward Portion of House of Taga Site, Tinian, Commonwealth of the Northern Mariana Islands. Report prepared for Historic Preservation Office, Commonwealth of the Northern Mariana Islands. Micronesian Area Research Center, University of Guam, Mangilao.

Carson, Mike T., and Hsiao-chun Hung, 2017. *Substantive Evidence of Initial Habitation in the Remote Pacific: Archaeological Discoveries at Unai Bapot in Saipan, Mariana Islands.* Access Archaeology. Archaeopress, Oxford.

Carson, Mike T., and Hsiao-chun Hung, 2020. Pacific Islands: Finding the First Sites. In *Encyclopedia of Global Archaeology*, edited by Claire Smith, pp. forthcoming. Springer International, Cham, Switzerland. https://doi.org/10.1007/978-3-319-51726-1_1520-2

Carson, Mike T., and Hsiao-chun Hung, 2021. Let's catch octopus for dinner: Ancient inventions of Octopus Lures in the Mariana Islands of the remote tropical Pacific. *World Archaeology* 53: signed copyright, scheduled to print in 2021.

Carson, Mike T., and Hiro Kurashina 2012. Re-envisioning long-distance Oceanic migration: Early dates in the Mariana Islands. *World Archaeology* 44: 409–435.

Craib, John L., 1993. Early occupation at Unai Chulu, Tinian, Commonwealth of the Northern Mariana Islands. *Bulletin of the Indo-Pacific Prehistory Association* 13: 116–134.

Craib, John L., 1999. Colonisation of Mariana Islands: New evidence and implications for human movements in the western Pacific. In *The Pacific from 5000 to 2000 BP: Colonisation and Transformations*, edited by Jean-Christophe Galiaud and Ian Lilley, pp. 477–485. Éditions de IRD, Paris.

Dickinson, William R., 2000. Hydro-isostatic and tectonic influences on emergent Holocene paleoshorelines in the Mariana Islands, western Pacific Ocean. *Journal of Coastal Research* 16: 735–746.

Dickinson, William R., 2003. Impact of Mid-Holocene hydro-isostatic highstand in regional sea level on habitability of islands n Pacific Oceania. *Journal of Coastal Research* 19: 489–502.

Dilli, Bradley J., Alan E. Haun, Susan T. Goodfellow, and Brian Deroo, 1998. Archaeological mitigation program, Mangilao Golf Course Project Area, Mangilao Municipality, Territory of Guam, Report prepared for Jetan Sahni. Paul H, Rosendahl, Ph.D., Inc., Hilo, Hawai'i.

Gifford, E. W., and Dick Shutler, Jr., 1956. *Archaeological Excavations in New Caledonia*. Anthropological Records Volume 18. University of California Press, Berkeley.

Hasebe, Kotondo, 1938. The natives of the South Seas Archipelago. *Lectures on Anthropology and Prehistory* 1: 1–35 (in Japanese).

Haun, Alan E., Joseph A. Jimenez, Melissa A. Kirkendall, and Susan T. Goodfellow, 1999. Archaeological investigations at Unai Chulu, Island of Tinian, Commonwealth of the Northern Mariana Islands. Report prepared for Commander, Pacific Division, US Naval Facilities Engineering Command. Paul H, Rosendahl, Ph.D., Inc., Hilo, Hawai'i.

Hornbostel, Hans. 1925. Unpublished field notes of archaeological research in Guam and the Mariana Islands, 1921–1924. Records on file. Bernice Pauahi Bishop Museum, Honolulu.

Horrocks, Mark, John A. Peterson, and Mike T. Carson, 2015. Pollen, starch, and biosilicate analysis of archaeological deposits on Guam and Saipan, Mariana Islands, northwest Pacific: Evidence for Chamorro subsistence crops and marine resources. *Journal of Island and Coastal Archaeology* 10: 97–110.

Hung, Hisao-chun. 2008. Migration and cultural interaction in southern coastal China, Taiwan and the northern Philippines, 3000 BC to AD 100: The early history of Austronesian-speaking populations. Unpublished doctoral thesis, the Australian National University, Canberra.

Hung, Hsiao-chun, Mike T. Carson, Peter Bellwood, et al., 2011. The first settlement of Remote Oceania: The Philippines to the Marianas. *Antiquity* 85: 909–926.

Kayanne, Hajime, Teruaki Isii, Eiji Matsumoto, and Nobuyuki Yonekura, 1993. *Quaternary Research* 40: 189–200.

Kirch, Patrick V., 1997: The *Lapita Peoples: Ancestors of the Oceanic World*. Wiley-Blackwell, London.

Kurashina, Hiro, and Russell N. Clayshulte, 1983.Site formation processes and cultural sequence at Tarague, Guam. *Bulletin of the Indo-Pacific Prehistory Association* 4: 114–122.

Kurashina, Hiro, Darlene Moore, Osamu Kataoka, Russell Clayshulte, and Erwin Ray, 1981. Prehistoric and protohistoric cultural occurrences at Tarague, Guam. *Asian Perspectives* 24: 57–68.

Laguana, Andrew, Hiro Kurashina, Mike T. Carson, John A. Peterson, James M, Bayman, Todd Ames, Rebecca A. Stephenson, John Aguon, and Harya Putra, 2012. Estorian I latte: A story of latte. *Micronesia* 42: 80–120.

Marche, Antoine-Alfred, 1889. Rapport general sur un emission aux Îles Mariannes. *Nouvelle Archives des Missions Scientifiques et Litéraires, Nouvelle Série* 1: 241–280.

Marche, Antoine-Alfred, 1982. *The Mariana Islands*, translated by Sylvia E. Cheng and edited by Robert D. Craig. Series Number 8. Micronesian Area Research Center, University of Guam, Mangilao.

Marck, Jeffrey, 1978. Interim report of the 1977 Laulau excavations, Saipan, MNI.

Report on file at the Commonwealth of the Northern Mariana Islands Division of Historic Preservation, Saipan.

Moore, Darlene R., Rosalind L. Hunter-Anderson, Judith R. Amesbury, and Eleanor F. Wells, 1992. Archaeology at Chalan Piao, Saipan, Report prepared for José Cabrera. Micronesian Archaeological Research Services, Mangilao. Guam.

Pellett, Marcian, and Alexander Spoehr, 1961. Marianas archaeology: Report on an excavation on Tinian. *Journal of the Polynesian Society* 70: 321–325.

Pregill, Gregory K., and David W. Steadman, 2009. The prehistory and biogeography of terrestrial vertebrates on Guam, Mariana Islands. *Diversity and Distributions* 15: 983–996.

Pugach, Irina, Alexander Hübner, Hsiao-chun Hung, Matthias Meyer, Mike T. Carson, and Mark Stoneking, 2021. Ancient DNA from Guam and the peopling of the Pacific. *Proceedings of the National Academy of Sciences of the United States of America* 118 (1): e2022112118; https://doi.org/10.1007/978-3-319-51726-1_152 0-2. 2017.1073/pnas.2022112118

Ray, Irwin R., 1981. The material culture of prehistoric Tarague Beach, Guam. Master of Arts thesis, Arizona State University, Tempe.

Reimer, Ron W., and Paula J. Reimer, 2017. An online application for ΔR application. *Radiocarbon* 59: 1623–1627.

Reimer, Paula J., Edouard Bard, Alex Bayliss, J. Warren Beck, Paul G. Blackwell, Christopher Bronk Ramsey, Caitlin E Buck, Hai Cheng, R. Lawrence Edwards, Michael Friedrich, Pieter M. Grootes, Thomas P. Guilderson, Haflidi Haflidason, Irka Hajdas, Christine Hatté, Timothy J. Heaton, Dirk L. Hoffmann, Alan G. Hogg, Konrad A. Hughen, K. Felix Kaiser, Bernd Kromer, Sturt W. Manning, Mu Niu, Ron W. Reimer, David A. Richards, E. Marian Scott, John R. Southon, Richard A. Staff, Christian S. M. Turney, and Johannes van der Plicht, 2013. IntCal13 and Marine13 radiocarbon age calibration curves 0–50,000 years cal BP. *Radiocarbon* 55: 1869–1887.

Spoehr, Alexander, 1957. *Marianas Prehistory: Archaeological Survey and Excavations on Saipan, Tinian and Rota*. Fieldiana Anthropology Volume 48. Chicago Natural History Museum, Chicago.

Steadman, David W., 1995. Prehistoric extinctions of Pacific Island birds: Biodiversity meets zooarchaeology. *Science* 267: 1123–1131.

Thompson, Dean, 1977. Archaeological surveys and test excavations along the leeward coast of Saipan, part 1: A summary of methods and procedures. Unpublished manuscript on file, Micronesian Area Research Center, University of Guam, Mangilao.

Thompson, Laura M., 1940. The function of latte in the Marianas. *Journal of the Polynesian Society* 49: 447–465.

13 What have palaeolandscapes revealed about the past and for the future?

Mike T. Carson

As was mentioned in the introductory chapter of this book (Chapter 1), the present concluding chapter considers how the case studies in this book can improve our knowledge about palaeolandscapes. Those ten case studies were contributed by 24 authors who, in total, have investigated diverse palaeolandscapes across the world. The authors presented their findings with the advantage of their first-hand datasets and their expertise to address numerous related research questions about palaeolandscapes. The case studies represented a sampling of different time periods, contexts, and topics that clarified various aspects of the past.

This concluding chapter offers an opportunity to assess the factors that made past landscape systems sustainable, unsustainable, or variable through time. Aligned with this goal, this chapter considers three major topics:

1. How did palaeolandscapes operate and change through time as complex and evolving systems?
2. How could the individual site-specific and region-specific case studies be applied in topics of global relevance?
3. What is the current outlook for the ongoing research contributions of palaeolandscape archaeology?

Palaeolandscapes as complex and evolving systems

This book has treated palaeolandscapes in the broadest sense of natural and cultural aspects that have interacted in variable ways, thereby constituting complex and evolving systems. Chapter 2 presented a general-utility framework about the functional operations of palaeolandscapes as complex systems, furthermore outlining how those systems can be variable through time and across geographic space. Following this introductory framework, Chapters 3 through 12 examined several examples of palaeoenvironments.

The preceding case studies situated the varied aspects of archaeological materials from specific sites or regions within the larger contexts of their environments of the past. All of the case studies considered landforms and habitat zones as a strong foundation for understanding ancient landscapes,

DOI: 10.4324/9781003139553-13

and then they added other lines of evidence when available or relevant for each given case study. The next most prevalent lines of evidence related to the roles of climate and sea level in shaping ancient landscapes and prompting cultural responses. Some chapters added other lines of evidence, such as water sources and hydrology, forest composition, traits of stone tools, and the roles of social perceptions of landscapes.

Given the broad range of natural and cultural components in palaeolandscapes, the case studies have traced how the numerous components of natural-cultural systems have changed concurrently but at different paces and magnitudes, thereby creating complex inter-relationships in the palaeolandscape records of each site or region of study. These issues of multiple concurrent chronologies may seem to be daunting, but in fact, they can be managed in a straightforward chronological narrative from an archaeological perspective. This book's focus in archaeology accordingly enabled descriptive narratives of how palaeolandscapes have changed through long spans of continuous chronological sequences.

A chronological approach, grounded in archaeological evidence, allowed recognition of whenever the diverse lines of evidence may have correlated, thus substantiating closer examination of how those correlations arose, how they functioned together, and how they created sustainable or non-sustainable results in their given landscape systems. This approach can be effective as an initial exploration of a region's research potential, for instance, when outlining the basic chronological narrative of a study region, and this approach indeed formed the starting basis of most of the case studies in this book. The studies of the longest time spans always offer the most potential for identifying more instances of change through time, especially when those time spans happen to include major transitions or fluctuations in large-scale regional or global environmental conditions. When the basic chronology of a palaeolandscape already is known, however, then research can focus more closely on reconstructing ancient landscapes during constrained parameters of the centuries that were relevant for studying specific topics, such as the initial migrations of people across the Bering Strait just prior to 14,000 years ago (studied by Buvit et al. in Chapter 2), the islanders who brought in pottery and other new cultural traits through eastern New Guinea and nearby archipelagos more than 3000 years ago (studied by Summerhayes in Chapter 10), and the response by people adapting to climate instability in Fiji after AD 1300 (studied by Nunn et al. in Chapter 11).

The case studies all shared an approach of starting with objective descriptions of ancient landscapes, and then next, they applied different forms of analytic methodology and interpretive methodology. Most of the chapters emphasised analytic methods to clarify the patterns and trends in the given evidence, such as when detailing what happened in a given site or region within a measured time interval or across a series of time intervals. Interpretive methodology overall was limited in terms of making inferences, and primary concerns of interpretive methods were for testing whether or not hypothetical models could be supported, such as hypotheses about cultural response to change in a palaeoenvironment in aspects of climate, sea level, inhabitable lands, access to water, available food resources, and social notions of how to live in a landscape.

All of the case studies in this book potentially may yet be refined with further evidence, but those possible future refinements should be recognised as categorically different from the future overturning of conclusions. For instance, the dating of the beginning and end of a palaeolandscape period could be refined through additional dating results, such as when narrowing a 1000-year-long period into two different 500-year-long periods. These refinements would not contradict the original datasets or interpretations of those datasets. Rather, the availability of more precise dating would enable finer details of examining a palaeolandscape sequence.

Linking case studies with global relevance

Increasingly during recent years, international concerns have revolved around how to cope with major global environmental change. These issues exist worldwide, although they may be manifest in variable forms in specific places or communities. In principle, the case-specific details in any single instance can be instructive toward globally relevant knowledge. In practice, singular examples always meet their limits of applicability in broader contexts beyond their given parameters, while more numerous and diverse examples can generate more realistic representation and understanding of the issues.

As a practical matter, this book's potential for globally relevant knowledge depended on the ability to gain primary datasets from the past, and therefore new advancements in visualising ancient sites and landscapes have been exceptionally valuable. In this regard, advancements may be noted in terms of visualising the physical shapes of ancient landscapes (as studied by Cochran et al. in Chapter 6) and in terms of visualising the social perceptions and functions of landscapes (as studied by Fulton and Mixter in Chapter 7). These advancements have allowed more productive identification and comprehension of palaeolandscape contexts, thereby generating greater amounts of useful primary datasets. Similar principles and research procedures were applied in the other case studies in this book, whenever illustrating what ancient landscapes looked like during specific time periods or through extended chronological sequences.

Overall, the examples in this book have shown that the most sustainable landscape systems were capable of adapting to change. For instance, generalist rather than specialist resource-use patterns were more effective when water sources became scarce or when climate and weather patterns created instability in basic food supplies. By comparison, the most unsustainable landscape systems occurred when people specialised in narrow ecological niches or intensive land-use patterns that simply could not tolerate a change in the system, such as when changing sea level resulted in loss or transformation of the targeted habitat niches, agricultural fields, or other factors on which people had relied.

This book's lessons from palaeolandscapes can pertain to situations of managing a broad range of modern issues, made possible by the wide-reaching scope of landscapes and palaeolandscapes as introduced in Chapters 1 and 2. As shown in Table 13.1, the specific issues in the case studies have involved:

- Contexts of migration and travel routes;
- Ability of people to cope with change in availability of water sources and hydrological conditions;
- Role of changing climate and sea level in shaping or re-shaping potential cultural use of landforms and resource habitats;
- Limits of specialising in given ecological niches or investing in rigid land-use patterns, and
- Ability of communities to adapt to changing social perceptions and structure as defining components of their landscapes.

As demonstrated in this book, individual palaeolandscape studies could address more than one topic of potential global relevance.

In terms of applying these lessons to the modern world, new planning for global change in climate and sea level will be most successful in a system-wide approach. As demonstrated in this book's palaeolandscape examples, the effects of changing climate and sea level often have extended into multiple aspects of natural and cultural landscape systems. In these situations of

Table 13.1 Globally relevant topics addressed in the case study chapters. "X" = the topic was addressed directly. "-" = the topic was addressed indirectly or in a limited fashion

Case study chapter	Migration and travel routes	Water sources and hydrology	Climate and sea level	Targeting of habitat niches	Social contexts
3 (Freeman)	X		-	X	
4 (Buvit et al.)	X		-	X	
5 (Maher et al.)		X	-	X	-
6 (Cochran et al.)			X	X	
7 (Fulton and Mixter)				-	X
8 (Zhuang and Du)		X	X	X	-
9 (Shaw)	-		X	-	
10 (Summerhayes)	-		X	X	-
11 (Nunn et al.)			X	X	-
12 (Carson and Hung)	-	-	X	X	

pervasive change in a holistic environment, the most successful cultural responses in the past have involved a change in multiple aspects, as noted in the cases of correlations among natural and cultural histories. The least successful reactions have been aimed at perpetuating the pre-existing *status quo* of land-use practice, economy, or social structure through intensified efforts that may have produced more food, trade goods, social status, tax or tribute, or other output for a short time but then inevitably met their limits and triggered a systemic collapse.

Currently, global climate change is causing overall warmer temperatures, among periods of unpredictable extremes of weather conditions, with variable outcomes of floods, landslides, droughts, forest fires, coastal erosion, and other concerns. People in some low-lying coastal regions may need to find higher ground for their homes, farmlands, and other activities in places where they had never lived previously, thus prompting substantially new natural-cultural landscape systems. In many places around the world, people will need to find new sources of water and new lands for food production. Several other issues will need to be considered about restructuring the positions and layouts of residential areas, roads within and connecting them, and businesses that support them.

Ongoing research and contributions

Global environmental change can be recognised as linked with several aspects of the world's landscapes, and therefore a future cultural response would need to occur at this scale. The possible next courses of action likely would include combinations of community-based initiatives, governmental policies, and international agreements that are beyond the scope of this book. Hopefully, though, future planning can recognise the potential values of palaeolandscape research.

New and continued research in palaeolandscape archaeology undoubtedly will proceed throughout the world, in cases of different time periods, and with increasing possibilities for cross-comparative analysis. In these respects, the research field appears to be growing. Ideally, the lessons from palaeolandscapes can be applied not only to learn about the past but also to plan for future change.

Index

Note: *Italicized* and **bold** page numbers refer to figures and tables, respectively.

For Product Safety Concerns and Information please contact our EU
representative GPSR@taylorandfrancis.com Taylor & Francis Verlag GmbH,
Kaufingerstraße 24, 80331 München, Germany

Printed and bound by CPI Group (UK) Ltd, Croydon, CR0 4YY
01/05/2025
01858469-0001